Digital, Analog, and Data Communication, Second Edition

WILLIAM SINNEMA

A Reston Book
Prentice-Hall, Inc.
Englewood Cliffs, New Jersey 07632

Library of Congress Cataloging-in-Publication Data
Sinnema, William
    Digital, analog, and data communication
    Includes index.
    1. Data transmission systems.   2. Digital commu-
nications.
TK5105.S558   1986   621.38   85-24481
ISBN 0-8359-1377-5

A Reston Book
Published by Prentice-Hall, Inc.
A Division of Simon & Schuster, Inc.
Englewood Cliffs, New Jersey 07632

Contents

Preface

Digital communication systems are being rapidly developed to provide reliable communication for digitized analog signals and data. Such analog signals as voice and television must first be translated into a digitized format for these systems, whereas data can be directly carried. Since data transmission is probably the fastest growing aspect of communications, chiefly because of the advent of the computer and microprocessor, it is important for technologists and engineers to become aware of the transmission techniques involved.

As the analog telephone system is still very much with us, we must of necessity consider some of its major transmission characteristics and how these are utilized to carry digital signals. At the same time, we should consider some of its limitations when carrying digitized signals. This book shall attempt to acquaint the reader with the widespread telephone network and with the principles and applications of the more recent digital communication schemes. It must be kept in mind that it is impossible, in a reasonably sized book, to cover each subject exhaustively. The reader can obtain more information on any particular area from one of a score of engineering texts.

A chapter-by-chapter breakdown of the material presented in this book is as follows. Chapter 1 is a general description of analog and digital techniques, which are all expanded upon in later chapters. It also defines some of the basic units used in telephony. The telephone network is described in Chapter 2, and Chapter 3 deals with some of the more technical details of the telephone network. This material is included to provide background information on the limitations of the network when attempting to transmit either analog or digital signals. Chapter 4 considers the most commonly used EIA (Electronic Industries Association) physical interface standards, covering the electric and physical connection procedures between DTE (Data Terminal Equipment) and DCE (Data Communication Equipment). These standards encompass voltage levels, transmission rates, slew rate, distance between the DTE and DCE and the data interchange and control circuits.

Chapter 5 is entirely devoted to modems.

Chapter 6 deals with the problem of aliasing when an analog signal is sampled. Chapter 7 considers the modulation methods most commonly employed when digitizing an analog signal, and also covers some practical aspects of PCM terminal equipment on the North American scene. Information theory is introduced as well. In Chapter 8, the effects of noise and other distortions on digital transmission are explored. Chapter 9 deals with the fundamental concepts of pulse transmission over a bandwidth-limited channel. The Nyquist criteria and various methods of reducing intersymbol interference, such as the raised-cosine channel response and partial response, are also considered. Chapter 10 covers many of the codes, as well as error detection and correction techniques, presently employed in the communication industry.

Chapter 11 is revamped around the ISO-OSI model and includes a section on asynchronous protocols used in personal computers. Chapter 12 complements Chapter 11's distance network emphasis by addressing Local Area Networks, a major trend in resource sharing within a local site.

This textbook is intended as an introductory text to digital and data transmission techniques. A student using this book should have a knowledge of algebra and be familiar with basic logic circuits. An acquaintance with differential and integral calculus would be helpful, although not essential. The description of the final equations should prove as an adequate interpretation.

When dealing with modern technology and with the computer in particular, it is imperative that engineers and technologists begin to reflect somewhat on the basic problems that have come to be associated with technology and its development. Towards this end, I would like to recommend the book *Technology and the Future: A Philosophical Challenge* by Egbert Schuurman (Toronto: Wedge Publishing Foundation, 1980). It gives particular attention to the basic dichotomies presently at work in the views of modern technology.

Acknowledgements

In the course of rewriting a textbook of this nature, one inevitably leans heavily upon the bits and pieces of materials developed by others and upon the various comments and criticisms given by students and fellow staff members. My efforts have chiefly consisted of organizing the various bits into a more consistent and logical manner and of expanding the sections of material that were difficult to comprehend. Thus I owe a debt of gratitude to the many authors in the electronic communication field and to the electronics students and staff at the Northern Alberta Institute of Technology.

I must thank particularly Mr. Tom McGovern for authoring the last two chapters on protocols and local area networks. Thanks also goes to Mr. Ken Clattenburg, who has rewritten and revised the original Telephone Network chapter, and to Mr. J. Want for his suggestions on partial response behavior; to Mr. D. Maruszeczka for his aid in discussing the Lenkurt PCM system and to Mr. Joseph Mah and Mr. Brian Krauss, both from Altel Data for their helpful advice in regard to modem evaluation and optioning.

Final appreciation must go to my wife Edith, who has been so willing to let me spend many evenings in my study, causing her to endure a rather solitary evening life.

William Sinnema

1

Introduction to Communication Systems

The purpose of any communication system is to convey a message from one point in space to another point in space. The originating input is frequently called a source whereas the terminating end is frequently referred to as the sink. If the message is understandable, then information has been conveyed from the source to the destination.

The nature of the message can take on a variety of forms. It can be a continuous time-varying quantity such as speech, music, motor bearing temperature, or pipeline pressure. It can be a discrete signal which only takes on fixed levels or values, such as a signal transmitted by a digital computer as a string of pulses or holes punched in a paper tape. It can be a continuous spatial varying quantity such as the image on a TV camera vidicon tube, or it can be some combination of the above.

Because most messages are not electrical in nature, to carry them over an electrical system the message must be converted to an electrical signal. At the receiver, the electrical signal must be reconverted into the appropriate form. These functions are performed by a transducer. Microphones, speakers, card readers, teletypewriters, and printers are all examples of transducers. With the exception of the discussion on the voice signals in a telephone hand-set, all the messages considered in this text are electrical in nature.

1-1.1   Block Diagram of a Communication System

After a message is converted to an electrical signal, the signal may directly modulate a carrier (high frequency signal), or it may be converted into another format by an encoder, for example, into a binary bit stream, before modulating a carrier wave. Encoding is rarely used in analog communication schemes such as AM and FM, but is frequently employed in digital modulation schemes such as delta modulation and pulse-code modulation (PCM). One of the chief advantages of employing the latter type of encoding is that the signal-to-noise ratio, or in binary terms, the error rate, is almost independent of the number of links or repeating amplifiers in a system. Modulation is employed in order to (a) more efficiently launch the radiated wave into space, (b) permit multiplexing, and (c) to improve signal-to-noise ratio. For efficient launching or reception of an electromagnetic wave or to obtain what is commonly called matching, the radiating or receiving device (antenna) must be a significant portion of a wavelength in size. This permits the resistive portion of an antenna impedance to be in the order of 10's of ohms. The larger the antenna in terms of wavelengths, the greater the antenna's "radiation resistance." The antenna resistance can thus closely approximate the driving generator impedance and associated transmission line; 50Ω, 75Ω, 300Ω, 600Ω, etc. not being uncommon representative values. The wavelength (λ) of an electromagnetic wave in free space is related to the velocity of light (C) by the following relation:

$$\lambda = \frac{C}{f} \qquad\qquad (1\text{-}1)$$

where   λ is the wavelength in m
   C is the velocity of light; 3×10^8 m/s
   f is the frequency of operation

For a signal frequency of 100 Hz, for instance, the wavelength would be $\lambda = 3 \times 10^8$ m/s $\div 100 = 3 \times 10^6$m. If the antenna must be at least one-tenth of the wavelength in size, this would result in an exceptionally large 300 km antenna. By employing modulation, this frequency can be translated upwards so that a more reasonably sized antenna can be used. Translating the various modulating or baseband signals upwards in different frequency bands also makes it possible to transmit many signals simultaneously. This is called frequency division multiplexing. It is not uncommon to carry 2,000 voice channels on one carrier with this technique.

As is done in FM, modulation can also be used to improve the demodulated signal-to-noise ratio. The price for this, however, is an increased transmission bandwidth for the same baseband signal—a severe penalty when the spectrum is already fully utilized.

The transmission medium in Fig. 1-1 is not restricted to a microwave link which transmits an electromagnetic wave in free space, but may include a pair

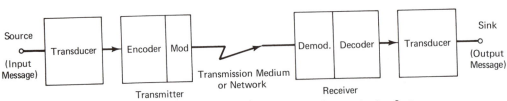

Figure 1-1   The Basic Blocks of a Communication System

of wires, a coaxial line, or an optical fiber. A loss or signal attenuation is experienced in each of these situations, which causes a reduction in the signal-to-noise ratio. For severe losses, repeater or regenerator stations must be inserted to compensate for some of the loss.

The receiver performs the inverse function of the transmitter. It amplifies the weak signal received and demodulates and decodes the signal. The output transducer then converts the electrical signal into its proper form.

Contaminants often affect the quality of a communication system. These contaminants include (1) signal distortion due to imperfect system response, (2) interferences from extraneous signals, and (3) thermal noise from random electron motions in a conducting medium. Within each block of Fig. 1-1, one or several contaminants can creep in to disrupt the system.

A full treatment of noise will be given in a separate chapter.

1-2   Voice Communication

Before considering the nature of voice in speech, first consider the decibel unit that is used to describe the amplitude and frequency response of amplifiers.

1-2.1   The Decibel (dB)

The decibel is a relative power measurement. It does not give the absolute signal level at some point in a system. It is symbolized by dB and defined as

$$dB = 10 \log \text{ (the power ratio)} \qquad (1\text{-}2)$$

Consider the examples of 1 milliwatt (mW) being applied to the inputs of two different amplifiers, resulting in 2 mW and 10 mW at their outputs. In these cases, the gains in dB of the amplifiers (10 log (output power/input power)) are, respectively:

$$10 \log \frac{2 \text{ mW}}{1 \text{ mW}} = 3 \text{ dB}$$

$$\text{and} \quad 10 \log \frac{10 \text{ mW}}{1 \text{ mW}} = 10 \text{ dB}$$

Please note from these examples that a doubling in power gives a 3 dB gain whereas a power increase of 10 gives a 10 dB gain. Similarly, power ratios of 100, 1000, and so forth result in gains of 20 dB, 30 dB, and so forth. If the power output is less than the power input, the gain is given in terms of negative decibels or often translated as a positive decibel loss. For instance, the gain of the circuit shown in Fig. 1-2 is 10 log (.05/1) = −13 dB. Alternatively, this can

Figure 1-2   A Negative Gain Network

be stated as having a loss of 13 dB. If the output and input resistances of a network are identical, the decibel gain can be restated as

$$dB = 10 \log \frac{(E_{output})^2/R}{(E_{input})2/R} = 20 \log \left(\frac{E_{output}}{E_{input}}\right) \qquad (1\text{-}3)$$

or $\qquad dB = 10 \log \frac{(I_{output})^2 \, R}{(I_{input})^2 \, R} = 20 \log \left(\frac{I_{output}}{I_{input}}\right) \qquad (1\text{-}4)$

Even if the resistance levels are not identical, Eqs. (*1-3* and *(1-4)* are often still used. The equations allow for ease in graphically plotting the gain of a frequency-sensitive network such as a filter. Note that ignoring the difference in resistance does not give a true power ratio.

In cascading networks, the gains in decibels of the individual networks can be added. This can be easily proven by observing Fig. 1-3 where the individual power gains are

$$g_1 = \frac{P_{out1}}{P_{in1}}, \text{ and } g_2 = \frac{P_{out2}}{P_{in2}}$$

where $\quad G_1 = \left(10 \log \frac{P_{out1}}{P_{in1}}\right) dB$

$$G_2 = \left(10 \log \frac{P_{out2}}{P_{in2}}\right) dB$$

$$G_{overall} = \left(10 \log \frac{P_{out2}}{P_{in1}}\right) dB$$

$$= 10 \log \left(\frac{P_{out2}}{P_{in2}} \quad \frac{P_{out1}}{P_{in1}}\right) dB$$

since $\quad P_{in2} = P_{out1}.$

Therefore

$$G_{overall} = \left(10 \log \frac{P_{out2}}{P_{in2}} + 10 \log \frac{P_{out1}}{P_{in1}}\right) dB$$

$$= (G_1 + G_2) \, dB$$

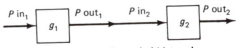

Figure 1-3 Cascaded Network

Without the use of a calculator, find the approximate power gain of a 13 dB amplifier.

This is equivalent to a 10 dB amplifier in cascade with a 3 dB amplifier. The overall power gain ratio is therefore $10 \times 2 = 20$.

Example 1-1 *Solution*

What is the dB loss of the Fig. 1-4 network?

The net overall gain is $-30 + 30 - 30 + 30 - 10 = -10$ dB or a loss of 10 dB.

Example 1-2 *Solution*

1-2.2 The dBm and dBW

The Decibel Referred to One Milliwatt (dBm) gives absolute measure of the signal level at some point in a system. It relates all power levels to a one milliwatt (mW) level.

$$\text{dBm} = 10 \log \frac{Power}{1 \text{ mW}} \qquad (1\text{-}5)$$

A power level of 1 mW for instance represents 0 dBm. An absolute power level of 2 mW represents 3 dBm. Similar to the dBm, the Decibel Referred to One Watt (dBW) is an absolute decibel unit using one watt as a reference. It is frequently used in microwave applications.

$$\text{dBW} = 10 \log \frac{Power}{1 \text{ W}} \qquad (1\text{-}6)$$

Thus, 1 mW represents O dBm, or

$$10 \log \frac{10^{-3}}{1} = -30 \text{ dBW}.$$

Similarly 1 W represents 0 dBW or 30 dBm.

When adding or subtracting two signals stated in dBm's or dBW's, it is important to convert back to milliwatts or watts and add the latter. For example, a 13 dBm (20 mW) signal added to a 10 dBm (10 mW) signal is equivalent to 30 mW, or 14.8 dBm. On the other hand, the addition of two equal power levels doubles the output power or, equivalently, exceeds either one or 3 dB. Thus, 15 dBm + 15 dBm = 18 dBm.

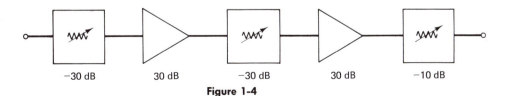

−30 dB   30 dB   −30 dB   30 dB   −10 dB

Figure 1-4

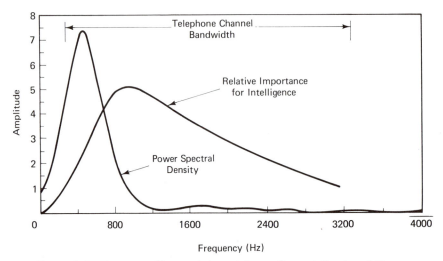

Figure 1-5 Frequency Characteristic and Power Spectral Density of Human Speech Adapted from ''A comparative study of various quantization schemes for speech encoding,'' P. Noll, **Bell System Tech. J.** vol. 54, no. 9 (Murray Hill, N.J.: Nov., 1975), pp. 1597–1614. © 1975, American Telephone and Telegraph Company. Reprinted with permission.

1-2.3 Characteristics of Human Speech

Human Speech consists of a complex assortment of frequency components ranging from about 100 Hz to 8000 Hz. The powers emitted vary from about 0.01 to 5000 μ W giving a dynamic range of around 57 dB. The major portion of energy content exists in the low-frequency range, as shown in Fig. 1-5. As far as understanding is concerned, however, the low frequencies (up to about 700 Hz) add very little intelligibility to the signal.

Intelligibility is a measure of how well a message is received over an imperfect communication medium. Studies have shown that the higher speech frequencies contribute more to intelligibility than the lower frequencies. A curve showing the relative importance for intelligibility is also shown in Fig. 1-5. According to the figure, bandwidth can be conserved and a good information transfer still obtained by limiting the bandwidth of the transmission medium. The standard CCITT (Comité Consultaif International de Téléphone et de Télégraphie—International Consultative Committee for Telephony and Telegraphy)* bandwidth of a voice channel is 300 to 3400 Hz. Many telephone subsets reduce this even further to about 500 to 3000 Hz.

1-2.4 Noise

Because the ear is not equally sensitive to all frequencies, any noise appearing in telephony does not have the same interfering effect for each frequency. In

* See Sec. 1-3 for more infotmation.

order to measure the relative interfering effect of noise, some form of frequency weighting is necessary to equalize the effects. When making noise measurements on a voice circuit, a filter is inserted that reduces the actual noise value in inverse ratio to its interfering effect.

By performing subjective experiments, many standard weighting curves have been adopted that depend upon the telephone subsets used in the experiments. The weighting adopted by CCITT, known as psophometric weighting, uses an 800 Hz reference frequency and is expressed in terms of pW (10^{-12} W). The units are expressed as pWp (p for psophometric weighting) or dBmp.

A uniform 300 Hz to 3400 Hz band of noise, such as white noise, results in a weighted power that is 2.5 dB less than the unweighted power. Thus, if the white noise power in a 300 Hz to 3400 Hz band is y dBm, the psophometric noise is $(y - 2.5)$ dBmp. For a 0-4 kHz band of white noise, the unweighted noise power is reduced by 3.6 dB.

In North America, voice circuit noise measurements are generally made using the C-message weighting filter characteristic shown in Fig. 1-6; it uses 1000 Hz as the 0 dB reference point. To maintain a positive number in the rela-

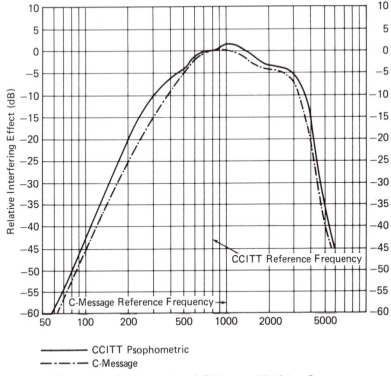

Figure 1-6   Psophometric and C-Message Weighting Curves

tive interfering effects, a 1 pW (−90 dBm) reference level is used. Any noise level is expressed in decibels above this reference noise as dBrn.

For example

$$0 \text{ dBm} = 90 \text{ dBrn.}$$
$$-1.5 \text{ dBm} = 88.5 \text{ dBrn.}$$
$$-60 \text{ dBm} = 30 \text{ dBrn.}$$
$$-90 \text{ dBm} = 0 \text{ dBrn.}$$

In general,

$$\text{dBm} = \text{dBrn} - 90 \qquad (1\text{-}7A)$$

or

$$\text{dBrn} = \text{dBm} + 90 \qquad (1\text{-}7B)$$

The notation dBrnc is used when the C-message weighting network is employed. 30 dBrnc indicates that the noise level is 30 dB above the reference noise level C-message weighted. White noise over a 300 to 3400 Hz bandwidth, when C-weighted, results in about a 1.5 dB reduction. Thus, a white noise signal at 0 dBm produces a −1.5 dBm, or 88.5 dBrn, C-weighted signal. If the white noise power is −60 dBm, the C-message-weighted noise is −60 + (90 − 1.5) = 28.5 dBrnc.

Conversion between the European and C-message North American units is given by dBmp = dBrnc − 90. Figure 1-6 gives both the CCITT psophometric and the C-message weighting curves.

1-2.5 Zero Transmission Level Point (OTLP) and dBmO

In dealing with telephone systems, it is convenient to define some point as the zero transmission level point (OTLP) and to measure the signal and noise levels at other points in the system relative to it. When referring signals back to the OTLP, the units are expressed in dBmO. The dBmO unit indicates what the power would have been at the zero transmission level point.

The exact location of the OTLP depends upon national practice. In North America, the output side of the Main Distribution Frame at the exchange (class 5 office) is usually considered as the OTLP. This is also the point where the local loop is connected to the exchange. A OTLP is a point at which the test tone level should be 0 dBm.

This 0 dBm test tone can often be automatically applied at the local exchange to the subscriber's loop by dialing a specified number. In Edmonton, Alberta, for instance, this number is 1200. The received signal at the subscriber's station in dBm's indicates the local loop loss for a 1000 Hz. tone. The difference in the power level of a signal (or noise) at some point in a system, compared to the OTLP level, in dB can be denoted by dBr (relative dB's). This difference in levels is also often called TLP. Thus

$$\text{dBr} = \text{TLP} = \text{Signal power (dBm)} - \text{dBm 0} \qquad (1\text{-}8)$$

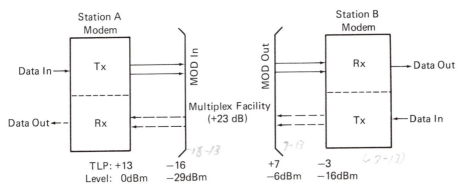

Figure 1-7 Data Transmission Levels on a Four-Wire Voice Circuit for Transmission from Station A to Station B.

The dBr or TLP value is given in dB like − 3 dBr or − 3 TLP. For instance, if the signal power level at the 0TLP is 6 dBmO, then at a relative level point of 7dBr (7TLP), the signal power will be 13 dBm. When the 0TLP is at − 13dBmO, the actual power level at the − 3dBr (− 3TLP) point is − 16dBm.

If we take a typical voice circuit, where the input is at + 13dBr (13 TLP), if the output power from a subscriber's apparatus is 0dBm (1mW), the 0TLP will be − 13dBmO.

In data circuits, the maximum allowable signal level is − 13dBmO. Fig. 1-7 shows typical transmission levels for a data circuit, where the modem output or station A transmit level is 0dBm, and located at a + 13dBr or TLP point.

1-2.6   Signal-to-Noise Ratio (S/N)

The term *signal-to-noise ratio* is frequently used to express in decibels the amount by which a signal level exceeds the noise level. If a signal level of 20 dBm experiences a noise level of 5 dBm, then the $(S/N)_{dB} = S_{dBm} − N_{dBm} = 20 − 5 = 15$ dB. For voice communication a S/N rate of 30 dB is considered satisfactory, whereas for video the minimum level should be 45 dB.

What is the signal-to-C-message-weighted-noise ratio at the 0TLP of a system when the signal power is 0 dBm (1 mW) and the unweighted white noise level is − 20 dBm spread over the 3.1 kHz voice bandwidth?

The signal power level is 0 dBm. The unweighted noise is − 20 dBm which, when C-message-weighted over a 3.1 kHz bandwidth, gives a weighted noise of − 20 dBm − 1.5 dB = − 21.5 dBm. Therefore $S/N_{C\text{-weighted}} = 0 − (−21.5) = 21.5$ dB.

The terms dBrnc0 and dBm0p refer to noise power measured in dBrnc's or dBmp's, respectively, at the 0TLP with C or phosphometric weightings.

Example 1-3

Solution

Example	What is the signal-to-C-message-weighted-noise ratio when the signal is at a
1-4	level of -9 dBm0 and the noise level is 48 dBrnc0?
Solution	Referring all levels to the 0 dB TLP, the signal has a level of -9 dBm0

and the C-weighted noise has a level $-90 + 48 = -42$ dBm0. Thus, the $S/N_{\text{C-weighted}} = -9 - (-42) = 33$ dB.

If the noise level was unweighted at 48 dBrn0, then the weighted noise would be decreased by about 1.5 dB, giving an $S/N_{\text{C-weighted}}$ ratio of 34.5 dB.

Example	A C-message noise reading of 30 dBrnc is obtained at the input to a modem
1-5	which is at a -3dBr or -3TLP level. What is the noise level at the 0TLP?
Solution	Using Equation 1-8, the noise level at the 0TLP will be

$$= \text{noise power} - \text{TLP}$$
$$= 30 - (-3) = 33 \text{ dBrnc0}.$$

1-2.7   Sampling Theorem

Any band-limited signal does not need to be continuously transmitted in order to retain all its information content. As will be more fully explained in Chap. 6, no information is lost if such a signal is sampled at regular intervals of time at a rate equal to or higher than twice the highest frequency component. The signal can be reconstructed by passing it through a perfect or ideal low-pass filter.

Take, for example, a voice signal that has an upper frequency component of 4 kHz. By sampling it at a rate of 8 kHz as shown in Fig. 1-8, the sampled signal can be transmitted and reconstructed without any loss of information. The reconstructed signal will naturally experience some delay.

1-3   The Telephone Network

Throughout North America, Europe, Australia, New Zealand, Japan, and several other nations, telephone hook-up is readily available to most households and offices. With almost 150 million telephones in North America and an equal number in the rest of the world, the telecommunications industry is intricately woven in the fabric of our society. The present world-wide switched public telephone system must be considered whenever a new communication service for the public-at-large is contemplated. Large, overlapping communications networks are extremely uneconomical.

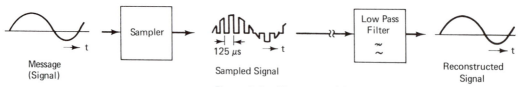

Figure 1-8   Illustration of Sampling

In addition, any changes or improvements in the existing network must be compatible with the overall system if they are to be introduced economically. Huge plant investments cannot be quickly written off, and plants cannot be replaced in a short time span. Just to replace telephone sets in every home would involve billions of dollars and many hours of installation time.

Thus, just as with monophonic FM and stereo FM, or black-and-white and color TV, improved communications technology must operate within the limits of old equipment. For example, one limitation of the present telephone set is the bandwidth range of 300 Hz to 3400 Hz. The present telephone sets cannot be used if operated below or above this frequency range.

Communication networks vary from country to country and from area to area within a country in terms of both the tariff structures and the type of ownership (private, public, or some mixture of the two). Public telephone networks, because of their monopolistic nature, are normally subjected to some form of regulation. In order to permit interconnection with networks located in different countries, each nation's network must meet some internationally set standards.

In Canada, the various companies that provide communication services signed a connecting agreement in 1931 to form the Trans Canada Telephone System (TCTS). Because significant changes have taken place in the Canadian telecommunications industry in recent years—i.e., services such as electronic messages, office communications, and data transmission have been added—the organization's name was changed in 1983 to Telecom Canada.

Telecom Canada compromises Canada's nine major telecommunication carriers and the satellite carrier, Telesat Canada. Each of the nine carriers provides a variety of telecommunication services for a particular geographically defined area such as a province. The major exceptions are Bell Canada, which operates in Ontario, Quebec, and part of the NWT, and AGT, which serves all of Alberta except the city of Edmonton which provides its own local telephone service.

Telecom Canada's nine members have varying ownership structures, regulatory conditions, and services unique to each company. The largest supplier of telecommunication products and services, Bell Canada, is privately owned and operates over 10 million telephones. It is federally incorporated and regulated by the Canadian Radio-Television and Telecommunications Commission (CRTC). The second largest telecommunications company is B.C. Tel. It is shareholder-owned and federally regulated by CRTC. The third largest telecommunication firm is AGT, a provincial crown corporation which is regulated by the Alberta Public Utilities Board.

Sask Tel and Manitoba Telephone System (MTS) are also provincial crown corporations. The New Brunswick Telephone Company (NB Tel), Maritime Telegraph and Telephone Company Ltd. (MT&T), Island Telephone Company Ltd. (Island Tel), and Newfoundland Telephone Company Limited are all investor-owned.

In the United States, the privately-owned American Telephone and Telegraph Company (AT&T) is the dominant telephone company. AT&T owns Western Electric Company and until 1982 owned most of the stock of 22 operating telephone companies. These companies are being divested from AT&T and provide services within their respective areas in the country. AT&T also owns one of the world's largest research organizations, the Bell Telephone Laboratories. It also has a Long Lines Department that is responsible for operating long distance interstate communication lines.

AT&T is beginning to enter the data processing, information distribution, and cable television markets, which was closed to them prior to 1982. In most countries, the communications industry is dominated by an organization commonly known as PTT (Postal, Telegraph and Telephone Administration). Frequently the term "common carrier" is used to denote the company that offers communication facilities to the public.

Telecommunications Standards Organizations

In order to promote and foster international telecommunications on a worldwide basis, to carry on studies of technical problems relating to the telecommunications network and to ensure proper interworking of the various national systems, the International Telecommunication Union (ITU) was founded. The ITU is a specialized agency of the United Nations, but predates it, having come into existence in 1932 with the merger of two older organizations, the International Telegraph Union (1865–1932) and the Radio Telegraph Union (1903–1932).

Within the ITU there are three major divisions:

1. *The Consultative Committee on International Telegraphy and Telephony (CCITT)* is responsible for making technical recommendations and studying tariff questions relating to telegraphy and telephony.

2. *The Consultative Committee on International Ratio (CCIR)* makes technical and operational recommendations relating to radio communications.

3. *The International Frequency Registration Board (IFRB)* registers and examines assignments for space and terrestrial services, records frequency assignments for orbital positions, and advises on the maximum number of radio channels in those portions of the spectrum where interference would occur.

The ITU also holds special conferences to consider new opportunities and particular issues in the light of technological advances. One such conference was the 1979 World Administrative Conference (WARC-79), which reviewed and modified the international regulations for spectrum use. An earlier WARC-63 conference was devoted to space communication and large frequency bands were allocated to space services.

The ITU operates under the ITU convention. This convention is like a treaty, to which the signatories are bound. It states the structure, goals, and responsibilities of the ITU and its members. Although most nations have ac-

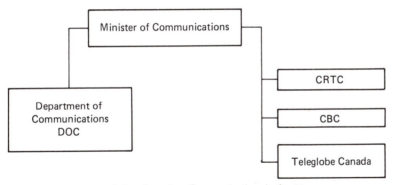

Figure 1-9 Canadian Communication Authorities

cepted the ITU standards, variations do occur, more in North America than in Europe. Each country has its own communication authorities that implement the decision of the ITU conferences. In Canada the federal Minister of Communications is responsible for all matters over which the Parliament of Canada has jurisdiction relating to telecommunications. The Department of Communications (DOC), the Canadian Radio Television and Telecommunications Commission (CRTC), the Canadian Broadcasting Corporation (CBC), and Teleglobe Canada—all independent organizations—report directly to the minister of communications (see Fig. 1-9).

The Department of Communications itself is composed of four sections:

1. The Policy sector formulates and implements policies relating to domestic and international telecommunications.
2. The Research sector which develops new communication systems and provides scientific advance for policy makers. Its principal in-house research faculty is located at Shirley Bay (near Ottawa), Ontario. It also distributes research contracts to various Canadian universities.
3. The Space sector which is responsible for the satellite communications operations such as the ANIK series.
4. The Spectrum Management and Government Telecommunications sector: the Spectrum Management service is responsible for the planning, allocating, and regulation of the frequency spectrum; the Government Telecommunications Agency plans and provides telecommunication services for the federal government.

The CRTC is the licensing and regulating authority for all broadcasting in Canada. It also has responsibilities in telecommunications in such areas as tariffs, regulations, and service complaints.

The CBC is responsible for the national broadcasting service, operating French and English TV and AM and FM radio networks. It must provide a balanced service of information and contributes to national unity and a continuing expression of Canadian identity.

Teleglobe Canada provides for all the external or overseas telecommunication services—telephone, telegraph, telex, video, and labs. It also participates in the deliberations of such organizations as Commonwealth Telecommunications Organization (CTO), International Telecommunications Satellite Organization (INTELSAT), ITU, and the Canadian Telecommunications Carriers Association (CTCA). It operates satellite earth stations at Lake Cowichan, B.C., Mill Village 1 and 2, N.S., and Des Laurentides, Weir, Quebec.

To supply the many communication services required across Canada, the Canadian Telecommunications Carriers and the federal government have formed Telesat Canada whose function is to establish a system of domestic satellite communications. Its aim is to give people anywhere in Canada an opportunity to communicate instantly and reliably with each other. It presently operates a series of ANIK (meaning "brother" in Inuit) domestic satellites operating in the $6/4$ and $14/12$ GHz bands.

In the United States, the 1934 Communications Act established the Federal Communication Commission (FCC) to regulate nongovernment use of frequencies and to regulate the interstate and foreign communications by wire and radio. It licenses transmitters within the United States and aboard U.S. registered ships and aircraft.

The regulating responsibility for frequency spectrum use by federal agencies has been delegated by the President to the National Telecommunications and Information Administration (N.T.I.A.) within the U.S. Department of Commerce. N.T.I.A. and FCC maintain a close liaison with one another. The decisions of the N.T.I.A. are based upon the recommendations of the Interdependent Radio Advisory Committee (I.R.A.C.).

Fig. 1-10 illustrates the U.S. frequency management structure.

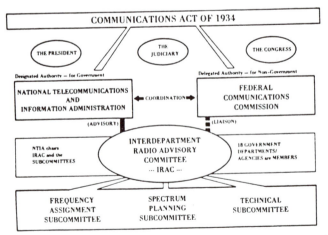

Figure 1-10 U.S. Frequency Management Structure (from: John J. Kellehar, "Regulatory Procedures and WARC-79," IEEE Transactions on Communications, Vol. Com-29, No. 8, Aug. 1981. Copyright © 1981 IEEE)

In North America the manufacturers of electronic components and equipment have formed a trade association known as the Electronics Industries Association (EIA). This association prepares comprehensive standards, many of which eventually become internationally recognized. For a listing of EIA standards and other engineering publications, write:

Electrical and Electronic Manufacturing Association
One Yonge Street, Suite 1608
Toronto, M5E 1R1

or

Electronic Industries Association
2001 Eye Street N.W.
Washington, D.C. 20006.

1-3.1 Modulation and Multiplexing

The common carrier most frequently transmits information through wire conductors (wire pairs and coaxial cable) and microwaves. The receiver and transmitter are usually hooked to wire carrier systems which in turn are interconnected by a microwave link. The message or signal to be transmitted is the baseband signal. The baseband signal can be a speech signal produced by a telephone, a series of pulses produced by a teletype, or, for wider bandwidth systems, a video signal produced by a television camera. The baseband signal is put on a carrier by a process called modulation. The modulation process in some ways modifies an otherwise continuous sinusoidal carrier signal. The three basic forms of carrier modulations are

    a. Amplitude modulation (AM)

    b. Frequency modulation (FM)

    c. Phase modulation (PM)

In amplitude modulation, the amplitude of the carrier varies in proportion to the amplitude of the baseband signal. In frequency and phase modulation, the frequency and phase departures of the carrier vary linearly with the amplitude of the baseband signal. Since frequency and phase are not independent quantities, there is always a phase shift with a frequency shift and vice-versa. Figure 1-11 illustrates these modulation waveforms.

The modulation process permits the message to be more efficiently transmitted and, in addition, through FDM allows several messages to be transmitted simultaneously. This can be more clearly understood by considering the frequency spectra of the various modulation schemes when a baseband signal ranging from a frequency of f_1 to f_2 modulates the carrier with a rest or unmodulated frequency of f_c. A table illustrating the various frequency spectra, as well as the bandwidth required in order to permit transmission of the modulated signal, is given in Table 1-1. Variations on the three systems are also included.

As illustrated, the bandwidth (BW) occupied by the modulated signal

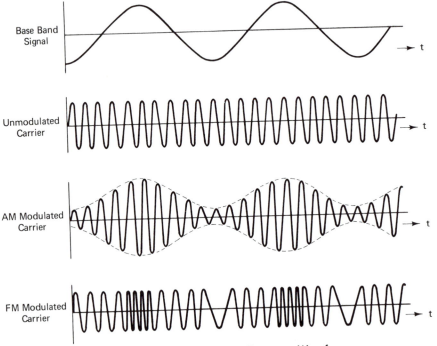

Figure 1-11   AM and FM Carrier Frequency Waveforms

depends upon the modulation scheme used. For AM, the bandwidth is twice the highest baseband frequency. For single sideband (SSB), the bandwidth is nominally equal to the baseband bandwidth. SSB suffers from poor low-frequency response since removal of one of the sidebands with real filters result in some erosion of the lower-frequency components due to the finite roll-off of the filter skirts. This will be discussed in more detail later when the low-frequency components in SSB generation are of significance, as in video and data transmission. The vestigial sideband can be used when a remnant of the removed sideband is retained.

Frequency modulation is very inefficient in its use of spectrum space, but has a greatly improved performance in a noisy environment. The bandwidth occupied is approximately by $2(f_2 + \Delta f)$, where Δf represents the peak frequency shift of the carrier from its rest frequency. This occurs when the baseband signal is at its maximum amplitude.

In order to increase the number of channels that can be carried over a single cable pair, several channels can be stacked so that each channel occupies a unique portion of the frequency spectrum. This process is referred to as frequency division multiplexing and is graphically illustrated in Fig. 1-12. As a result, several channels can be simultaneously transmitted over one physical circuit, with each signal occupying a different frequency slot, or portion of the total available bandwidth.

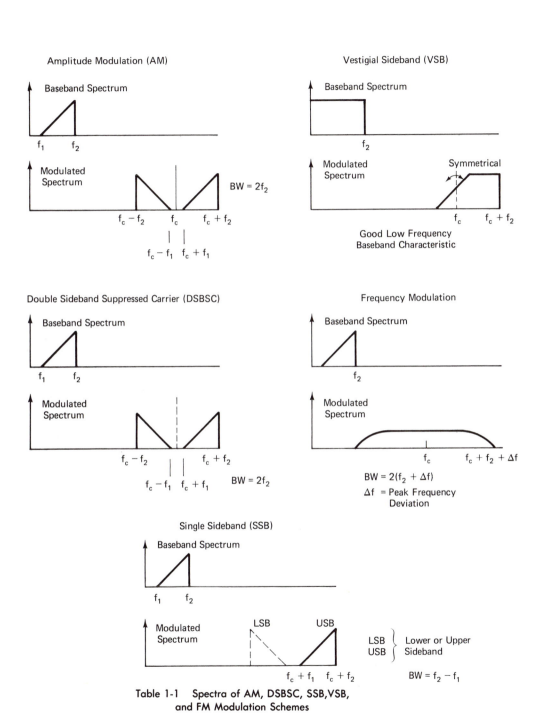

Table 1-1   Spectra of AM, DSBSC, SSB, VSB, and FM Modulation Schemes

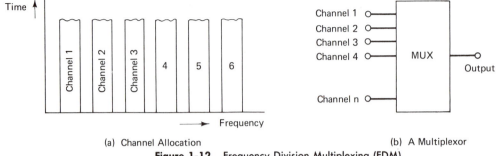

(a) Channel Allocation

(b) A Multiplexor

Figure 1-12 Frequency Division Multiplexing (FDM)

To illustrate how telegraph signals, voice signals, and television signals can be simultaneously transported from one point to another, consider a simplified but not untypical FDM system as shown in Fig. 1-13.

Because telegraph machines transmit at very low speeds, the bandwidth requirement is minimal for each telegraph channel. OFF and ON pulses, transmitted at 75 or 150 b/s, must first be converted to a narrow band voice frequency signal by the VFCT block. These narrow band signals are then stacked to allow up to 18 telegraph channels on one voice frequency channel.

The voice channel can be stacked with 23 other voice frequency channels to obtain a grouping of 24 channels. This grouping can be transmitted by an N-carrier system. Arranging 25 or more such groups of 24 channels results in 600-plus channels which can be transmitted by the L carrier. Since a single TV channel requires a large bandwidth, television can take up a bandwidth equivalent to 1200 voice channels. A TV channel can readily be carried on an L carrier. The Bell L3 coaxial cable system, for instance, can carry 660 voice channels and one TV channel. By further upward modulation in frequency, thousands of voice channels can be transmitted over a microwave or satellite link.

To recoup a single voice or telegraph channel, the process illustrated in Fig. 1-13 must be reversed.

FDM does have some distinct disadvantages, namely:

a. Bandwidth is wasted because of the guardbands required between adjacent channels to minimize channel interference (crosstalk between channels).

b. The system is inflexible to bandwidth changes. If for instance the BW of one channel is increased, the center frequencies of all the other channels may need to be redefined or the number of channels reduced.

1-3.2 Dedicated/Dial-Up Circuits

The common public switched network is a complex network of switching stations. Depending upon the number "dialed up," a call can be routed to many

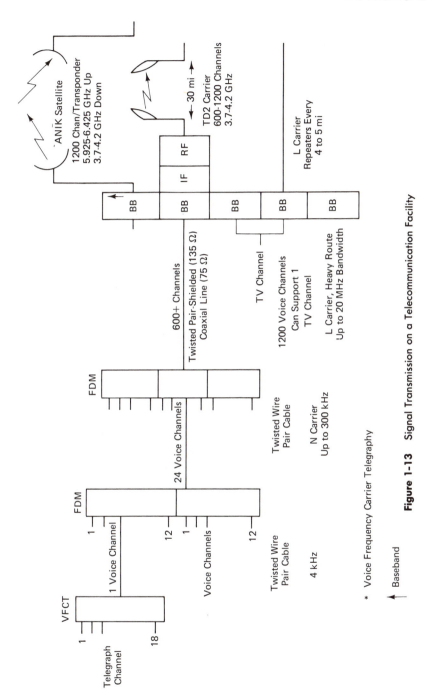

Figure 1-13   Signal Transmission on a Telecommunication Facility

* Voice Frequency Carrier Telegraphy

← Baseband

phones via the switching stations. Depending upon the routing, the line qualities may vary. Switching equipment tends to introduce a great deal of impulse noise in the channel.

A dedicated or private line provides a private full time connection between two locations. It may be routed through a central office, but the line is physically connected to the dedicated channel and bypasses the switching network and its associated signaling equipment.

Private lines are always available and we just have to pick up a telephone set or throw a switch to make a connection. They are also generally less noisy than switched circuits.

1-3.3   Two-Wire/Four-Wire Transmission

When communicating by voice or when transferring data, transmission in both directions is desirable. If both directions are carried by a pair of wires, we have two-wire or one-loop transmission. Two wires are used by the millions in connecting the telephone set to the local office.

Four-wire circuits or two-loop transmission has a pair of wires for each direction, one pair for the transmit side and another pair for the receive side. Four-wire is used pretty well in all sections of the telephone network, other than that between the subscriber and the local office, usually called the subscriber or local loop. The transmit and receive signals are physically separated.

To convert from two-wire to four-wire, a terminating set or hybrid is used. This is illustrated in Fig. 3-2.

Lines can be either balanced or unbalanced, depending upon the termination used. If both wires of a line have equal impedances with respect to ground, the line is said to be balanced. The twisted wire pair used in private line applications for the subscriber's loop is usually balanced as illustrated in Fig. 1-14 (b). On the other hand, most dial-up lines use an unbalanced line as shown in Fig. 1-14 (a), with one wire grounded at the central office.

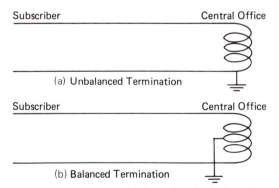

Subscriber   Central Office

(a) Unbalanced Termination

Subscriber   Central Office

(b) Balanced Termination

Figure 1-14   Illustration of Balanced and Unbalanced Lines

Lines also have a "characteristic impedence," which depends upon the size and spacing as well as upon the type of insulating material. Telephone sets and modems should be matched to the telephone line circuit which is normally 900 ohms for the switched telephone network and 600 ohms for private lines.

1-4   Digital Communication

The previous section dealt with the transmission of a continuously variable analog signal. Such a signal has an infinite number of possible levels since it can assume any value between the "rails." This section will briefly consider the digital transmission of a signal when discrete separate pulses have a finite number of levels are transmitted. Most commonly, a binary system of pulses is used. Only one of two conditions may exist at any particular time, as exemplified by the various nomenclatures in Table 1-2. The two conditions represented by the digits 1 and 0 are termed binary digits, or bits.

Table 1-2   Nomenclature of Binary States

Application	Boolean Nomenclature	
Binary numbers	1	0
Binary numbers	HI	LO
Transistor	MARK	SPACE
Signal	OFF	ON
Transistor	Minus	Plus
Paper tape	Hole	No hole

1-4.1   The Binary, Octal, and Hexadecimal Numbering Systems

The base of the binary number system is 2, and the system consists of the two digits 0 and 1. The placement or position of the binary number within a sequence of digits determines the weighting that the bit represents. In general, the sequence $a_3a_2a_1a_0 \bullet a_{-1}a_{-2}a_{-3}$, where a_n represents a 1 or a 0, has a weighting of:

$$2^0 \text{ for the } a_0 \text{ bit}$$
$$2^1 \text{ for the } a_1 \text{ bit}$$
$$2^2 \text{ for the } a_2 \text{ bit}$$
$$2^3 \text{ for the } a_3 \text{ bit}$$
$$\text{and} \quad 2^{-1} \text{ for the } a_{-1} \text{ bit}$$
$$2^{-2} \text{ for the } a_{-2} \text{ bit} \quad \text{etc.}$$

For any other numbering system, the base value of 2 changes to the new base. For the octal numbering system where the base is 8, the weightings change to 8^1, 8^2, and so on.

Example 1-6	Convert the binary number 1011.01 to its decimal equivalent.
Solution	$1 \bullet 2^3 + 0 \bullet 2^2 + 1 \bullet 2^1 + 1 \bullet 2^0 + 1 \bullet 2^{-2}$ $= 8 + 2 + 1 + .25 = 11.25$
Example 1-7	Convert the octal number 652.7 to its decimal equivalent.
Solution	$6 \bullet 8^2 + 5 \bullet 8^1 + 2 \bullet 8^0 + 7 \bullet 8^{-1}$ $= 426.875$

Table 1-3 gives the decimal, binary, octal and hexadecimal numbers up to $20_{\text{base 10}}$.

As can be observed from Table 1-3, a grouping of three binary digits can be represented by an octal number. Thus, to convert a binary number to its octal equivalent, break it up into groups of three digits starting at the base point and working in both directions. Leading zeros can be added or trailing zeros can be added after the base point to make a group of three.

Table 1-3 Binary, Octal, and Hexadecimal
Numbering Systems

Decimal Base = 10	Binary Base = 2	Octal Base = 8	Hexadecimal Base = 16
0	0	0	0
1	1	1	1
2	10	2	2
3	11	3	3
4	100	4	4
5	101	5	5
6	110	6	6
7	111	7	7
8	1000	10	8
9	1001	11	9
10	1010	12	A
11	1011	13	B
12	1100	14	C
13	1101	15	D
14	1110	16	E
15	1111	17	F
16	10000	20	10
17	10001	21	11
18	10010	22	12
19	10011	23	13
20	10100	24	14

Convert 10011.0111 to octal grouping.	**Example** **1-8** *Solution*

$$10|011.011|1$$
adding leading or trailing zeros
$$010|011.011|100$$
octal equivalent 2   3   . 3   4

Similarly, hexadecimal numbers can be converted to binary and vice-versa by breaking the binary number up into groups of four.

Convert 10011.0111 to hexadecimal grouping.	**Example** **1-9** *Solution*

$$1|0011.0111$$
hexadecimal equivalent 1   3   . 7

Conversion between octal and hexadecimal requires the use of binary as an intermediary.

Convert 372.3 octal to hexadecimal	**Example** **1-10** *Solution*

$$3 \quad 7 \quad 2 \, . \, 3$$

Step 1   372.3 octal = 011|111|010.011
Step 2   0|1111|1010.0110 = FA.6

When dealing with microprocessors or computers it is much easier to read the binary numbers off in octal or hexadecimal form than to use binary throughout. The PDP-11 minicomputer, for instance, uses octal notation, and the 8080 microprocessor uses hexadecimal.

1-4.2   Encoding of an Analog Signal into a Digital Signal

An analog signal can be converted to a digital signal by converting the signal level to a binary equipment. For example, a signal varying from 0 to 8 V can be converted to a binary number using Table 1-3. All voltages between 0 and 1 can be represented by 000, all voltages between 1 and 2 by 001, and so on to a maximum of 111. In turn, the binary digits can represent the presence of a pulse in the case of a 1 digit and the presence of no pulse in the case of a 0 digit. This digital technique is known as pulse code modulation (PCM). By this technique only the transmission of pulses or the absence of pulses takes place. The amplitudes of the pulses are not critical, since only their presence or absence needs to be noted.

In analog transmission, any noise on the system is amplified along with the signal, and the noise accumulates as it passes through many amplifiers, or repeaters. In digital transmission, however, each amplifier or regenerator first acknowledges whether a 1 or a 0 is received and then retransmits a cleaned-up pulse. Thus, any noise on a pulse is removed and the distortion does not accumulate. Of course, if the noise is excessive, an error can occur in the detection of a pulse, and the error will continue as the pulse train moves down the system.

1-4.3   Time Division Multiplexing (TDM)

As illustrated in Fig. 1-12, frequency division multiplexing assigns each of several signals to a narrow bandwidth slot and transmits all channels simultaneously. In time division multiplexing, each of the several signals is transmitted sequentially.

At any instant, only one signal is transmitted. As illustrated in Fig. 1-15, Channel 1 occupies the output channel during the first time slot. During the second, Channel 2 occupies the entire bandwidth of the output channel. During the thirteenth time slot, Channel 1 may be transmitted again, and so on. Sampling must be done at a rate high enough to prevent losing any significant information of the input signal. For an analog signal, each signal must be sampled at least twice as fast as its highest significant frequency component. As a result, the bandwidth requirements of the higher speed circuits can become quite large.

1-4.4   A Digital AM System

In a very simple digital modulation system the analog signal is initially sampled to generate a sequence of pulse amplitudes, as illustrated in Fig. 1-16. These amplitudes are then quantized, or represented by a level that most nearly corresponds with the actual level. These quantized levels are next encoded into a suitable digital signal which in turn amplitude modulates a carrier that can be

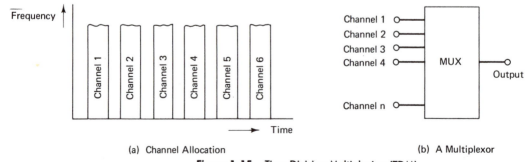

(a) Channel Allocation   (b) A Multiplexor

Figure 1-15   Time Division Multiplexing (TDM)

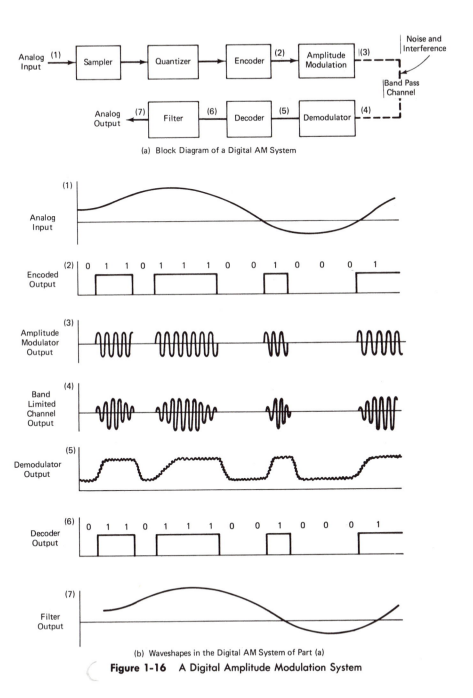

(a) Block Diagram of a Digital AM System

(b) Waveshapes in the Digital AM System of Part (a)

Figure 1-16 A Digital Amplitude Modulation System

transmitted over a bandlimited transmission path. At the receiver, the signal is demodulated and hopefully decoded to a good approximation of the original message. Poor filtering plus quantization noise, and interference entering the system can cause degradation of the original message.

As can be observed from the sketched waveshapes in Fig. 1-16(b), significant signal distortion can occur before errors are introduced. If errors are kept to a minimum, the overall performance of a system can be made virtually independent of the number of multisections or regenerators in the system. That is, if errors in a system are kept within reasonable limits over a given section, the overall performance of a link is almost independent of the distance between the terminals. Noise does not accumulate since the regenerators at the end of each section clean up the signal. This nonaccumulation of noise makes digital systems more immune to noise than analog systems and gives them a great advantage in long-distance transmission, which requires many repeaters.

1-4.5 Considerations in the Usage of Digital Communications

Digital communications have several distinct advantages over analog transmission systems, but there are definite disadvantages to this system as well. Because of the heavy investments in analog facilities, it will be many years before digital becomes the major technique in communication systems. In fact, if it were not for the many analog earth stations already in existence, satellite communications would probably now be almost wholly digital. The following lists provide a good summary of the advantages and disadvantages of digital communication.

Advantages of Digital Communications

1. Transmission quality is almost independent of the distance between terminals. The error rate is virtually unaffected by distance due to the regeneration, or cleaning up, of the signal at the regenerators. In analog transmission, the noise accumulates with the number of links between terminals. For a 24-channel FM carrier telephone system, the S/N ratio after five repeaters is reduced by about 3dB; after 10 repeaters it is reduced by 7dB.

2. A mixture of the traffic ranging from telephony and telegraphy to data and video information can be easily carried.

3. The capacity of certain existing transmission systems can be increased. For instance, the capacity of a single channel over a cable pair can be increased by employing digital multiplexing equipment whereby several signals are combined into a higher rate signal without any significant impairment. Similarly, the capacity of a satellite communication system can be increased by going to Time Division Multiple Access (TDMA). In either case, the intermodulation problems encountered in a frequency division multiplexed (FDM) system become non-existent, as only one channel in a digital system is being amplified at any particular time. Thus, the satellite amplifier can be heavily driven without

the danger of introducing intermodulation noise. In the case of FDM, all channels are simultaneously transmitted, each taking up a limited amount of bandwidth space. In time division multiplexing (TDM), only one channel is transmitted at any particular time, but it takes up the entire bandwidth of the system.

4. Digital cable systems are more economical for 15 to 40 km distances, particularly if cable capacity is nearly exhausted. Rather than laying new cable or providing new ductways, it is more economical to convert to digital on the old cable.

5. Digital communications lends itself to such novel facilities as cryptography, storage, and other forms of digital processing.

6. Digital communications is more suitable for the new types of transmission media such as light beams in optical fibers and circular or helical waveguides operating around 30 to 100 GHz. By providing a great deal of redundancy and employing special processing techniques, digital transmission also make it possible to communicate over long distances—from space vehicles, for example.

7. Digital signal characteristics are convenient for electronic switching in which groups of digits are selected to be switched in turn onto various highways. This is called packet switching.

Disadvantages of Digital Communications

1. One of the major disadvantages of a digital system is its large bandwidth requirement. An audio channel which may normally require an analog bandwidth of 4 kHz will require 64 kHz when transmitted by a PCM system. This can be reduced by signal processing and the employment of special modulation techniques but at the expense of increased costs and complexity.

2. Time division digital transmission is not compatible with the frequency division analog transmission in current use. Both cannot be carried simultaneously as can be done, for instance, with monaural and stereo FM. It is possible to transform analog into digital, and vice versa, by using converters. This is done on digital trunks that are presently in operation.

   Because of the basic incompatibility, however, it is very difficult to gradually change over from analog to digital transmission. Unless digital systems become highly economical or demand becomes high, it will be some time before an international satellite digital system becomes viable. Over 100 nations cannot be expected to replace all the earth stations, which cost about $20 million each, with more efficient digital equipment.

In the future, the flow of traffic in communication systems can be expected to become increasingly more digital. Because of the heavy capital investment in the present analog system, the existing telephone network facilities will continue to be used as the main transmission medium. This limits the signal spectrum to the audio frequency range of the telephone channel, or 200 to 3400 Hz. Future discussions in this text will be restricted to this constraint when considering practical applications of digital techniques.

1-5   Data Communications

Data communications is primarily concerned with the movement of data from one location to another. The term *data* refers to alphabetical, numerical, or special purpose characters which, when appropriately grouped, constitute some word or message. This data can be taken from, or delivered to, various types of equipment. It can be sent to some instrument, such as a temperature controller. It can come from a source document, such as an inventory or payroll or from a storage medium, such as paper or magnetic tape or punched cards. It can also come from some register within a computer. Pulses, or digital transmissions, are used in the transmission of such signals because most digital computers can directly handle them. By using communication facilities that can handle digital data, interconnected digital computers can directly process the data.

The devices most frequently encountered by a data user are the Central Processing Unit (CPU) and the terminal. The "CPU," "main frame," or "host" is capable of receiving input information, manipulating the input information based on a set of instructions called a "program," making decisions based on the result of some computation and making the results available to the user.

The terminal is a device which can input data or request the services of a CPU and can receive the output from a CPU. Some terminals function as input or output devices only.

Data communications is rapidly expanding as more and more businesses and agencies employ computer-based communication systems. Typical business operations that use data communications are

1. reservation services; for example, airlines, trains, and the like
2. mail service
3. inventory control
4. industrial control
5. computer time-sharing
6. automatic utility metering
7. packet switching

1-5.1   Communication Codes

To transfer or exchange information between machines, certain codes have been developed to represent graphic and control characters. Graphic characters are the "printable" alpha-numeric, punctuational, and other symbolic characters. Control characters provide format codes for business machines such as Back Space, Line Feed, Form Feed, Carriage Return, and communications codes to maintain the discipline in communication formatting, i.e., Start Of Header (SOH), End of text (ETX) CANcel etc.

Two of the most common codes sets are American Standard Code for Information Interchange (ASCII) and Extended Binary Coded Decimal Interchange Code (EBCDIC). These are outlined in Tables 10-1 and 10-2 respectively.

ASCII characters are seven bits in length with an optional eighth bit often added for parity or error checking. Seven bits allow for 2^7 or 128 codeable characters.

To transmit an ASCII character, a "1" or MARK could be represented by a positive voltage or current pulse and a "0" or SPACE by zero voltage or no-current condition. In this manner, characters can be transmitted from one point to another over a single pair of conductors.

1-5.2 Parallel and Serial Data Transmission

A code character can be transmitted by the current pulse in either parallel or serial fashion. In parallel data transmission, each code element is transmitted simultaneously. For a 5 bit code, five pairs of lines must interconnect the receiver and transmitter, as illustrated in Fig. 1-17.

In serial data transmission, each code element is sent sequentially as shown in Fig. 1-18. Only one pair of wire conductors is required for connecting the receiver to the transmitter.

For long distance transmission, the serial method is preferred because the line costs are more reasonable. However, serial transmission is slower than parallel transmission since the several code elements must be transmitted in sequential order instead of simultaneously. Parallel transmission is usually employed when computers are located in the same proximity.

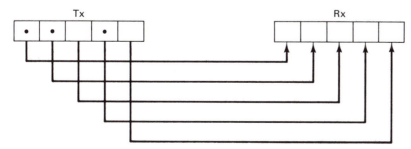

Figure 1-17 Parallel Data Transmission

Figure 1-18   Serial Data Transmission

1-5.3   Asynchronous and Synchronous Transmission

Serial data transmission can be broken down into two separate classes of transmission: synchronous and asynchronous.

In asynchronous serial data transmission, each character has its own synchronizing information to inform the detector of the first and last bit location of the character and therefore can be sent at varying rates. Each information character is synchronized by the use of stop-and-start elements, and the length of the time gap between characters is usually not fixed. Asynchronous transmission can be easily generated and detected by mechanical equipment. However, it is slower than synchronous transmission and is not as efficient in code transmission because of the added synchronizing stop-and-start elements.

As an example of asynchronous serial transmission, consider the transmission of the letter, or character J in the seven-bit ASCII format (refer to Fig. 1-19). When the character is transmitted, it is preceded by a start bit which is always a "space" (binary 0) and followed by an optional parity bit and one or more stop bits. The stop bit(s) is always a "mark" or a "1."

The receiver detects the start bit by noting the transition from a mark to a space, and then decodes the next seven bits as a character. If more characters are presented, the process is repeated. Both the receiver and transmitter provide their own clocks, but they must be running at close to the same rate. Asynchronous transmission allows for a variable interval between the transmitted characters, as occurs for instance with a teletype machine that transmits a character every time an operator depresses a key. Many devices, however, buffer their messages in preparation for transmission. Upon transmission a steady stream of characters are sent.

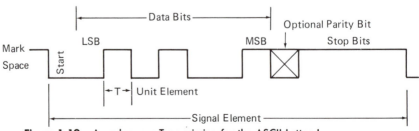

Figure 1-19   Asynchronous Transmission for the ASCII Letter J.

A block diagram of a typical asynchronous serial receiver is shown in Fig. 1-20. The purpose of the receiver logic is to assure that incoming pulses are sampled at or near each pulse center. This sampling tends to minimize any errors due to pulse distortion.

The clock, which is generally a crystal oscillator, is divided down to provide pulses at 8 times the input pulse rate. By initially setting the divide-by-8 counter to 4 (100), the overflow count, 8 (000), can be used to strobe the incoming pulses into the shift register character buffer. At the end of a character, the start pulse will overflow and set the character flag. This flag can be used to reinitiate the system. At the beginning of a character, the incoming start pulse sets off the code detector flip-flop, which starts the clock.

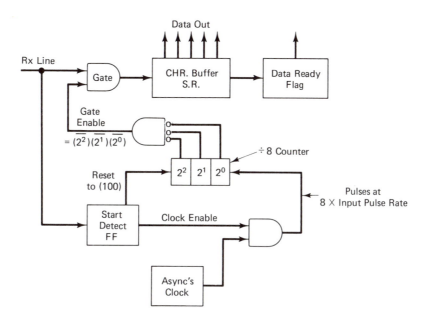

(a) Block Diagram of an Asynchronous Receiver

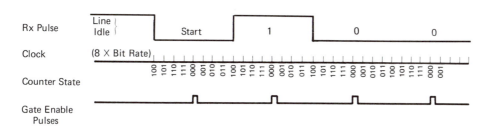

(b) Generation of Gate Enable Pulses

Figure 1-20   A Serial Asynchronous Receiver

In the above system, the "end-of-code" is detected by *counting* incoming pulses. This can result in the Rx falling out of step with the Tx if an extra bit, such as noise, gets into the system. In such a case, all the following information would be erroneous. The system should therefore include some pulse-width-testing logic which can separate the extra stop bits from the others and assure re-synchronization at the end of each signal element.

Synchronous transmission is not made up of start-and-stop pulses, but uses a common clock pulse at the transmitting and receiving end (and a few framing pulses) to achieve synchronization and character pulse identification. The receiver can recognize a unique code in the received bit stream which allows it to lock in on an incoming bit stream and to identify the characters. The receiving device is thereby clocked at exactly the same rate as the transmitter. This clock synchronization is referred to as "bit synchronization."

With synchronous transmission, synchronization is dealt with on a message basis rather than on a character basis. It does not allow for an interval between characters and thus is limited to devices that have the ability to buffer messages.

One binary synchronous protocol is the character-oriented IBM's Binary Synchronous Data Transmission or BISYNC, shown in Fig. 1-21. Each of the blocks represents an eight-bit EBCDIC character, the Hex code for the various characters are written above the EBCDIC characters. The PAD character, which is just an alternating sequence of 1's and 0's, allows the clocks to regain synchronism. The SYN characters provide the specialized bit pattern for the receiver to sense when an eight-bit character has been captured. The receiving device can then begin to decode the special control characters and the messages.

Synchronous operation can be characterized by the following:

a. There are no start-stop bits to synchronize each character.

b. Information is sent in an entire block, which can consist of many signal elements with no separations.

c. The entire block is framed by unique codes which indicate to the Rx the beginning, end, and so forth of blocks of information.

d. The Rx must know the code length and be able to recognize the unique codes which are used to control the RX.

e. Every bit in the Tx and the Rx must be synchronized to a common clock.

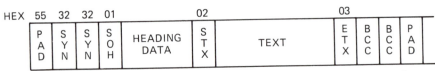

Figure 1-21 Typical BISYNC (Synchronous) Format (a) with Brief Description of Control Characters (b).

Table 1-4   Advantages and Disadvantages of Synchronous and Asynchronous Serial Data Transmission

Asynchronous	Synchronous
1. Easily generated mechanically	1. High efficiency for long messages as start-stop bits are not required (block framing is required)
2. Can easily drive mechanical equipment	
3. Suitable for manual operation, such as pecking on a typewriter	2. Not as sensitive to distortion because timing is included along with data
4. Requires two tracking clocks	
5. Sensitive to distortion	3. Bit loss can cause whole message to be erroneous
6. Inefficient due to start-stop bits for long messages	4. Special codes must be detected for timing and logic control
7. Speed limited due to above and due to the fact that pulse-width margin must be allowed to accommodate distortions and timing error	5. Not suitable for mechanical equipment
	6. Tends to be costly in terms of equipment because of the more complex circuit required

The advantages and disadvantages of serial asynchronous and synchronous transmission are shown in Table 1-4.

1-5.4 USART, 20mA Loop and Modem

To connect a digital computer with a parallel *n*-bit output (where *n* is typically a multiple of 8) to a single communications line, a Universal Synchronous Asynchronous Receiver Transmitter (USART) is employed. Often, a UART is the only available option. This device either transforms the parallel *n*-bits into a serial bit stream or converts the serial bit stream into *n*-parallel bits. (The USART is often considered part and parcel of a computer facility, as shown by the dotted enclosure in Fig. 1-22.) Some of the options available on a commercial LSI USART will be considered in a later chapter.

In order to directly transmit serial data over a limited distance of up to several hundred meters, sources capable of delivering around twenty milliamperes of current are traditionally used. This is known as a 20 mA loop. Many of the common teletypewriters, for instance, switch the 20 mA currents by sweeping a carbon brush over commutator bars. If at least 18 mA of current is present, the contacting surfaces are kept reasonably clean. Fig. 1-22 illustrates, in block diagram form, the use of a 20 mA loop when connecting a UART to a teletype (TTY).

For much longer distances, when the telephone system could be most conveniently employed, the modem is used to transform the binary bit streams into a signal which conforms to the telephone voice bandwidth. Modem is a contraction of the terms modulator and demodulator. A modem thus permits

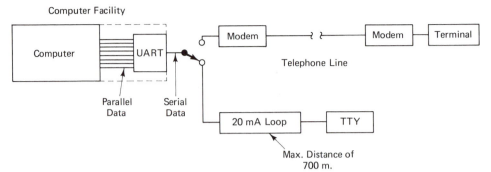

Figure 1-22 Transmission of Data to and from a Computer to a Teletypewriter or over a Telephone Line

digital signals to be sent over the telephone analog communications facilities. Modems are frequently referred to as data sets by telephone companies. The interface between the modem and the computer/terminal facility is internationally standardized. When connecting a modem to data processing equipment in North America, the EIA standard RS-232-C is followed. This conforms to the CCITT's Recommendation V.24, concerning definitions of the interchange circuits. These definitions will be discussed in much more detail in a later chapter.

1-5.5 Simplex, Half-Duplex and Full-Duplex Channels

A *channel* may be defined as a single path on a line through which the electrical signal flows, where *line* is defined as the component part of the system extending between the stations. Transmission on a channel can be characterized in one of three ways: simplex, half-duplex, and full-duplex. In North America, the terms have the following meanings (also refer to Fig. 1-23).

Simplex operation indicates transmission in one direction only. In this mode, a terminal can transmit but not receive, or receive and not transmit. A two-wire dial-up or private line can be used.

Half-duplex operation indicates transmission in either direction, but only in one direction at a time. Modems, for instance, may switch from transmit to receive and vice versa. Two- or four-wire circuits are required, with the latter used like two simplex circuits. On a 2 W circuit, the line must be turned around to reverse the direction of transmission. The 4 W circuit eliminates the line turn-around delay.

Full-duplex operation indicates transmission in both directions simultaneously. It is encountered less frequently than half-duplex. A 4 W circuit is normally required, although by splitting the frequency spectrum, a 2 W circuit can be used.

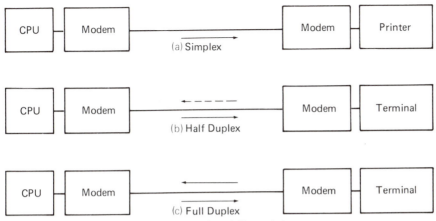

Figure 1-23 Modes of Channel Operation

Some modems feature a reverse Channel that can be used on a 2 W circuit. In this application a narrow band of frequencies is reserved for transmission of an acknowledgment signal, indicating to the transmitter whether the signal was received error-free or not. The major portion of the frequency spectrum is still used for the main data flow. This is still considered half-duplex operation and is a method of immediately indicating an error to the transmitter without turning the channel around. On full-duplex channels, echoplex can be used to echo the character back from the far end so that it can appear on a printer or display located near the operator's keyboard. Echoplex cannot be used with half-duplex channel as immediate transmission is not available in the reverse direction.

1-5.6 Line Structure

In interconnecting terminals to computers or to other terminals, the communication link falls into either the point-to-point or the multipoint categories. In the point-to-point structure of Fig. 1-24, the communication line connects only two points using either a private line or the switched network. In this configuration one of the machines is usually assigned "primary" status, giving it to the authority to transmit first. The other machine has "secondary status," permitting it to transmit only at the request or permission of the primary station. If transmission is FDX, both machines can enjoy equal status, as both may transmit at the same time.

In the multipoint configuration (see Fig. 1-25) two or more terminals/computers share a common 2 W or 4 W private line. Although two or more terminals may receive a message at the same time, only one terminal at a time can transmit data. As in the case of the point-to-point discipline, one station is designated as the "primary," with all the others designated to "secondary" status. Contention, a term used to describe the possibility if more than one

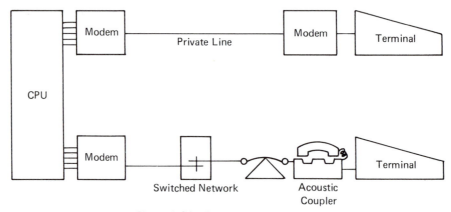

Figure 1-24 Point-to-point Network

machine attempts to take control or transmit at the same time, is avoided by employing a technique known as poll and select. Transmission can occur in either the FDX or HDX mode.

Each station on the line is allocated a unique address. When the primary station polls a terminal, it asks each terminal in a predefined sequence if the terminal has data to transmit. Each secondary terminal receives the poll message and examines the identification address. One station will recognize the proper identification and must then respond to the poll. If the terminal has no data, it so informs the primary station, and the poll sequence continues to the next secondary. If the terminal has data to send, it sends a positive response to the primary station. The secondary may either begin transmitting data at this time, or wait until further instructions are received from the primary station, as

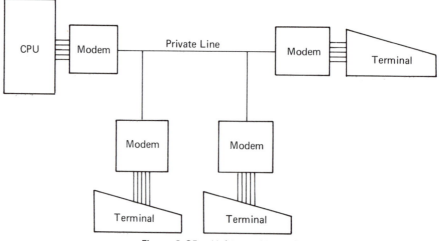

Figure 1-25 Multipoint Network

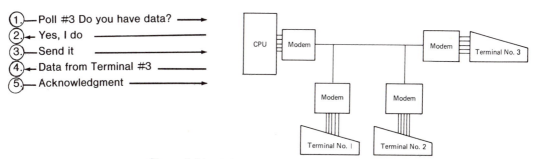

Figure 1-26 Polling

illustrated in Fig. 1-26. Upon receipt of data from the secondary, the primary responds with an acknowledgement that the data was correctly or incorrectly received. Upon the correct receipt of data, the poll sequence can resume.

In addition to polling, selection can be used. In this case, the primary selects a particular secondary (terminal) address to which data is to be transferred and requests permission to send this data to that terminal. The terminal identified by the select address responds with a positive or negative reply. Upon the receipt of a positive acknowledgement, the primary begins to transmit data to the secondary station. If the reply from the secondary was negative, the primary may, at a later time, attempt to select the second again.

After the secondary has correctly received the data, it so informs the primary which is now free to resume the selection of another secondary or convert to polling. Fig. 1-27 illustrates a typical select sequence.

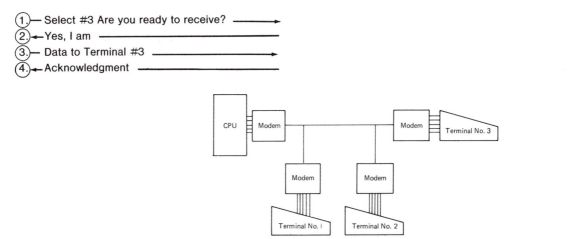

Figure 1-27 Selection

In 4 W multidrop networks, a four-wire bridge such as that shown in Fig. 1-28 is inserted into the lines. A separate amplifier is used for each direction of transmission. The number of 4 W drops is not limited by the feed through coupling problems experienced in 2 W bridges, but only on system factors such as data reliability and throughput.

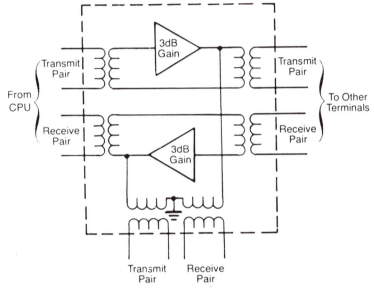

Figure 1-28 A Four-Wire Bridge (Courtesy Racal-Vadic)

Problems

1. a. State three purposes for modulation.
   b. Describe the three basic forms of carrier modulation; that is, AM, FM and PM.

2. Calculate the physical length of a half wavelength antenna ($\lambda/2$ dipole) at frequencies of 1 MHz, 100 MHz and 10 GHz.

3. a. Calculate the power and voltage gains of a +3 dB and a +10 dB amplifier.
   b. If the power level at the input of a 13 dB attenuator is 2 mW, what is the power level at the attenuator output? Convert the input and output power levels to dBm.

4. Individually, two noise sources produce 6 dBm and 3 dBm of noise. What will the accumulative noise level be in dBm?

5. a. Describe the frequency characteristics of human speech.
   b. Give the rationale for employing weighting networks when monitoring noise levels.

6. a. What is the level in dBm of a 1000 Hz 0 dBrnc tone? Of a 90 dBrnc tone?

   b. What is the noise level in dBrnc of a 0 dBm, 300-3400 kHz white noise signal?

7. a. What is the zero transmission level point in a system?

   b. What is the definition of dBm0?

   c. What is the level in dBm0, of a + 35 dBm tone at 18 dB relative transmission level point?

   d. What is the definition of dBm0p?

8. Fill in the blanks for the circuit shown.

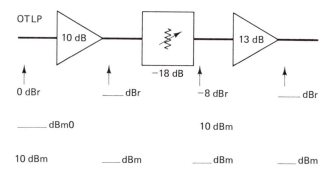

9. a. Complete the following table:

$$dBm = 10 \log \frac{P(\text{watt})}{10^{-3}}$$

$$dBrn = 10 \log \frac{P(\text{watt})}{10^{-12}}$$

Power (watts)	dBm	dBrn
10^1	40	130
10^0		
10^{-3}		
10^{-6}		
10^{-12}		

   b. If the power represents white noise levels over a 300 to 3.4 kHz bandwidth, what are the dBrnc levels in (a)?

10. If the output level of a repeater is at −5 dBr, what is the noise level at the zero transmission level point if the noise from the system is −60 dBm?

11. a. Convert 50,000 pWOp to dBm0p.

   b. Convert 2000 pW0p to dBm0P.

12. A C-message noise reading of 25dBrnc is obtained at the input to a modem which is at a −3 dBr or −3 TLP level. What is the noise level at the 0 TLP?

13. a. State the Sampling Theorem.
    b. What is the minimum sampling rate required to retain all the information within the baseband of 300 to 3.3 kHz?

14. a. Why are international standards in the telecommunications industry desirable? Are there undesirable aspects as well? What are they?
    a. Why are international standards in the telecommunications industry desirable? Are there undesirable aspects as well? What are they?
    b. What are the functions of the CCITT and EIA organizations? What do these mnemonics represent?

15. a. Describe the bandwidth employed by the following modulation schemes:
    1) AM
    2) SSB
    3) VSB
    4) FM
    b. Describe the difference between TDM and FDM.

16. What particular frequencies are not weighted in the psophometric and C-message weighting networks?

17. a. What is a bit?
    b. Convert 1101.101 to its decimal equivalent.
    c. Convert 63 octal to its binary equivalent.
    d. The flag bits in the High Level Data Link Control (HDLC) are given by 01111110. Convert this to hexadecimal.

18. Explain why overall system performance in a properly designed digital system is almost independent of the number of regenerator stations.

19. Draw a simple block diagram of a digital SSB modulation system.

20. List the advantages and disadvantages of converting an analog communications system to a digital communications system.

21. a. What is data communications?
    b. How many unique combinations can a seven-element code have?

22. a. Compare the speed of parallel and serial transmission.
    b. Based on the answer to (a), why is serial transmission employed?

23. Describe the difference between synchronous and asynchronous transmission.

24. a. What is the function of a modem?
    b. What is the function of a USART?
    c. What is a 20 mA loop?

25. Describe simplex, half-duplex, and full-duplex transmission.

26. Describe the purpose of a hybrid or terminating set.

27. Sketch the asynchronous transmission bit pattern for the ASCII letter F. Assume one stop bit and even parity.

28. Sketch a typical BISYNC format.

29. Describe the difference between "polling" and "selection."

30. Frequency weighting of noise is used in telephony to
   a. Mask the intelligence of the received signal
   b. Determine the frequency response of an "average" ear.
   c. Take into account the relative interfering effect of noise on the human ear
   d. Attenuate the interfering effects of echo on the human ear

2

The Telephone Network*

2-1   Introduction

From the first significant long-distance telephone call placed between Brantford, Ontario and Paris, Ontario in 1876, to the present, the telephone network has evolved to carry electrical signals which represent the human voice. Until recent decades, the network has had to carry only relatively low frequency signals corresponding to those audible frequencies generated by the human vocal cords. New requirements for transmitting data at very high speeds has necessitated upgrading the frequency response of the existing telephone network and installing special data transmission networks to handle the greater transmission speeds needed.

This chapter explores the transmission of the human voice and the North American telephone network.

2-2   The dc Signal Path

The most efficient means of communication between individual humans is talking—that is, causing air to flow across the vocal cords, making them vibrate at various frequencies. The vibration of the vocal cords causes air molecules to vibrate at frequencies from approximately 100 Hz to 4000 Hz which in turn, the human ear perceives as sound.

Sound energy is limited to the distance it may travel, so to converse over great distances we must be able to convert sound energy into another form of energy which will travel long distances quickly.

Any device which converts one form of energy to another is called a *transducer*. To convert sound energy into electrical energy, sound must cause a variation directly in either resistance, voltage, or current in an electrical circuit.

* This chapter was revised by Ken Clattenburg, Coordinator of the Telephone Systems Program at Sir Sandford Fleming College in Peterborough, Ontario.

The device used in a telephone is called a telephone *transmitter* and accomplishes the conversion by directly varying resistance in proportion to the amplitude of the sound wave.

Referring to Fig. 2-1, sound energy striking a flexible diaphragm causes carbon particles to be alternately compressed and expanded. When compressed, more surface area of each particle is forced into contact with the surfaces of adjacent particles. If there is a larger cross-sectional area through which electrons may pass, the effective resistance of the transmitter is reduced.

When the carbon particles are allowed to expand, less surface area of each particle comes into contact with its neighbor, thereby increasing the effective resistance of the transmitter.

The varying resistance of the transmitter causes the current in the circuit to vary inversely with the change in resistance. Therefore we have electrical energy which varies in response to applied sound energy.

To complete a basic one-way communication system, it is necessary to have a transducer which will convert electrical energy back to sound energy, so that the human ear and brain may decode the information.

The transducer must be able to cause air molecules to vibrate; it must create a physical motion. The telephone receiver utilizes the phenomenon of magnetism which can cause the motion of a magnetic material placed in close proximity to a magnetic field.

Current through a conductor creates a magnetic field which varies in direct proportion to the amount of current applied. If the conductor is formed into a coil of many turns, the strength of the field is increased significantly. Fig. 2-2 depicts a basic one-way voice transmission system in which sound energy applied to the transmitter causes the resistance of the transmitter to vary, re-

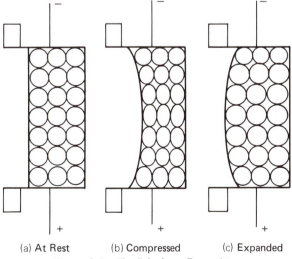

(a) At Rest   (b) Compressed   (c) Expanded

Figure 2-1   The Telephone Transmitter

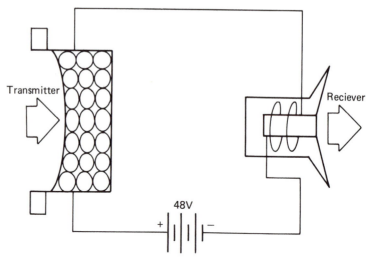

Figure 2-2 One-way Transmitter

sulting in a fluctuating DC current. Current applied to the coil causes a magnetic field which attempts to draw the magnet toward the center of the coil. The natural resilience of the diaphragm causes it to return to rest position when the magnetic field weakens. The rapid movement of the diaphragm causes air molecules to vibrate, creating a reproduction of the sound applied to the transmitter. The coil, magnet, and diaphragm form the basic components of the telephone receiver, which functions much in the same way as the speakers of a home stereo system.

However, for economic reasons, telephone transmitters and receivers are constructed to respond to a relatively narrow frequency band of 300 to 3400 Hz. Most information in the human voice is contained in this band, but some high and low frequencies are eliminated during transmission. This sometimes results in the listener's failure to recognize a caller's voice because certain qualities of that voice are not reproduced.

The power source used to provide the talk battery is nominally −48 Vdc. A minus potential with reference to ground tends to reduce oxidation of the copper conductors of the telephone line and the 48-volt potential is considered the highest DC level that can be handled safely.

2-3   The ac Signal Path

In order to have a practical communication system, a means of signaling the receiving end must be provided. In telephony, an AC voltage of approximately 90 Vac at 30 Hz is used to drive an electromagnetic device, the ringer or bell, as shown in Fig. 2-3. AC current through the coils alternately causes the clapper

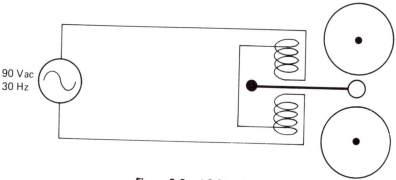

Figure 2-3 AC Signal Path

to be drawn toward one gong and then the other. Each gong is made of a different metal; this creates the familiar ringing sound of the standard telephone.

2-4 Battery Systems

Only one pair of wires is used to interconnect two normal telephones. Therefore, DC talk battery and AC signaling must use the same pair. DC battery and AC generator may be provided by the telephone set as in a local battery system or by the central office as in a common battery system.

2-4.1 Local Battery Systems

The term *Local Battery* refers to a type of telephone system used widely in the first decades of this century. Fig. 2-4 depicts one model in use at the time. Fig. 2-5 is a simplified schematic of a local battery system.

When both telephones are in the idle condition, the ringers of both sets and the magnetos are connected to the line. If set A wishes to call set B, the crank on the magneto is turned rapidly to generate an AC signal, which causes the ringers in both sets to ring. The caller at set A removes the handset at their set and thus places their receiver and transmitter in the circuit. The induction coil in each set isolates each battery from the other, yet allows both parties to hear themselves and the other party. When the called party lifts the handset, the caller stops cranking and contact is established.

One obvious drawback of this system is the costs associated with maintaining the batteries and the possibility of not being able to transmit because of low batteries.

2-4.2 Common Battery System

The solution to the inherent problems of local battery systems is to construct a centralized power source or common battery installation which can easily be maintained. Fig. 2-6 illustrates how this is accomplished.

Figure 2-4 The Blake Magneto Wall Telephone (Copyright Bell Canada, Courtesy Bell Canada Telephone Historical Collection)

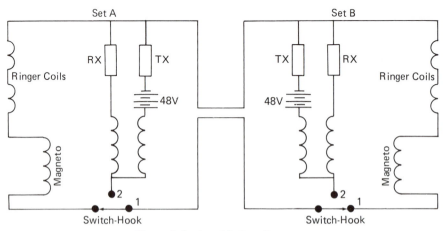

Figure 2-5 Local Battery System

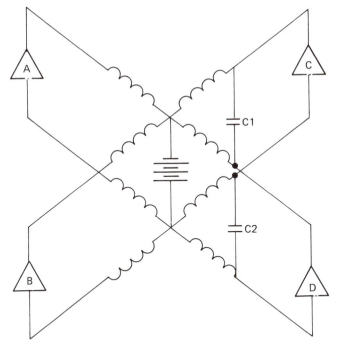

Figure 2-6   Common Battery System

The coils in series with each set perform two functions. They are designed to inhibit voice frequencies from spilling over into other conversations through the battery and are actually relay coils which detect *off-hook* conditions. Additional equipment provides contacts to establish connection between two parties of a conversation. Connection is represented by capacitors C1 and C2 which allow voice frequencies to pass but block DC talk battery.

2-5 Instruments

The telephone set is by far the most familiar appliance in the transmission path. It is a moderately complex piece of equipment and must perform the following functions:

1. Inform the subscriber of an incoming call by ringing.
2. Inform the local office that a call has been originated, answered, or completed.
3. Relay to the local office the number that has been called.
4. Provide the proper amount of sidetone (the sound in the speaker's ear due to his or her own voice) to the receiver. A slight amount of sidetone is desirable since without it the telephone sounds dead. Too much

sidetone causes the speaker to lower the voice, produces ear fatigue, and reduces the intelligibility of the signal due to the reduced signal-to-noise ratio.

The telephone is most commonly connected to the end office through a single pair of wires called the subscriber's loop. Current is supplied to the telephone over the loop pair from the common battery source.

2-5.1 The 500-Type Telephone

The schematic diagram of a model 500-type telephone set (Fig. 2-7) will permit a greater appreciation of the operation of a telephone. The 500-type telephone has been an industry standard in North America for over 30 years and although gradually being replaced by new technology, is still the instrument most people have in their homes.

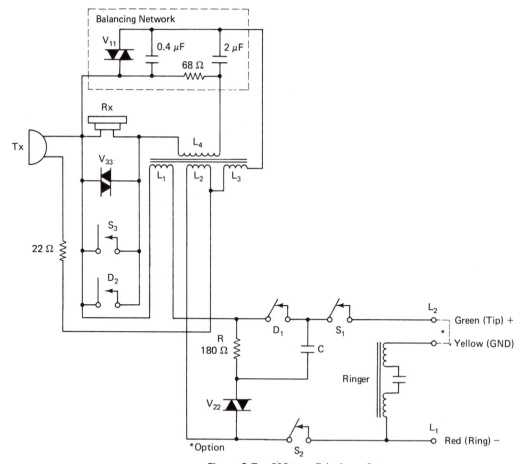

Figure 2-7   500-type Telephone Set

The telephone transmitter (TX) picks up sound and converts it to electrical signals while the telephone receiver (RX) picks up electrical signals and converts them to sound energy, as discussed in Section 2-2.

The varistors V11 and V22 in Fig. 2-7 are non-linear silicon carbide devices whose resistance decreases with increase in current. This tends to provide automatic regulation of the signal. Because loop lengths between the end office and each subscriber may differ, the transmission loss will also vary. To maintain almost constant transmit and receive levels, these varistors tend to reduce the signal levels on short loop lengths by producing a larger shunt loss at the telephone terminals. When the loops are long, the lower line current will cause the varistors to have greater resistances and therefore less shunt loss. The impedance of this type of telephone normally varies between 600 ohms and 900 ohms, depending on the dc line current. The varistor V33 is inserted to suppress dial clicks in the receiver.

In order to keep sidetone to a reasonable level, (or to keep the signal in the receiver while transmitting), a hybrid transformer arrangement is used. It has the ability to divert power from the transmitter to the line, and at the same time to receive power from the line which drives the receiver. The hybrid consists of the inductor coils L1 to L4 and the balancing network. While transmitting, the voltage induced in L4 cancels most of the voltage across the balancing network. The small voltage that remains at the receiver is used as sidetone. Capacitor C and resistor R suppress the high frequency interference components of the dial pulses to nearby radio receivers.

Usually, the L2 lead is at dc ground and the L1 lead is at a nominal −48 Vdc. The polarity of a telephone line has traditionally been identified by the designations of Tip for the ground side of the line and Ring for the battery or −48 Vdc side of the line. This convention originates from the telephone plug and jack used in the manual switchboard as illustrated in Fig. 2-8.

By applying a 20-Hz, 85-to-100 Vac signal to the line, the telephone rings. The capacitor in series with the ringer coils allows the flow of the 20-Hz signal through the coils but blocks dc current, preventing the ringer from shorting the line and causing false off-hook conditions.

Going off-hook (lifting the receiver from its cradle) causes switches S1 and S2 to close and S3 to open. The normal open circuit line voltage of 48 volts drops to about 5 volts due to the line impedance experienced by the dc current now flowing. The telephone is now ready to receive or transmit a message (the transmitter now experiences a dc bias) or to dial a number.

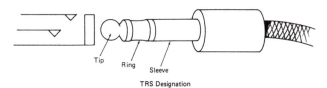

TRS Designation

Figure 2-8   TRS Designation

With a standard rotary type dial, the action of selecting a digit and rotating the finger wheel to the finger stop winds a spring within the dial. At this point, no signal has been sent but a normally open contact within the dial (D2) closes to short the receiver, thus reducing the clicking noise that the dial pulses will create in the receiver.

Releasing the finger wheel allows it to return to rest position during which time a cam within the dial causes normally closed contact (D1) to open as many times as the digit dialed. In other words, dialing the digit 4 causes contact D1 to open four times. The pulsing rate is 10 Hz. When the finger wheel reaches rest position, contact D1 is closed and contact D2 opens until the next digit is dialed. A dial pulse train is shown in Fig. 2-9.

The time delay between digits, called the interdigit time, must be of sufficient duration to allow the end office equipment to determine that the digit is completed. In the Step by Step (SXS) office, where selecting electromechanical switches are directly and sequentially controlled by the dialed digits, about 600 ms of interdigit time is required. In common control offices, where the digits are stored in a register for a brief period of time, the minimum interdigit time is cut in half to about 300 ms.

The bridged capacitor circuit at the central office shown in Fig. 2-10 (a) supplies pattern current to the subscriber loop through the high impedance relay windings which prevent grounding of the voice frequencies and prevent the voice signals from passing to other subscriber lines through the common dc source. The capacitor C permits voice signals and ringing generator to pass through the network while blocking the dc current. Repeating coils (Fig. 2-10 (b)) can also be employed to perform the same functions. These use magnetically coupled transformers to pass the voice and ringing signals through the network and also provide dc to the subscriber loop. The ballast resistors limit the current flow in short loops. Capacitor C bypasses the voice signal around the relay. The relay or dial pulse detector in the central office repeats the signaling or dialing pulses and passes them on to the switching machine.

The break intervals of dialed pulses exceed the make intervals because of the initial charging of the bridged ringing capacitors (C) in the central office

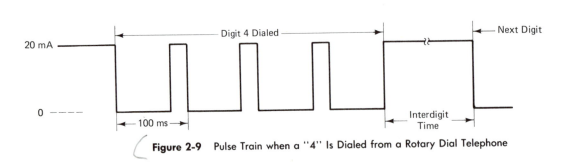

Figure 2-9 Pulse Train when a "4" Is Dialed from a Rotary Dial Telephone

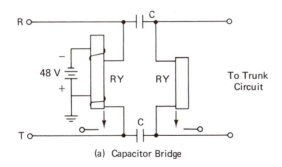

(a) Capacitor Bridge

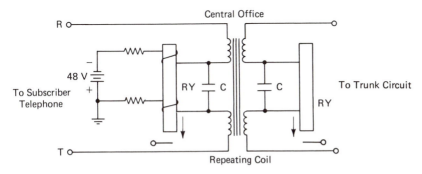

(b) A Central Office Repeating Coil

Figure 2-10

when the dialed contacts open. The charging circuit causes a slight delay in the pulsing relay in the central office. The break-to-make ratio is in the order of 60/40.

2-5.2 DTMF Signaling

As with the Step by Step central office, which allows the user to control the speed at which the connection is made, the rotary dial makes inefficient use of common equipment in a central office, in that equipment is tied up for the entire dialing sequence. In an office processing high call volumes, this can increase equipment level requirements significantly.

The Dual Tone Multi-Frequency (DTMF) dial, commonly known as Touch Tone, greatly reduces dialing time as well as providing more convenient dialing for the user.

DTMF signaling consists of sending a pair of frequencies concurrently to the central office. One high and one low frequency is generated by the dial and the combination is decoded by the central office as a particular digit. DTMF signaling is about ten times faster than rotary or dial pulsing and is less prone to errors since there is little distortion of voice frequencies even on the longer subscriber lines. The high resistance and capacitance of the longer local loops severely distort the dc pulsing.

2-5.3   The Touch Tone Telephone

The Touch Tone (or DTMF) telephone differs from a dial pulse set only in that a Touch Tone dial is used instead of a rotary one. The Touch Tone dial consists of an oscillator connected through a 4-by-4 matrix to an inductor or coil. As shown in Fig. 2-11, the pressing of a button corresponding to the digit desired causes the selection of coils which generate two particular frequencies. For example, pressing the digit 5 causes 770 Hz and 1336 Hz to be transmitted to the central office. Pressing 6 causes 770 Hz and 1477 Hz to be transmitted.

Fig. 2-12 shows a Touch Tone telephone set. Notice that power for the oscillator in the dial is drawn from the −48 Vdc of the telephone line. A diode bridge prevents a line reversal from causing improper operation. Earlier models experienced the symptom where dial tone could be heard but the subscriber could not call out. A reversal of the line caused the wrong polarity to be applied to the oscillator.

Seizure and line disconnect still remain under the control of the hookswitch. Damping of the tones is accomplished in the same manner as in the rotary dial, in that a normally open contact closes when a button is pushed, shorting out the receiver. Thus a subscriber complaint of not being able to hear could be caused by a defective dial.

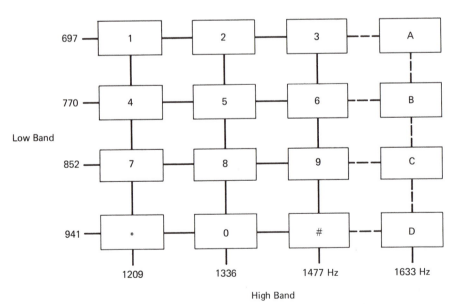

Figure 2-11   Touch-tone Dialing Frequency Assignments. The Dashed Tones are reserved for Special Purposes such as Military and Security

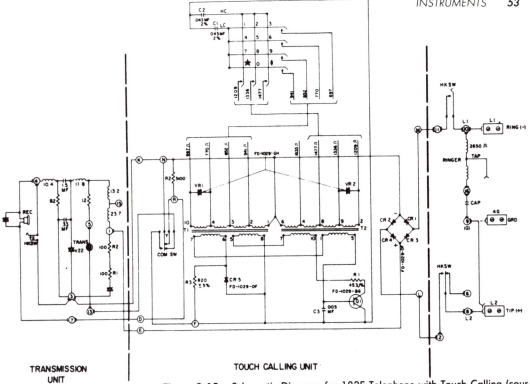

Figure 2-12   Schematic Diagram for 182E Telephone with Touch Calling (courtesy GTE Automatic Electric Incorporated).

2-5.4   The Touch Tone Decoder

Touch Tone Decoding may be accomplished by circuits integral to the central office or by stand-alone units. Fig. 2-13 depicts a decoder which utilizes tuned circuits to detect the presence of the respective frequencies. The presence of 770 Hz and 1336 Hz will cause an output on gate 5 only. This output will be interpreted by the C.O. as digit 5. The other gates will respond only to the particular two frequencies assigned to their respective digits.

2-5.5   The Electronic Telephone

The 500-type telephone is a hardy and dependable device but the number of separate parts and the time to construct it has made it increasingly expensive to manufacture. Advances in microchip technology have enabled the development of a new generation of telephone instruments containing fewer separate parts and simpler to manufacture. Telephones are now available in shapes ranging from cartoon characters to cigar boxes. Fig. 2-14 depicts one such example of the newer models available.

Integrated circuits have also enabled the creation of a third type of dial which allows subscribers to have push-button dialing even if the local office is not equipped with DTMF receivers or the subscriber does not wish to pay the additional charge for Touch Tone. Known as Digi-Pulse, Digit Pulse, Touch

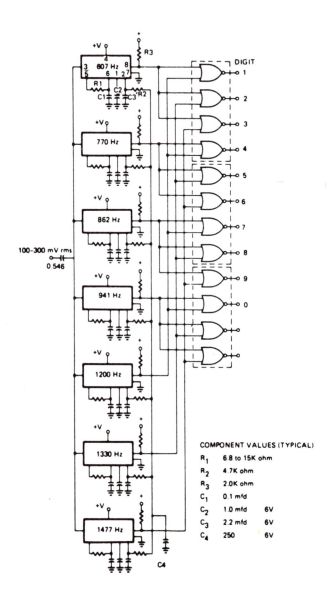

COMPONENT VALUES (TYPICAL)

R_1	6.8 to 15K ohm	
R_2	4.7K ohm	
R_3	2.0K ohm	
C_1	0.1 mfd	
C_2	1.0 mfd	6V
C_3	2.2 mfd	6V
C_4	250	6V

Figure 2-13 Touch-Tone Decoder (Courtesy of EXAR Integrated Systems, Inc.)

Figure 2-14 The Harmony Telephone, manufactured by Northern Telecom Canada Ltd. 1984 (copyright Bell Canada, courtesy Bell Canada Telephone Historical Collection).

Pulse, and other names, the basic concept is that the pressing of a button causes the respective number to be stored temporarily in a chip within the set which then causes dial pulses similar to those generated by a rotary dial to be transmitted to the local office. These dials are not much faster than a rotary dial since the digit pulse train is still of the same duration. A user can identify Digit Pulse dials because after all digits have been dialed, the dial pulses can be heard still transmitting out over the line.

However, the act of dialing is simplified and the dial pad is similiar to a Touch Tone pad, allowing the design of more compact and decorative telephone sets.

2-6   Concepts of Centralized Switching

Connecting several telephones could conceivably involve directly linking every telephone to each of the others. No switching would be required, and there would be no blocking of calls because of a busy line. The disadvantage of such an interconnecting scheme would be the extraordinarily large number of lines required for the number of telephones in existence.

For example, Fig. 2-15 shows that 10 lines would be required for the direct interconnection of 5 telephones, and 15 lines would be required to connect 6 telephones. In general, for N telephones, $N + N - 1 + N - 2 \ldots + 1 = N (N-1) /2$ lines are required. In other words, for a typical community with 10,000 telephones, nearly 50 million lines would be required. These numbers become all the more unrealistic when considering the direct interconnection of all the telephones in the world.

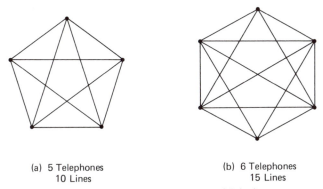

(a) 5 Telephones
10 Lines

(b) 6 Telephones
15 Lines

Figure 2-15   Direct Interconnection of Telephones

2-6.1   The Manual Central Office

To establish a temporary connection between two telephones, a means of switching between telephone sets must be provided. The first centralized switching was accomplished by terminating each line at a *switchboard* such as the one shown in Fig. 2-16. The telephone operator was signaled by the caller originally by means of cranking the magneto in the subscriber's set, which acti-

Figure 2-16   No. 1 Standard Magneto switchboard, introduced by Bell Canada in the 1880s (Courtesy Bell Canada Telephone Historical Collection

vated a supervisory light or flag corresponding to the caller's line and an audible signal such as a bell or buzzer. The operator then inserted the rear cord of the cord circuit providing talk battery to the caller and inquired what party the caller wished to contact. If the caller wished to reach another party connected to the same switchboard, then the operator simply placed the front cord of the cord circuit in the jack corresponding to the called party and caused the bell of the called party to ring by throwing a ringing key on the cord circuit. The availability of the called party's line was obvious by the lack of another cord in the jack. The operator monitored the line occasionally to determine when the call was terminated. Later enhancements to the switchboard included elimination of the need for magnetos in the sets and automatic dialing of the called party by the operator.

The need to contact parties in centers other than a local community led to the connection, by means of *tie trunks,* of two or more switchboards in a particular geographic area. Callers wishing to reach someone in another community would be connected to a jack terminating a tie trunk to that community, and completion of the call would be handled by an operator at the distant end. Callers were usually assessed an additional charge or *toll* for calls of this nature. Fig. 2-17 shows the evolution of switching networks, the dotted lines representing tie trunks that were added as demand grew for service between *central offices* or exchanges.

2-6.2 The Hierarchy of Exchanges

The natural evolution of the telephone network led to a situation similar to what would occur if all telephones were directly interconnected to each other. The number of tie trunks would become unmanageable. The solution was to configure the network so that local exchanges are connected through another switching center, called a Toll Center, and to interconnect the Toll Centers through other switching centers, and so on.

First, consider a very elementary but typical system with only two levels of switching: an End Office (EO), or local Central office (CO), and a Toll Center (TC). As illustrated in Fig. 2-17, a call on such a switching system could

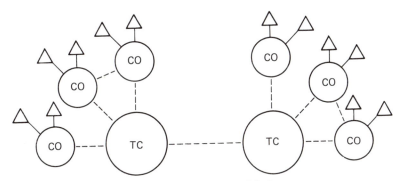

Figure 2-17   Interconnection of Exchanges

take one of several paths. It might be routed from one central office directly to the other central office, which would be the shortest or preferred route. If that line or trunk was busy or not available, it could go to a higher level office or the Toll Center(s).

By employing wide band trunk lines that accommodate hundreds of voice channels with suitable switching stations, a large number of subscribers can be interconnected. The drawback is the possibility that blocking will occur. Blocking is the result of design factors for switching networks which take into account that not all telephones will be in use at the same time. Therefore, less than a full complement of switches is provided. In peak calling periods such as Christmas, switching facilities become fully utilized and some calls cannot be completed.

In the telephone network, up to five levels of switching centers are employed. Fig. 2-18 illustrates this hierarchy. The volume of calls handled by each End Office and the number of End Offices is determined by the population density of the geographical area served. In the business district of a large city, the call volume and the number of individual lines required is quite large within a relatively small geographic area. Rural areas have need of fewer lines and generate lower call volumes, therefore requiring fewer End Offices.

Within cities and nearby towns or suburbs, many telephone companies provide local or free calling. Monthly tariffs are determined by the number of other telephone numbers that can be called toll-free. Calls to outside these areas are charged to the subscriber on a time-and-distance basis.

In rural areas, the density of subscribers and the distance between small towns and villages makes it uneconomical to offer free calling on the scale available in metropolitan areas. However, monthly tariffs are usually reduced accordingly.

In North America as well as other parts of the world, each end office can handle up to 10,000 telephones. Thus each subscriber can be identified by a unique four-digit code, or the last four digits of a telephone number. Once subscription to telephone service approaches the maximum capacity of an end office, additional end offices must be established to meet increased demand. Distances between population centers also makes it economical to establish local central offices to avoid excessively long subscriber loops.

Each end office is identified by a three-digit number or office code. In most places a subscriber dials a seven-digit number to reach another subscriber

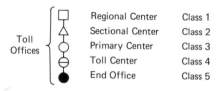

Figure 2-18 Switching Center Hierarchy

within the local calling (toll-free) area. Therefore, there may be up to 1,000 end offices (10 x 10 x 10) within an Area Code.

An Area Code is a three-digit number used to identify a large geograhic area or a metropolitan area requiring up to 1,000 end offices. It may serve an area the size of Alberta or a section of New York City.

Calls to subscribers outside the local calling area are subject to a toll charge and are routed through a Class 4 or toll center. If the call is to an end office within the subscriber's area code, the subscriber must dial 1 plus seven digits. If the call is to an end office outside the subscriber's area code, 1 plus area code plus seven digits must be dialed. To reach Boston from Canada, for instance, a subscriber would dial

1	617	458	6132
Station to Station	Area Code	End Office Code	Subscriber's Number

For overseas calls, an international access code, 011, is dialed instead of 1. The area code and office code are then dialed, using a numbering plan dependent upon the country. To reach Bonn, West Germany from Canada, a subscriber might dial

011	49	2221	613243
Station to Station	Country (Germany) Code	Bonn	Subscriber's Number

To keep tie trunks between switching centers to a minimum, toll centers may be routed between each other via *Primary Centers,* Primary Centers via *Sectional Centers* and Sectional Centers via *Regional Centers.* The traffic volume between the highest level centers is extraordinarily high, requiring high speed trunks utilizing multiplexing techniques to increase efficiency. Fig. 2-19 indicates the locations of the higher control switching centers in the United States and the southern part of Canada.

The longest route any signal can take is to go all the way up and back down the hierarchy. Some of the many possible routes are indicated in Fig. 2-20 by the dotted trunk lines. The actual route taken is determined by the availability of trunks between switching centers. The best route is the shortest one.

The tandem office in Fig. 2-20 is employed in areas of high population density to reduce the number of directly interconnected trunks between end offices within a toll-free or local calling area.

The configuration of the North American telephone network is dictated by the need to provide service to population centers whether they have a high or low population density and are proximate or remote from one another.

2-6.3 The Step by Step Exchange (SXS)

As the dependence upon telephone service increased, it became necessary for telephone companies to man the switchboards 24 hours a day. Assuming

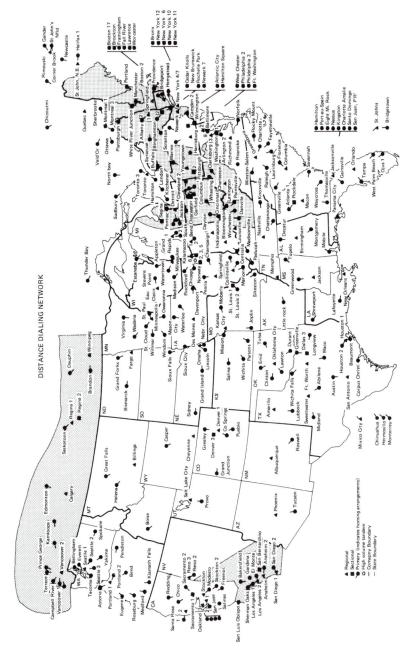

Figure 2-19 Control Switching Points in the United States (Reproduced with permission of AT&T Co.)

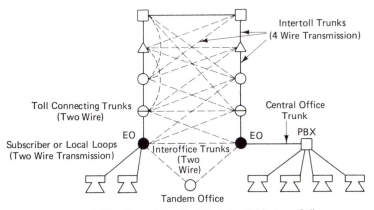

Figure 2-20   Typical Routing Plans Available to a Call

eight-hour shifts, each switchboard position required three people to operate it each day. In large urban areas many switchboards were required, meaning a large staff of operators had to be employed.

People subscribing to telephone service were never assured of privacy or equitable treatment in business matters. For example, a Kansas City undertaker, A. B. Strowger, believed that the operator in the local exchange was routing calls for his services to a competitor. He was so convinced of this mistreatment that he developed an automatic switch that was the basis for the development of the automatic dial exchange of today (see Fig. 2-21). Mr. Strowger became a millionaire from the sale of his patents, but his original suspicions about the operator were later found to be unjustified.

Figure 2-21   Strowger Automatic Switch—1893 (Courtesy Bell Canada Telephone Historical Collection)

Figure 2-22 Step-By-Step Dial Switching Equipment, introduced by Bell Canada in 1924 (copyright Bell Canada, courtesy Bell Canada Telephone Historical Collection).

The descendant of the Strowger switch is shown in Fig. 2-22 and is still in use in many exchanges today. An exchange which uses Strowger switches is commonly called a step-by-step (SXS) central office.

2-6.4   The Step by Step Switch Train

Fig. 2-23 graphically represents the SXS switch train of a call placed from 363-5674 to 363-6349 or, in other words, from one subscriber in the 363 exchange to another in the same exchange. The sequence of operation of the various Strowger switches is outlined below.

1. Caller goes off hook. Line relays operate.
2. Line Finder steps up to the seventh level and sweeps to the fourth set of contacts. The line finder is one of several serving the 5600 group; it

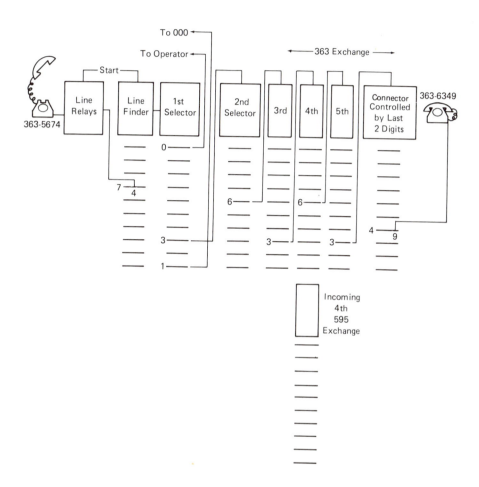

Figure 2-23 Step-by-Step Switch Train

therefore locates 5674 by its physical location on the switch.

3. Dial Tone is connected through the line finder to assure the subscriber that dialing may commence and the first selector associated with this line finder goes from rest position to ready.

4. Caller dials first digit (3). Dial tone is removed from line and First Selector steps up three levels and sweeps across to contacts associated with the first available second selector.

5. Caller dials second digit (6). Second selector steps up six levels and sweeps across to contacts associated with the first available third selector.

6. Caller dials third digit (3). Third selector steps up three levels and sweeps across to contacts associated with the first available fourth selector.

7. Caller dials fourth digit (6). Fourth selector steps up six levels and sweeps across to contacts associated with the first available fifth selector.

8. Caller dials fifth digit (3). Fifth selector steps up three levels and sweeps across to contacts associated with the first available connector.

9. Caller dials sixth digit (4). Connector steps up to the fourth level and waits for seventh digit.

10. Caller dials seventh digit (9). Connector sweeps across to ninth set of contacts. Dialing is complete and ringing generator is placed on the called subscriber's line.

Note that a subscriber's line is connected in parallel to contacts on each line finder in a group of 100 subscribers and contacts on each connector in a group of 100 subscribers. The number of Line Finders and Connectors provided for each 100 subscribers is determined by calling patterns of the local community. An exchange serving a downtown business area would statistically place more calls at the same time than a rural farm community; therefore, it would require a high number of line finders and connectors.

If our caller wished to call a number in another exchange, say 595-3478, the first three digits would be handled by the 363 exchange. On the third digit (5), the third selector would connect to a tie trunk to the 595 exchange. The last four digits would be handled by the 595 exchange. In some cases it is necessary to trunk on the first digit. For example, dialing 0 for an operator causes the first selector to step to the tenth level and sweep to the first available operator or attendant trunk. Dialing 1 immediately trunks the subscriber to a toll office via the first selector, which steps to the first level and sweeps to the first available toll trunk. Remaining digits are handled by the Toll Network.

The Strowger switches actually provide the talk path for the call and therefore must be held up for the entire duration of the call. A simple intra-office call could hold up seven switches while an inter-office call could hold up as many as three switches in the office. Additionally, the speed at which a call is processed is determined by the caller. This is an inefficient use of expensive equipment. For the relative merits of SXS Switching, see Table 2-1.

If, at any point during dialing, there is no available next selector or connector, the caller receives a busy tone and the call must be retried. If there are no line finders available, the caller simply must wait for dial tone. If the called party is off-hook for any reason, a busy signal is returned to the caller.

Table 2-1   Relative Merits of Direct Progressive Control Switching

Advantages	Disadvantages
• Signaling is relatively simple and easy to understand.	• Speed controlled by subscriber and limited by the characteristics of the switching machine signal repertory.
• Inexpensive for small number of subscribers.	
• Post-dialing delay is negligible.	
• Easy to add more switches for expansion.	
	• Number scheme controls routing of the call.
	• High level of impulse noise due to current switching.
	• High level of impulse noise generated by the physical vibrations in neighboring equipment due to mechanical movement of wiper arms.
	• Maintenance is costly.

2-7 Common Control Switching

The most serious drawback of step by step switching is that the caller controls how long the relatively expensive switching equipment is in use. Considering that once a call is established all that is needed for a talk path is basically a battery and a pair of wires, step by step is indeed an inefficient use of equipment.

Equipment can be used more effectively if the circuits used to carry the conversation are kept relatively simple and the equipment used for call control is used only as long as is necessary. This concept is called common control switching.

2-7.1   The Telephone Matrix

A telephone matrix is simply a configuration of wires and contacts which is easily understood by visualizing a grid of vertical and horizontal pairs of wires such as shown in Fig. 2-24 (a). If a set of normally open contacts are situated at the points where the verticals and the horizontals cross, a talk path may be established between any two sets attached to the grid's horizontals by closing two sets of contacts on the same vertical as represented by the points shown as X. In this example, a talk path has been established between set 1 and set 3. This matrix is a 4 by 2 matrix which means there can be four inputs and two conversations or talk paths. Usually there are fewer talks paths than terminations because not every one is on the phone at the same time. If a larger matrix is desired we can either construct it to size or connect two matrices together as in Fig. 2-24 (b).

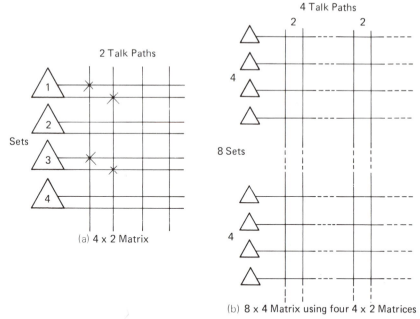

Figure 2-24 The Telephone Matrix

2-7.2 Common Control Switching

The telephone matrix is simply a grid of wire pairs. To control the completion of a call, there must be equipment to determine line status (on or off hook), decode and store dialed digits, set up talk paths, and so on. In step by step switching, the talk path is setup as digits are dialed. In common control switching, the dialed digits are collected and stored until all digits are dialed. The digits necessary to determine talk path connections within the caller's central office are handled locally and remaining digits are retransmitted to the next office involved in the call. Talk paths are not usually connected until it is verified that the call can be completed. This prevents unnecessary use of existing office capacity and represents a significant saving in facilities in a busy office.

2-8 The Crossbar Central Office (X-Bar)

The Cross Bar central office is a common control central office that utilizes electromechanical switching, but in a much more efficient way than SXS offices. The X-Bar switch is the basic matrix in an X-Bar office and is shown in Fig. 2-25.

Figure 2-25   No. 5 Crossbar Switches, first used for local service in Canada in 1956 (Copyright Bell Canada, Courtesy Bell Canada Telephone Historical Collection)

The X-Bar switch consists of a grid of horizontal selecting bars and vertical holding bars activated by magnets, causing the closure of matrix cross points. The switch depicted provides a 10 x 20 matrix.

Fig. 2-26 shows the individual crosspoint mechanism in detail. Operation of a horizontal selecting bar places a selecting finger into the upper or lower notch of the contact card. Operation of the vertical holding bar forces the closure of the contact selected. Up to 20 independent contact closures may be maintained on a 10×20 switch. In addition, each one of the inputs to the switch may be to any output provided that output is not in use.

Fig. 2-27(a) graphically represents a simple 4 x 2 matrix, that is, four inputs and two outputs. By itself, this switch can provide two talk paths or up to two conversations may take place at any one time. However, Fig. 27(b) shows how by using additional switches in different configurations, an increase in the number of possible talk paths or an increase in the number of inputs may be obtained.

In X-Bar central office, 10 x 20 matrixes are used most commonly, giving the subscriber better odds at being able to complete a call.

The number of talk paths provided in a Central Office does not allow for each and every subscriber to place a call at the same time. To do so would not be economical use of facilities. Statistically, only a small percentage of subscribers attempt calls at the same time.

Fig. 2-28 illustrates, in block diagram form, the major components of a #1 X-Bar office.

Common Control Equipment

Originating Marker — Handles the completion of all calls originating in this office. Other devices used in the completion of a call placed by a local subscriber operate under the direction of this Marker.

Terminating Marker — Handles the completion of calls received from other central offices.

Originating Register — Stores digits dialed by the local subscriber. Checks that the dialed number is valid. Notifies the originating marker when all digits have been dialed.

Outgoing Register — Stores digits to be retransmitted to another central office.

Incoming Register — Stores digits received from another office.

Line and Trunk Equipment

Line Circuit — Subscriber is hard-wired to this circuit. Notifies originating marker of change of state (on/off hook).

Line Link Frame — Connects subscriber to central office. X-Bar switch.

Junctor — Provides connection between line link frame and trunk link frame.

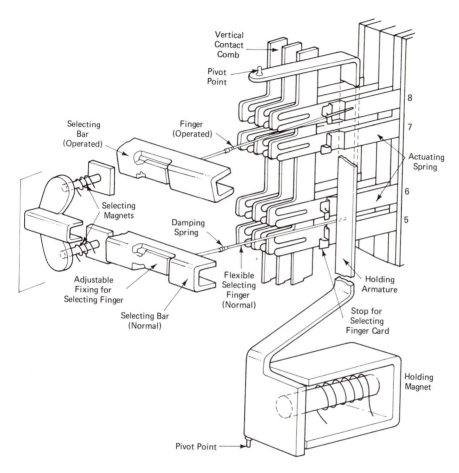

(a) Principle of Operation of Crossbar Switch

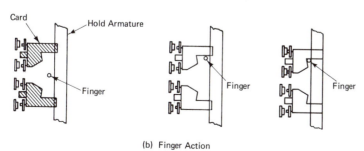

(b) Finger Action

Figure 2-26   A Cross-bar Switch (Courtesy of Northern Telecom)

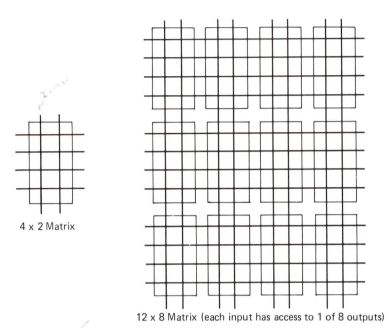

4 x 2 Matrix

12 x 8 Matrix (each input has access to 1 of 8 outputs)

Figure 2-27 Cross-bar Switch Configurations

Trunk Link Frame   Connects incoming and outgoing trunks to the central office. X-Bar switch.

Trunk Circuit   Incoming, outgoing and intra-office trunks are hard-wired to this circuit. Advises the markers of seizure incoming and of disconnects on termination of a call.

Completion of an Intra-Office Call

1. Local subscriber goes off-hook.

2. Line circuit senses off-hook and advises the originating marker of request for dial tone.

3. Originating marker sets up transmission path via line link frame to idle originating register. This is a temporary connection to enable the transmission of dial tone to the subscriber and the digits dialed by the subscriber to be transmitted to the originating register.

4. Originating Register provides dial tone to subscriber. Dial tone assures subscriber of proper connection of central office and the subscriber begins the dialing sequence. The dialed digits are stored in the originating register and decoded by the originating marker.

5. The originating marker determines the nature of the call (intra-office) and tests the called subscriber for idle condition via auxiliary control circuits connected to the subscriber's line circuits. If the called subscriber is idle, a talk path is set up internally via the calling sub-

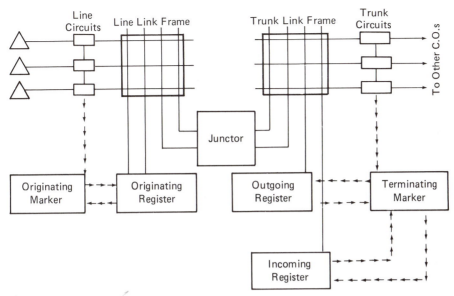

Figure 2-28   Cross-bar Central Office Diagram

scriber's line link frame and the called subscriber's line link frame through a junctor and intra-office trunks. Ringing generator is applied to the called subscriber's line. If the called subscriber's line had not been idle, a busy tone would have been returned to the calling subscriber.

6. The talk path is held for the duration of the call but the common control equipment is free to handle other calls. On termination of the call by either party, the line and trunk circuits, link frames and junctor release to an idle condition.

Completion of an Interoffice Call

1. If, after decoding the dialed digits, the originating marker determines the call is to another office, an idle trunk to the required office is selected. The number of digits stored in an outgoing register is determined by the type of call; i.e. 4 digits if the call is to another local C.O., 7 digits if to a tandem office, and 10 digits if to a toll office.

2. An internal talk path is set up via the calliing subscriber's line link frame, a junctor and a trunk link frame. Digits stored in the outgoing register are transmitted to the next office in the network.

3. If the receiving office is also an X-Bar office, the seizure of an incoming trunk is sensed by the associated trunk circuit which advises the terminating marker.

4. The terminating marker selects a terminating register and sets up a transmission path via the trunk link frame. When the register has received all the digits from the transmitting office, the terminating marker decodes the digits.

5. The called subscriber's line is checked for an idle condition. If idle, an internal talk path is set up by the terminating marker via the trunk link frame, junctor, and line link frame of the called subscriber. Ringing generator is applied to the called party's line.

6. The common equipment is free to handle other calls.

In #5 type X-Bar the functions performed by the terminating and originating markers are combined into one device called a completing marker. Additionally, a separate marker called a dial tone marker handles strictly requests for dial tone. In urban areas where call volumes tend to be greater, a more efficient use of equipment is possible with this configuration.

For the relative merits of common control cross-bar switching, refer to Table 2-2.

Table 2-2   Relative Merits of Common Control Cross-bar Switching

Advantages	Disadvantages
• Faster switching than S × S. • Only a few common control circuits are needed for connection set-up by a given office—typically six for a 10,000 line office. • Lower probability of blocking than with S × S because switches look ahead. • Elementary diagnostic techniques are available for fault-finding.	• A fault in a common control circuit causes the removal of several circuits. • Relatively high start-up cost. • More complex to understand than S × S.

2-9   The ESS Central Office

In order to handle the increasing volume of traffic on the telephone system, it is necessary to move away from the slow switching speeds of the electromechanical Strowger and cross-bar switches to the more modern electronic switch, which uses computer technology with Stored Program Control (SPC). A system such as the Bell Laboratories No. 1 ESS can handle up to 100,000 subscriber lines. In this and similar ESS's a set of software programs instructs a high-speed processor on establishing a connection. A stored program resides in a semipermanent memory that cannot be erased by circuit errors or equipment malfunction. It contains an instruction set which informs the processor what actions to take under a given system condition. It also contains such information as the location of each subscriber and trunk line, class of service, alternate trunk routes, and so forth. In order to change this information, which is typically stored as magnetized or demagnetized spots (1's and 0's) on an aluminum sheet, the memory card can be temprarily removed and altered by a separate machine.

In addition to the semipermanant memory, a temporary ferrite sheet read/write memory keeps track of such transient actions as the state of the subscriber and trunk lines, the digits dialed, and so forth. It is wiped clean after a call is completed.

A very elementary block diagram of an ESS is shown in Fig. 2-29.

To determine whether a subscriber is ON or OFF HOOK, a saturable-core ferrod sensor is used in the line circuit. It consists of two identically wound solenoid coils around ferrite rod. When current flows in the line and through the solenoid coils, the ferrite saturates. In addition, a pair of sensing wires are threaded through two holes drilled in the ferrite rod. When an interrogate pulse is sent down one of the sensing wires, with no dc current flowing in the line, a readout pulse is detected in the other sensing wire. If a dc is flowing in the line, the ferrod is saturated, no transformer action is possible, and no readout pulse is detected. The scanners scan these ferrods to determine the line and trunk sta-

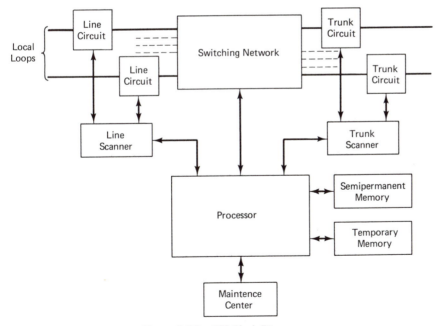

Figure 2-29   ESS Block Diagram

tus, and this information is stored in the temporary memory. Each ferrod sensor is scanned about 10 times per second or every 100 ms.

The crosspoint switches employed in the switching network are similar in character to the cross-bar switch, except that they employ much faster reed switches called ferreeds. One such sealed reed switch is shown in Fig. 2-30.

The two reeds, made from a magnetic material, are firmly supported at the opposite ends of a nitrogen-filled glass tube. A small gap is normally maintained at the free ends. The glass tube is surrounded with a coil winding which, when energized, causes an induced N-pole at the free end of one reed, and an induced S-pole at the free end of the other reed. The free ends are therefore magnetically attracted, causing closure of the relay contacts. They can operate in the order of a fraction of a millisecond. Rows of free-end switches are stacked vertically to form a switching block in the switching network.

The central processor or control is much faster than any of the external switches in the network since it must translate all the information given to it by the line and trunk scanners, memories, and maintenance center into a program of action. It then forwards this to the same equipment. Most of the instructions given to the processor are related to maintenance rather than call processing in order to assure a high standard of reliability. To achieve the generally accepted standard of no more than two hours outage in a 40-year time span, much time is spent on diagnosing faults. As soon as a fault occurs, an alternative working configuration must be found and the faulty unit taken out of service. A diagnostic routine is used to pinpoint the exact fault location. The technician makes the repair and rechecks the unit with a diagnostic routine. After passing the test, the unit can be put back into service.

Some electronic manufacturers such as Northern Telecom and Nippon Electric Company, have combined the electromechanical cross-bar switch with

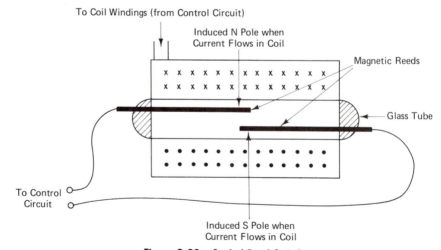

Figure 2-30 Sealed Reed Switch

a stored program common control system. A central processor still controls the system and determines the actions to be taken. The implementation of these actions is carried out by the electromechanical line and trunk link networks, originating and terminating junctors, markers, and so forth. The marker operates and releases the network crosspoints under the order of the central processor unit. It does not make any decisions of its own as in the common control crossbar system.

The block diagram of the Northern Telecom's SP-1 ESS system, which employs this techinque, is illustrated in Fig. 2-31. The signal distributor is similar in function to the line scanner. It takes orders from the central processor to perform such actions as operating a trunk relay and informing the processor of

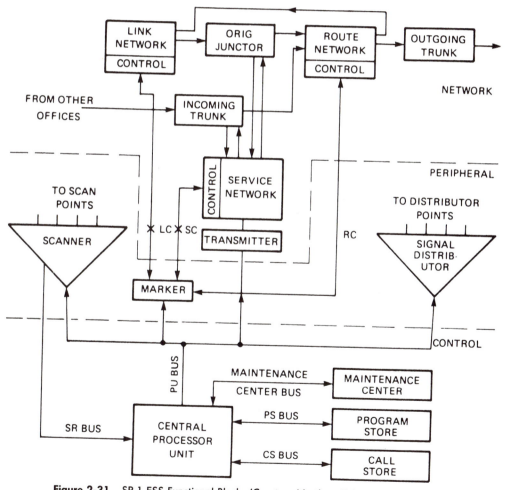

Figure 2-31   SP-1 ESS Functional Blocks (Courtesy Northern Telecom)

Table 2-3   Advantages and Disadvantages of ESS

Advantages	Disadvantages

1. Up to 1000 times faster to establish a connection than by electromechanical system. Once digits have been dialed, digital switches and control logic operate in fractions of microsends.
2. Automatic routing and fault-record producing facilities.
3. Extensive diagnostic routines.
4. Offers many classes of service, e.g.:
   a. Rerouting of a dialed call to another number (call forwarding).
   b. Modification of the way in which a call is processed; e.g., coin box call collection and restriction on PBX lines to local calls.
   c. Modification of the way in which a call is charged; e.g., free calls and wide area dialing.
   d. Special dialing codes; e.g. abbreviated dialing, call forwarding, permission of toll calls.
5. Small space requirements
6. High reliability.
	1. Expensive for small system.
	2. Catastrophic in case of failure.

any originating trunk calls. The link network is a cross-bar switching matrix which concentrates the incoming traffic via the originating junction to the route network cross-bar switching matrix. The originating junction provides the access to the route network and to the service network for calls originating from subscriber lines via the link network.

For the relative merits of electronic switching refer to Table 2-3.

2-10   The Digital Central Office

In an analog switch such as a cross-bar, a physical connection or path is established between the input and output for the duration of a call. The advent of solid-state switches which can make connections at rates exceeding millions of times per second has made it feasible to interleave connections in both space and time.

Figure 2-32   ESS No. 1 Display and Master Control Panel, Plymouth, Michigan.
(Reproduced by permission of AT&T Co.)

If a message is sampled often enough, which according to the Sampling Theorem must be at a rate of at least twice the highest frequency component in the message, no information is lost in the sampled signal and the original signal can be accurately reconstructed from the sample. In an elementary pulse code modulation (PCM) system, such as that illustrated in Fig. 2-33, an analog-to-digital converter translates each sampled amplitude into a binary coded eight-bit word. These eight bits are transmitted in a small time interval called a time slot. For voice transmission in which the bandwidth is limited to a range of 300

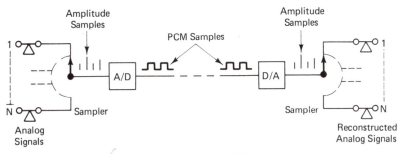

Figure 2-33   A Basic PCM System

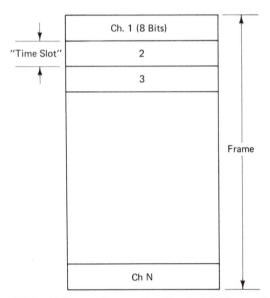

Figure 2-34   N-Channel Frame with a Nominal 8 Bits per Channel

Hz to 3.4 kHz, a sampled rate of 8 kHz is used, which slightly exceeds the minimum sampling rate of $2 \times 3.4\,k = 6.8$ kHz. The time taken to sample all of the N inputs, or the collection time for all N inputs, is called a frame, as illustrated in Fig. 2-34. This will be dealt with in much more detail in Chapter 7, Sec. 7-8.

Two types of switches are used in digital switching—space and time. In space switching, each input time slot is switched to the correct output. For the brief time slot interval, the proper connection between the input and output is held. Such a space switch could well be called a time-multiplexed space switch. In time switching, the eight bits within each time slot are stored in memory and read out at some later time. Time switching allows for time delays so that the processor can find appropriate empty time slots, particularly when transferring calls between two buses or highways.

A *bus* or *highway* is a single line carrying a number of time interleaved channels (TDM), as illustrated in Fig. 2-35. The channels are cyclically sampled by the time division switch and placed on the highway. Figure 2-35(b) shows a typical consecutive sampling arrangement.

To understand the need for both space (S) and time (T) switches, consider the requirement to connect a subscriber on one 24-channel single bus system, called a module, to a subscriber on another 24-channel module. As an example, see Fig. 2-36 and assume that Channel 6 in Bus 1 must be interconnected with Channel 9 in Bus 2.

To accomplish this connection, it is necessary to find an empty or idle pair of matching time slots on the two buses. If, for instance, Channel 6 on Bus 2 is busy, connecting Bus 1 to Bus 2 during the Channel 6 time slot interval would cause interference between the Channel 6's in the two modules. For this

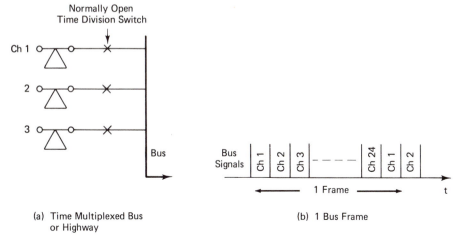

(a) Time Multiplexed Bus or Highway

(b) 1 Bus Frame

Figure 2-35 Illustration of Bus or Highway

example, assume that Channel 20 is idle in both modules. The following operations could then be followed to establish a suitable linage.

1. Delay Channel 6 in Bus 1 for 14 channels, so information resides in Time Slot 20 (time switching).
2. Connect Bus 1 to Bus 2 in Channel 20 (space switching).
3. Delay Channel 20 for 13 channels; $20 + 13 = 33 - 24 = 9$ in a 24-channel system (time switching). This assures that information emanating from Bus 1 Channel 6 is coincident in time with Bus 2 Channel 9. The arrows in Fig. 2-36 illustrate the switching procedure.

Since with this method both idle slots must occur at the same time during the same frame, it is easy for a blockage to occur. A more flexible approach is to add a buffer to each bus so that an intermediate message can be processed and delayed. This can be illustrated by using an example similar to the previous one, in which Channel 6 in Bus 1 is trying to communicate with Channel 9 in Bus 2. See Fig. 2-37.

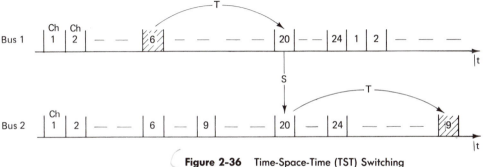

Figure 2-36 Time-Space-Time (TST) Switching

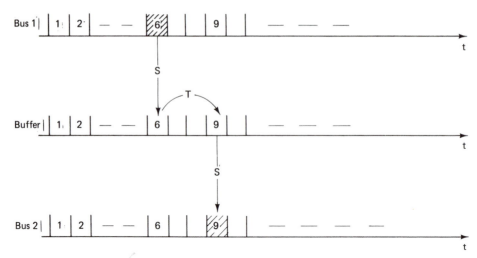

Figure 2-37 Space-Time-Space (STS) Switching

1. Connect Channel 6 in Bus 1 to Buffer Time Slot 6 (space switching).
2. Delay information in Buffer Time Slot 6 to Time Slot 9 (time switching).
3. Connect Buffer Time Slot 9 to Bus 2. This is also where Channel 9 resides in time.

To visualize the operation of the buffer store and to understand how time delay can be accomplished, consider a buffer with 24 time slots, each of which can hold 8 bits. Even though the incoming messages are stored in a certain cyclic sequence, they can be read out in a different cyclic sequence, depending upon the switching instructions. Figure 2-38 shows such an example, in which the incoming Message 1 is written into Time Slot 1, Message X is written into Time Slot X, and so on. The contents can be read out in a different cyclic sequence which need not be consecutive and may appear to be quite random. Delays can be readily obtained by reading out a channel on a delayed time slot, such as reading Channel X out in Time Slot Y.

For simplification purposes, we will omit including the connection store memories when symbolizing the time switch. The following is a more detailed block diagram of a TST configuration, which includes both T and S switches.

In the block diagram shown in Fig. 2-39, assume Line 15 in Module 1 wants to communicate with Line 23 in Module N. Also assume that Line 15 in Module 1 is assigned to Time Slot 3 and Line 23 in Module N is assigned to Time Slot 13. As illustrated on Time Switch 1, Line 15 is read into Time Slot 3. If Time Slot 3 and 13 are busy on network Bus BN, Line 15 must be transferred to an idle slot such as 19. The steps are written as follows:

Module 1 Line 15   *to*   Module N Line 23
(assigned to Time Slot 3)   (assigned to Time Slot 13)

1. Write Line 15 into Time Slot 3.
2. Read Time Slot 3 out at Time Slot 19 (T).
3. Write Time Slot 19 Module 1 into Time Slot 19 Module N (S).
4. Read Time Slot 19 out at Time Slot 13 (T).

The digital-to-analog converter sends the message to Telephone 23 in Module N.

The input signal may also be digital. In such cases, no conversion circuitry is needed. Synchronization is required to permit separation of the control signals from the information signals. It is also necessary to assure proper data rates for multiplexing to a higher order bus structure that has an integral multiple of the 24-channel format.

The advantages of digital switching are impressive, particularly when a system goes totally digital. Some of these advantages are listed in Table 2-4.

Table 2-4   Advantages of a Digital Switching System

Easier to maintain than analog; self diagnostics.
Requires less floor space.
Dissipates less power.
Handles more traffic per equipment frame.
Can multiplex a variety of sources; e.g., video, data, audio, and facsimile.
Allows for ready cryptography.
Eliminates crosstalk.

2-11   Office-to-Office Signaling

In North America, signaling is relayed from office to office in an independent fashion. In such types of signaling, Office A sends routing information to Office B which in turn sends this information, possibly in a different form, to C and so on. This technique is known as point-to-point signaling.

Point-to-point signaling can be performed by in-band or out-of-band frequencies. In-band signaling, in which signaling frequencies occur within the voice bandwidth, is in general use in North America. It runs the danger of being fooled by loud sounds which occur at the signaling frequencies. Out-of-band signaling, which employs a signal outside the voice bandwidth (at 3825 Hz, for example), has the disadvantage of requiring extra bandwidth. However, it is not susceptible to voice interference.

Although dc or dial pulse signaling similar to that generated by a rotary dial is occasionally used in inter-office signaling, Dual Tone Multi-Frequency (DTMF) is more common. As shown in Table 2-5, the frequency pairs used in

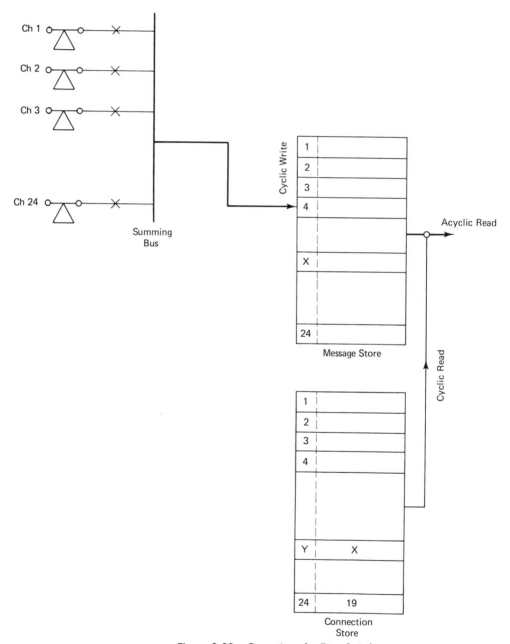

Figure 2-38 Operation of a Time Switch

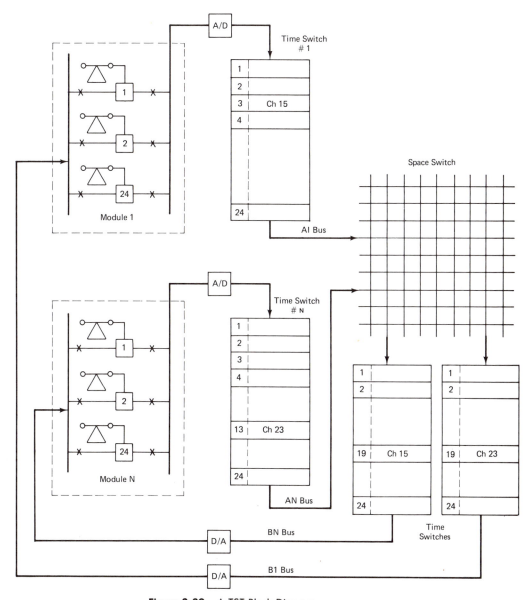

Figure 2-39   A TST Block Diagram

inter-office signaling are different from those generated by the DTMF dial of a telephone set.

When a toll channel is not in use in North America, a 2.6 kHz signal is continuously present in the channel.

Table 2-5   North American Office-to-Office Signaling Frequencies

Signal	Frequency Pairs
Start of digit transmission	1100 and 1700
1	700 and 900
2	700 and 1100
3	900 and 1100
4	700 and 1300
5	900 and 1300
6	1100 and 1300
7	700 and 1500
8	900 and 1500
9	1100 and 1500
0	1300 and 1500
End of digit Transmission	1500 and 1700

2-12   Transmission Line Fundamentals

The connection between any voice terminal (i.e., telephone sets and key telephone systems) and any switching device (i.e., exchange) is called a *telephone line*. The connection between any two switching devices such as exchange to exchange, exchange to PBX, PBX to PBX, etc. is called a *trunk*.

The medium used as a transmission path may be fiber optics, broadband (COAX), or microwave transmission; still, however, the most common medium is wire cable pairs.

In order to properly understand the behavior of electrical signals when transmitted on cable pairs, a brief overview of transmission line theory is necessary. For a full development of the theory, refer to an introductory text on the subject.

Any transmission over wire cable experiences losses and distortion due to four distributed constants present in the wire cable: a distributed series inductance, a distributed shunt capacitance, a distributed series resistance, and a distributed shunt conductance. These constants are present in any circuit, but become significant when considering the distances involved in telephone transmission. The four constants are illustrated in Fig. 2-40. The four constants are represented on a unit length basis by the following symbols:

L-series inductance   H/unit length

R-series resistance Ω/unit length

C-shunt capacitance   F/unit length

G-shunt conductance   S/unit length

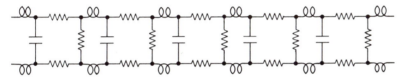

Figure 2-40 Schematic Representation of the Distributed Constants of a Transmission Line

R represents the resistance or the imperfection of the conductor and is frequency-dependent due to the fact that at higher frequencies electrons tend to travel at the surface of the conductor, thus reducing the effective cross-sectional area through which they may pass. This is referred to as skin effect.

G represents the imperfection or dielectric loss of the insulation around each conductor which allows some leakage current to pass between conductors. For modern dielectrics, it is often assumed to be negligible.

L represents the opposition to change in current levels which results from the collapsing and expanding magnetic fields created due to fluctuating current levels. Inductive reactance is expressed as $X = 2\pi \, fL$. Thus an increase in frequency increases the opposition to current flow.

C represents the capacitance created by placing two conductive materials in close proximity separated by a dielectric. Higher frequencies applied to these two plates will cause current flow between the plates, effecting a short circuit of these frequencies. Capacitive reactance is expressed as $X = \dfrac{1}{2\pi fC}$. Thus an increase in frequency decreases the opposition to current flow.

The overall effect of the distributed constants is called the characteristic impedance of the line and is expressed in rectangular form as:

$$Z_o = \sqrt{\frac{R + j2\pi fL}{G + j2\pi fC}} \qquad (2\text{-}1)$$

This is the impedance experienced by a wave or signal, commonly known as an incident wave if traveling away from the source and a reflected wave if traveling away from the load along a transmission path. If a line is terminated with its characteristic impedance, or "matched," no reflected wave is present. In general, the ratio of the reflected voltage to the incident voltage, known as the voltage reflection coefficient Γ, is given by,

$$\Gamma = \frac{V \text{ reflected}}{V \text{ incident}} = \frac{Z_R - Z_o}{Z_R + Z_o} \qquad (2\text{-}2)$$

where Z_R is the load impedance.

With the line matched, no reflected wave or echo is present, but the signal does experience some attenuation due to line resistance and the dielectric conductance. The ratio at which a wave travels along a line and the amount of attenuation depends upon the propagation constant γ of the line. This propagation constant, when separated into its real and imaginary components, is symbolized by $\alpha + j\beta$ where α is known as the attenuation constant and β is known as the phase constant. The rate of attenuation is determined by α, which has a unit of nepers per unit length. To convert nepers to decibels, multiply by 8.686. The phase velocity of the signal is determined by β, where $V_p = \omega/\beta$. In general,

$$\gamma = \alpha + j\beta = \sqrt{(R + j\omega L)(G + j\omega C)}$$

In practice, variations in temperature, moisture content of media, and other extraneous factors cause a change in the characteristic impedance of the line. Therefore, maintaining a perfectly matched condition is impossible, leading to some echo present to varying degrees on any line.

The telephone line, electrically, is a low pass filter because the series resistance and series inductance block high frequencies and the shunt capacitance shorts high frequencies.

The telephone network is designed primarily to transmit voice frequencies. To minimize the attenuation of the higher frequencies of the audible band, it is possible to increase the inductance of the line which tends to offset the shunt capacitance and flatten the frequency response within the 300 Hz to 3400 Hz range. This ensures that the quality of reproduction of the talker's voice is maintained.

The addition of inductance to the line is called loading and may be one of two forms. Distributed loading involves wrapping a highly permeable material, such as iron wire, helically around the conducting core of a medium such as submarine cable, where the cost is justified.

On land line systems, lumped inductances, as shown in Fig. 2-41, are added to the line at regular intervals. The effect of lumped loading is similar to the distributed loading up to a critical cutoff frequency. Beyond the cutoff, the attenuation increases sharply, as shown in Fig. 2-42. Using 88 mH loading coils at 1.6 Km spacing results in a cutoff frequency of about 3.5 kHz.

The propagation velocity of the signal on a loaded line also tends to be more constant than for an unloaded line, but has the disadvantage of being much lower. This flatness results in very little delay distortion, but the longer

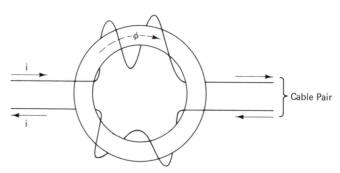

Figure 2-41 Loading Coil Winding

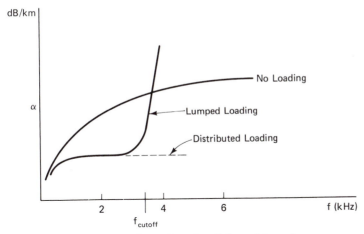

Figure 2-42   Effect of Loading Coils on Attenuation

delay is of real concern when reflections or echoes are present on the line. The latter tend to be annoying if the delay is in the order of several milliseconds. Typical transmission velocities along a loaded wire pair are around 30,000 Km/sec at voice frequencies. This results in a delay of 0.03 ms per Km or 30 ms for a 1000 Km run.

A low amplitude signal may be amplified or reconstructed if it is not allowed to deteriorate below a certain level. The sharp cutoff of high frequencies on lumped loaded cable precludes the transmission of the higher frequencies necessary for multiplexing or data transmission. Therefore, loading coils must be removed from a line if these higher frequencies are to be transmitted and amplifiers used at regular intervals to prevent signal degradation on long lines. Baseband data transmission rates of 300 Kbits/sec. are generally obtainable on subscriber lines as long as all loading coils are absent.

2-12.1 Subscribers' Loop

The subscriber loop connects the telephone handset to the local switching office. It is generally a two-wire circuit which carries signals in both directions simultaneously, i.e., full-duplex. Its length and minimum conductor size are limited by the maximum loop resistance permitted for either supervisory signaling or transmission attenuation of the voice signal, depending upon which exceeds the allowed degradation first. For step-by-step switching stations, the loop resistance is determined by the signaling and dial pulsing requirements and is limited to 1200 ohms. For crossbar and ESS offices, the resistance is limited to 1300 ohms. If the resistance of a telephone handset is 50 ohms, 1250 ohms are left for the actual subscriber line.

The amount of signal attenuation allowed in the local loop depends on the overall national transmission plan in use. The maximum allowable loss is about 8 dB, with about 3.5 dB the average.

Table 2-6 shows the typical resistances of copper conductor employed in local loops. Calculating the maximum loop length that can be used if the resistance is to be kept below 1250 Ω is quite straightforward. For any increase of three gage numbers, the cross-sectional area of the conductor is halved, which causes a doubling in resistance.

For an AWG #22 loop, the maximum loop length determined by the signaling limit would be:

$$l = \frac{1250}{105.96} = 11.8 \text{ km.} \qquad (2\text{-}4)$$

As discussed earlier, the attenuation experienced by a signal on a transmission line is determined by the electrical characteristics of the line as well as the signal frequency. In addition, the resistance of the line is slightly temperature-dependent. Any moisture in a coaxial line has a very detrimental effect on the transmitted signal.

Table 2-7 gives losses of standard types of paper-insulated cable circuits. In order to obtain longer loops for a given total attenuation, the conductor diameter can be increased, which would involve going to a lower AWG gage; or inductive loading can be used. Going from an unloaded ASM AWG #24 wire to an unloaded DNB AWG #19 wire, one can increase the loop length by a factor of 2.15/1.12 or 1.9; loading the line by adding series-lumped inductances at fixed intervals can further increase the loop length. Inserting 88-mH inductors at 6000-ft intervals increases the loop length of loaded versus unloaded DNB cable by a factor of 1.12/.38 or 2.9. As can be further seen from Table 2-7, loading greatly decreases the velocity of propagation, which results in greater signal delays. Loading also has a tendency to increase the characteristic impedance of the line. For local loops, 19H88 and 22D66 cable is most frequently used. The first two digits represent the wire gage, the letters H and D indicate loading-coil spaces of 6000 and 4500 ft, and the last two digits indicate the inductance in millihenries.

Table 2-6

American Wire Gage AWG#	Ω/km of Loop at 20°C
12	10.420
14	16.568
16	26.352
18	41.896
19	52.828
20	66.601
21	83.990
22	105.960
23	133.596
24	168.438
25	212.402
26	267.782

Table 2-7 Approximate Characteristics of Standard Types of Paper-Insulated Exchange Telephone Cable Circuits. (Courtesy of Howard W. Sams Inc., Indianapolis, Indiana. From *Reference Data for Radio Engineers*, 6th Edition, p. 35–12.)

Wire Gauge (AWG)	Code No.	Type of Loading	Loop Mile Constants C (μF)	Loop Mile Constants G (μmho)	Propagation Constant Polar Mag	Polar Angle (deg)	Rectangular α	Rectangular β	Char. Impedance Mid-Section Polar Mag	Polar Angle (deg)	Rectangular Z_{01}	Rectangular Z_{02}	Wavelength (miles)	Velocity (miles per second)	Cutoff Freq (hertz)	Attenuation (dB per mile)
26	BST	NL	0.083	1.6	—	—	—	—	910	—	—	—	—	—	—	2.9
	ST	NL	0.069	1.6	0.439	45.30	0.307	0.310	1007	44.5	719	706	20.4	20 400	—	2.67
24	DSM	NL	0.085	1.9	—	—	—	—	725	—	—	—	—	—	—	2.3
	ASM	NL	0.075	1.9	0.355	45.53	0.247	0.251	778	44.2	558	543	25.0	25 000	—	2.15
		M88	0.075	1.9	0.448	70.25	0.151	0.421	987	23.7	904	396	14.9	14 900	3100	1.31
		H188	0.075	1.9	0.512	75.28	0.130	0.495	1160	14.6	1122	292	12.7	12 700	3700	1.13
		B88	0.075	1.9	0.684	81.70	0.099	0.677	1532	8.1	1515	215	9.3	9 270	5300	0.86
22	CSA	NL	0.083	2.1	0.297	45.92	0.207	0.213	576	43.8	416	399	29.4	29 400	—	1.80
		M88	0.083	2.1	0.447	76.27	0.106	0.434	905	13.7	880	214	14.5	14 500	2900	0.92
		H88	0.083	2.1	0.526	80.11	0.0904	0.519	1051	9.7	1040	177	12.1	12 100	3500	0.79
		H135	0.083	2.1	0.644	83.50	0.0729	0.640	1306	6.3	1300	144	9.8	9 800	2800	0.63
		B88	0.083	2.1	0.718	84.50	0.0689	0.718	1420	5.3	1410	130	8.75	8 750	5000	0.60
		B135	0.083	2.1	0.890	86.50	0.0549	0.890	1765	3.3	1770	102	7.05	7 050	4000	0.48
19	CNB	NL	0.085	1.6	—	—	—	—	400	—	—	—	—	—	—	1.23
	DNB	NL	0.066	1.6	0.188	47.00	0.128	0.138	453	42.8	333	308	45.7	45 700	—	1.12
		M88	0.066	1.6	0.383	82.42	0.0505	0.380	950	8.9	939	146	16.6	16 600	3200	0.44
		I188	0.066	1.6	0.459	84.00	0.0432	0.459	1137	5.2	1130	103	13.7	13 700	3900	0.38
		H135	0.066	1.6	0.569	86.53	0.0345	0.570	1413	4.0	1410	99	11.0	11 000	3200	0.30
		H175	0.066	1.6	0.651	87.23	0.0315	0.651	1643	3.3	1640	95	9.7	9 700	2800	0.27
		B88	0.066	1.6	0.641	86.94	0.0342	0.641	1565	2.8	1560	77	9.8	9 800	5500	0.30
16	NH	NL	0.064	1.5	0.133	49.10	0.0868	0.1004	320	40.6	243	208	62.6	62 600	—	0.76
		M88	0.064	1.5	0.377	85.88	0.0271	0.377	937	4.6	934	76	16.7	16 700	3200	0.24
		H88	0.064	1.5	0.458	87.14	0.0238	0.458	1130	2.8	1130	55	13.7	13 700	3900	0.21

1000 hertz.
In the third column of the above table the letters M, H, and B indicate loading-coil spacings of 9000 feet, 6000 feet, and 3000 feet, respectively, and the figures show the inductance in millihenries of the loading coils used. NL indicates no loading.

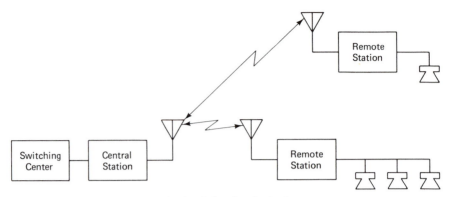

Figure 2-43 Subscriber Radio System

In order to obtain even longer subscriber loops, amplifiers or range extenders can be added. Range extenders increase the battery voltage on the loop to extend the signaling range. In addition, they often contain amplifiers to extend the transmission loss limits.

For very long loop lengths, the exact method used to connect the individual subscriber to the switching center depends upon the distance between the two and the number of subscribers. For a few subscribers at distances up to 50 km from the exchange, a distributed carrier system such as radio can be used. In this case, several subscribers on a multiparty line or on private lines are connected to the remote station. The central station is located at the exchange as shown in Fig. 2-43.

When there are a small number of subscribers along a rather long route, it can be convenient to allocate a slightly different carrier frequency to each channel. Up to six channels can be placed on a single wire pair, as shown in Fig. 2-44. Such a system is often referred to as a distributed carrier.

When there are clusters of subscribers in a remote region, it is more convenient to go to a lumped carrier system. In such a system, up to a few hundred subscribers are digitally multiplexed onto a common highway to the central station. Such a system is shown in Fig. 2-45.

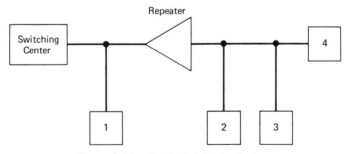

Figure 2-44 Distributed Carrier System

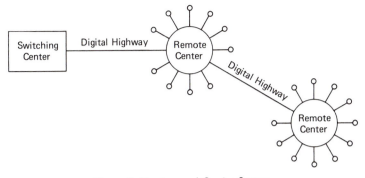

Figure 2-45 Lumped Carrier System

Figure 2-46 graphically illustrates what type of subscriber loop system would be most applicable in a given situation. For a large number of subscribers near a switching center, standard cable pairs can be used. For slightly longer distances, loaded cables or amplifiers or some similar device needs to be used. For a combination of long loop lengths and a large number of subscribers, remote centers using digital lines are the most attractive. The boundaries illustrated by the dotted lines are, in practice, rather fuzzy.

2-12.2 Trunk Circuits

In urban areas there may be more than one central office located in the same building and more than one telephone exchange building within a relatively small geographic area. Interoffice trunks are required to interconnect these of-

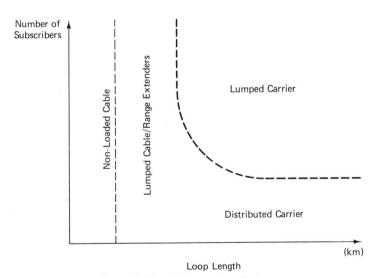

Figure 2-46 Optimum Loop System

fices to each other and to other switches such as toll centers or private branch exchanges for large business and institutional subscribers. Additionally, toll offices must be interconnected by intertoll trunks. There are several different types of trunk circuit configurations. Some of these are listed in Table 2-8.

Instead of placing a short on a line as is done on a loop start line, a ground start trunk operates by placing a momentary ground on the tip side of the line which pulls in control relays to make connection. The advantage of ground start over loop start is that glare or the simultaneous seizure of the same trunk at opposite ends is avoided. Glare occurs in loop start operation when a trunk circuit is seized outbound during the first silent period of the ringing cycle from an inbound call. E&M trunks provide a more definite seizure and release supervision. Four-wire circuits allow for the insertion of echo suppressors and amplifiers to improve transmission quality.

The intertoll trunks and many of the toll-connecting trunks, however, must handle large call volumes. Therefore, four-wire circuits are used, which enables multiplexing several voice channels on the same circuit and makes more efficient use of facilities.

Table 2-9 uses single diagrams and brief descriptions to illustrate the various types of media used to provide trunk circuits between switching centers. Each channel is assumed to cover a bandwidth of 300 to 3400 Hz. The crosstalk refers to a message on one wire pair being coupled to an adjacent pair.

Because of the heavy investment in both local loops and trunk circuit cable, any new or alternative communication modulation scheme will have to employ cable systems already in service. Severe demands have been placed on cables by implementation of the more recent schemes with their higher frequency response requirements. Newer transmission media such as fiber optics are increasingly being installed to provide the necessary transmission speeds. Also, new methods of employing the properties of light in digital transmission modes have enhanced the capacity of fiber optics.

In most locales, wire cable must still be used at some point along the transmission path. One major concern with cable, and also with optical fibers, is the increased attenuation caused by moisture. Water vapor diffusing through a cable sheath can cause a 50% increase in signal attenuation. In more severe cases, it can result in the loss of wire pairs due to corrosion.

Moisture can creep into cables through sheath damage due to incorrect cable handling procedures, lightning-caused pin-holes, improperly encapsulated splices, rodents, and so forth. Some penetration occurs even with the best protected jackets, so that cable moisture gradually increases over the years. The moisture problem is reduced by using polyethylene jackets bonded to aluminum, as shown in Fig. 2-47, or, in the case of air core cable, by pressurizing the cable with dry air or by filling the cable with a petroleum jelly compound. As the latter tends to flow out of cables that are placed in the sunlight, pulverized polyethylene is added to give a more putty-like consistency.

Fig. 2-48 shows additional cable faults that can cause signal degradation.

Table 2-8   Trunk Circuit Configurations

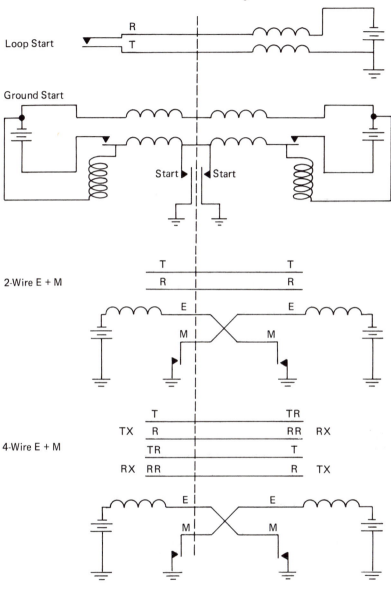

Table 2-9 Types of Trunk Circuits

Transmission Media	Construction	Distance Between Repeaters	No. of Channels on a Single Line or Carrier	Operating Frequency Range	Notes
Open Wire	25 cm; ≈ 3.25 mm Diameter	40 km	12	Up to 160 kHz	Bulky and Unsightly. Affected by Weather Conditions, i.e. Large Leakage with Wet Insulators. Severe Crosstalk. High Radiation Losses at the Higher Frequencies. Resistance Increases with Frequency Due to Skin Effect. Large Distance Between Amplifiers.
Twisted Wire Pair Cable	Polyethylene	3 to 6.5 km	12–120	Up to 1 MHz	Subject to Severe Crosstalk Because of Nearness of Conductors. Resistance Increases with Frequency Due to Skin Effect.
Coaxial (Tube)		3 to 65 km	L5 Carrier 10,800 per Tube 108,000 per Cable	3 to 60 MHz	No Radiation Losses. Negligible Crosstalk. Large Number of Channels can be Sent Over a Single Tube.
Microwave	Parabolic Dish (1° Beamwidth); Feed Horn	30–50 km	TD-2 Carrier 600–1200 per Radio Channel 10,800 per Route	3.7 to 4.2 GHz 5.925 to 6.425 GHz etc.	Fewer Amplifiers Required than Coaxial System. Affected by Rain when Frequency Exceeds 10 GHz. Large Number of Channels can be Sent Over a Single Antenna. High Velocity of Propagation, Minimizing Delay Times. Can Cause Radio Interference.

Table 2-9 (continued)

Transmission Media	Construction	Distance Between Repeaters	No. of Channels on a Single Line or Carrier	Operating Frequency Range	Notes
Geosynchronous Satellite	Satellite / Earth	Direct	600 per Transponder (12 Transponders per Satellite)	Intel Sat IV Uplink 5.925 to 6.425 GHz Downlink 3.7 to 4.2 GHz	Subject to Interference from Terrestial Link and to Interfere with Terrestial Link Long Delay Times – Up to 270 ms Large and Thus Costly Earth Transmitting Antennas Required Large Distance and Large Area Coverage. Economical Receive Only Earth Stations. Much Lower Cost per Channel than Submarine Cable for Transatlantic Communications.
Submarine Cable	Copper (Inner Conductor) / Steel Strenth Member / Copper (Outer Conductor) / Polyethylene	18 km	4000	28 MHz max. Frequency	Employ Time Assignment. Speech Interpolation (TASI) While a User is Silent, Even Between Words, the Channel is Transferred to Another Talker.
Optical Fibre	Core / Protective Coat / Cladding / Graded Index Fibre	4 to 30 km	Several Hundred Voice Channels	1 micron Wavelength 1 Ghz–km	Difficult to Splice in the Field. Cannot be Used to Carry DC Power. Not Subject to Interference or Tapping.

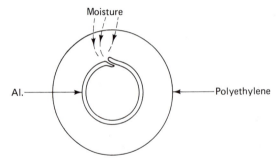

Figure 2-47 A Bonded Jacket

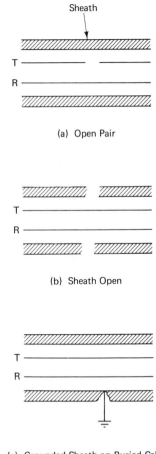

(a) Open Pair

(b) Sheath Open

(c) Grounded Sheath on Buried Cable
(Due to Sheath Perforation)

Figure 2-48 Common Cable Faults

2-12.3   Frequency Division Multiplexing

In order to increase the utilization of a transmission line, several telephone conversations can be sent down the line simultaneously. By stacking several low-frequency signals into a higher spectrum location through repetitive modulation, multiplexing is achieved. Multiplexing is the name given to the technique which allows a communication circuit to carry more than one channel.

For example, since open-wire transmission lines can be used for frequencies up to approximately 300 kHz and twisted wire parts for frequencies up to 120 kHz, several voice channels, each 4 kHz wide, can be frequency-spaced so that all channels can be carried together. This 4-kHz bandwidth includes the nominal 300 to 3400 Hz audio bandwidth along with the guardband, which is required due to the finite roll-off of the bandpass filters. A common form of frequency division multiplexing (FDM) is the standard CCITT group which contains 12 voice channels over the 60 to 108 kHz frequency range, as shown in Fig. 2-49.

2-12.4 CCITT Groups

The assignment of a frequency bandwidth to a particular channel must be standardized to enable interfacing between switching centers. Fig. 2-50 shows the block diagram of the equipment employed to transmit and receive the frequency division multiplexed signal of a Standard CCITT Group.

The low-pass filters are inserted at the sending end to ensure that each channel is limited to well within the 4 kHz band, so as not to interfere with adjacent channels when all the channels are combined. A guard band is also inserted between the channels to minimize the interference between channels. Overall, about 3.3 kHz of useable bandwidth is allocated to each channel.

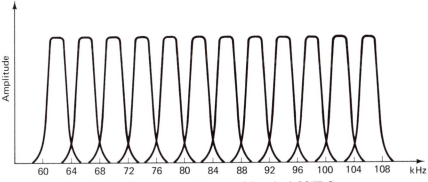

Figure 2-49   Frequency Division of Standard CCITT Group

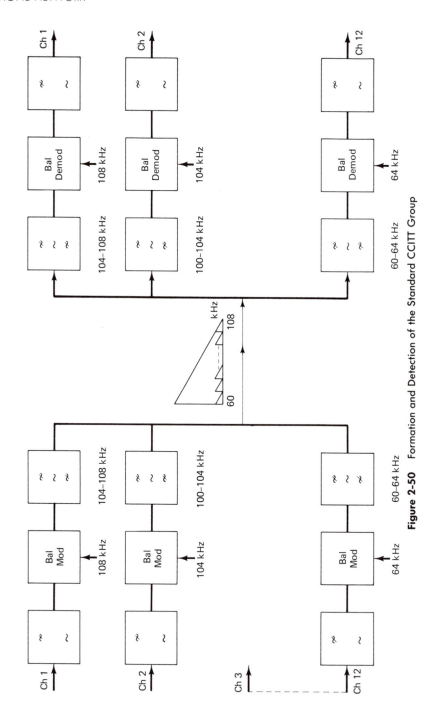

Figure 2-50 Formation and Detection of the Standard CCITT Group

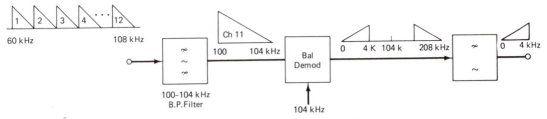

Figure 2-51 Frequency Spectrum for Channel II in the Receiver

The balanced modulators form a double sideband signal with no carrier. The bandpass filters reject the upper sidebands, which results in a neatly packaged transmitted signal of 12 channels within the 60 to 108 kHz range. On the receiver end, the various bandpass filters allow only their respective channels to pass through.

The balanced demodulators, which are identical to the balanced modulators, cause the sum and difference frequencies to appear at the output. The difference frequencies are chosen by the final low-pass filter, causing an inversion of the received channel spectrum. This is illustrated in Fig. 2-51 where Channel II is extracted.

In order to accommodate more traffic, these 12-channel groups can be further combined to form a supergroup. A supergroup can be combined to form a mastergroup and so on. The CCITT hierarchy is given in Table 2-10. Such standard combinations allow modularity of equipment as well as common international modulation and demodulation schemes. Multiple modulation has the disadvantage of introducing more noise and distortion each time modulation and demodulation takes place.

In North America, a variation is made above the supergroup level.

10 Supergroups = 1 Mastergroup (600 Channels)

6 Mastergroups = 1 Jumbogroup (3600 Channels)

3 Jumbogroups = 1 Jumbogroup Multiplex (10,800 Channels)

Twelve super-mastergroups can also be combined to give 10,800 channels.

Table 2-10   CCITT Standard for Multiplex Groups

	CCITT Standard	Number of Voice Channels
12 telegraph channels or 24 telex channels	= 1 channel	1
12 channels	= 1 group	12
5 groups	= 1 supergroup	60
5 supergroups	= 1 mastergroup	300
3 mastergroups	= 1 super-mastergroup	900

The maximum shift in frequency permitted by CCITT is + 2 Hz. To maintain such frequency accuracy, a pilot signal is often transmitted along with the speech signal which is used by the receiver for frequency synchronization. The receiver oscillator is phase-locked to the incoming pilot tone. If the pilot is lost for some reason, the system continues to operate, but without synchronization.

To obtain the various local oscillator frequencies illustrated in Fig. 2-50, a master frequency generator can provide a 4 kHz signal to drive a harmonic generator. Various narrow-band filters select the desired harmonics to act as the local oscillator signal.

On international circuits, the end-to-end frequency tolerance should be better than 2 Hz. A frequency synchronizing pilot is normally used to maintain this accuracy. With some of the recent master oscillators, stabilities are such that synchronizing pilots are not required. In order to maintain signal level regulation, level regulating pilots are initiated on each group, super group, master group, and so on. These pilot signal amplitudes are monitored when manually adjusting amplifier and attenuator gains. To maintain signal levels on a dynamic basis because of variations in amplifier gains and line losses as a result of aging and temperature changes, automatic gain control circuits also monitor pilot signal levels. Signal levels must be tightly controlled to prevent poor signal-to-noise ratios at reduced signal levels and excessive intermodulation noise from overdriving the circuits. The latter can be more fully appreciated after circuit loading is considered.

Incoming pilot levels in the multiplex-receive equipment are also monitored to detect equipment failure. If the pilot drops outside the level regulating range (usually about +5 dB), an alarm is triggered. In the absence of a pilot, the channel is assumed to have failed and a standby channel unit is brought into operation.

2-13   Special Telephone Services

To meet the many demands of voice, television, data, facsimile, and telemetry transmission, the common carriers have developed many communication services. The following tree shows some of the services offered by Telecom Canada, a consortium of the major common carriers in Canada. Although other systems may use slightly different descriptive terms, the services listed are representative of those offered by other communication companies.* It should be noted that data transmission over the switched voice network could be subject to high error rates due to switching noise and narrow bandpass characteristics inherent in a system designed for the lower frequencies of the voice bandwidth.

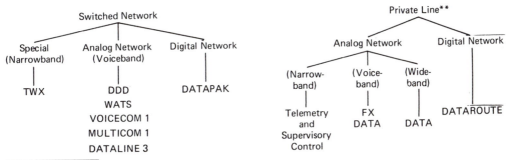

* For more information about communication services offered by common carriers, contact personnel in the computer communications group in your area.
** A private line consists of a channel or circuit supplied to a subscriber for exclusive use.

Teletype Exchange Services (TELEX and TWX)

Teletype Exchange Service provides direct-dial point-to-point, or conference, connections between teletypewriters. No voice transmission is involved. Data is transmitted at a 110 b/s rate. Either a Model 33 or 35 terminal is used—the Model 33 for light to medium usage, and the 35 unit for heavy usage (multi-copy forms). The eight-level ASCII code is used, which has seven information bits and one parity bit.

Direct Distance Dialing (DDD)

Direct Distance Dialing is the most common network offering. It is used primarily for voice transmission. Through this method, any telephone subscriber can be connected to any other telephone on the network by direct dialing.

Data sets allow for the transmission of data over two-wire half-duplex facilities at speeds up to 1200 b/s, asynchronous and 2400 b/s, synchronous.

Wide Area Telephone Service (WATS)

WATS operates over the regular DDD network. It gives the subscriber access to any dialable telephone within a specified or restricted geographical area identified as a WATS zone (see Fig. 2-52). The user can select inward WATS (InWATS) which can receive incoming calls only, or outward WATS (OutWATS) which allows outgoing calls only. Two-way service on the same WATS trunk is not available. A flat rate is charged based on the zone size and whether it is a full time service or a measured time service: e.g., 5, 10, or 160 hours per month.

Voicecom 1

Voicecom 1 provides voice-only service between specified locations on the DDD network. Connection is established by a preprogrammed automatic

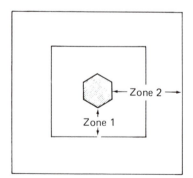

Figure 2-52   Areas Served

dialer. Only one digit needs to be dialed. An exclusion feature can be added to prevent access to the line by any other extensions.

Multicom 1

Multicom 1 provides voice and data communication between specified locations on the DDD network. As in the case of Voicecom 1, connection is established by a preprogrammed automatic dialer. Data sets are required for data (up to 2400 b/s). The voice feature is generally used to ensure the proper establishment of a connection.

Dataline 3

Dataline 3 provides data transmission from remote originating data stations to a centralized data terminal. The central terminal, such as a time-shared computer, may process the data and return it to the originating station. Calling is permitted only from the remote originating data station. Automatic dialing is provided. Bit rates of 2400 b/s, synchronous and 1200 b/s, asynchronous are permitted.

Datapak

Datapak is a nationwide, public, packet-switching network. It consists of switching nodes at various centers interconnected with high speed 56 Kbits/s digital transmission trunks. Transmission facilities are allocated only after a data or control package has been assembled. Normal service handles data packets of up to 256 octets (2048 bits) and provides an average network transit delay of 0.49 s.

Non-packet-mode terminals such as teletypewriters are connected to Datapak by Network Interface Machines (NIMs). Packet-mode terminals use the Standard-Network-Access Protocol (SNAP). SNAP establishes and clears calls and, in general, manages the flow of data along the network.

Telemetering and Supervisory Control

These channels are designed to transmit meter readings from remote locations to a central point (telemetering). If, in addition, a control signal can be sent out to the remote station from the central station to change some operation, supervisory control is possible. Typical pulse rates of 30 and 55 baud-dc pulse transmission are employed.

Foreign Exchange (FX)

Foreign exchange provides a dedicated line between a customer location and a remote telephone exchange. For example, a subscriber in Montreal could have an FX service to Toronto which allows the subscriber to call any telephone within the local calling area of Toronto toll-free or to make a toll call at a reduced rate to Buffalo because of the closer distance. A flat rate per month is charged for this service based on distance.

Voice and data or both can be transmitted at a rate of up to 1200 b/s, asynchronous and 2400 b/s, synchronous. This is a very economical method of transferring information when a customer has many calls to make to subscribers in a remote exchange area.

Data

Voice-grade data channels on a private line can transmit up to 9600 b/s, depending upon the conditioning of the channel. Two- or four-wire circuits can be made available and data sets are required. More than two stations located in different exchanges can be interconnected (multipoint), or more than two stations within the same exchange can be interconnected (multidrop).

Wide-band data channels can transmit at a rate of up to 50k b/s and higher. Full-duplex operation is used and a separate voice-grade circuit is used for voice contact, supervision, and so forth.

Dataroute

Dataroute on a private line provides full-duplex data transmission at rates of up to 50k b/s. It can provide multipoint and multidrop service (see preceding section on Data). Voices cannot be carried. For customers outside the Dataroute serving area, access can be made through a dialed-up or private line. Such would be the case if the subscriber was located outside Victoria or Vancouver in British Columbia.

Problems

1. a. How does a transmitter convert sound energy into electrical energy?
   b. How does a receiver convert electrical energy into sound energy?
   c. Explain the operation of a bell or ringer in electrical terms.
2. Name two types of battery systems and explain the difference between them.
3. State two possible causes for the following subscriber trouble reports.
   a. Can't hear very well.
   b. Can't be heard.
   c. Get dial tone but can't call out.
   d. No dial tone.
   e. Loud clicks in the ear when dialing out.
4. Name three types of dials available for telephone instruments and briefly describe their operation.
5. Sketch out the switching center hierarchy.
6. What are the chief differences between loop and trunk circuits?
7. Draw the switch train for a call from 876-4357 to 876-7952 through a Step by Step office. Label all switches.
8. Briefly explain the theory of common control switching.
9. State the major components of a No. 1 X-bar office and their functions.
10. State the major components of an ESS office and their functions.
11. What is in-band and out-of-band signaling and what is the disadvantage of each?

12. What characteristic of wire cable causes the electrical properties listed below?
   a. series resistance
   b. shunt capacitance
   c. shunt conductance
   d. series inductance

13. How does transmission line loss vary with frequency?

14. a. How is a line physically loaded?
   b. State two significant advantages of line loading.
   c. What effect does line loading have upon propagation velocity?
   d. What effect does lumped loading have on the attenuation characteristics of a line?

15. a. If a subscriber's loop length is limited by a maximum signaling resistance of 1300 ohms, what maximum length (in Km) can be achieved using AWG #24 and #19 lines? Assume the internal resistance of a telephone is 50 ohms.
   b. If a subscriber's loop is limited by the signal attenuation, say 8 dBm at 1000 Hz, what maximum length (in Km), can be achieved with #24 ASM and # 19 DNB non-loaded wire?
   c. If both conditions in (a) and (b) must be met, what condition limits the maximum loop length?
   d. If H88 loading were added to the cable circuit of (b), what maximum length (in Km) could be achieved when 8 dBm of signal attenuation is permitted at 1000 Hz.

16. State four types of trunk circuit configurations.

17. What is the advantage of multiplexing?

18. Referring to Table 2-9, state the maximum frequency response of the following types of transmission media.
   a. Wire cable
   b. Coaxial cable
   c. Microwave
   d. Geosynchronous satellite transmission

19. a. What is a standard CCITT group size?
   b. How many channels does a Jumbo Group carry?
   c. What is the maximum oscillation frequency drift permitted by the CCITT in carrier transmission?
   d. Give three uses of a pilot signal in carrier transmission.

20. Explain what the following telephone services are.
   a. FX
   b. WATS
   c. DDD
   d. Private Line

3

Telephone System Performance Requirements and Impairments

3-1 Introduction

In order for the telephone systems to satisfy customers, certain performance criteria must be met. This chapter will consider some of the fundamental transmission objectives that must be met for the systems to operate properly. It will also consider some reasons for signal-to-noise degradation as the signal passes through the system.

Most telephones are connected to the local switching center by means of a two-wire circuit. Both the received and transmitted signals travel over the same electrical transmission path within the same frequency band. If any gain needs to be added to compensate for transmission loss on the cable, it is not feasible to use a bidirectional amplifier, as shown in Fig. 3-1, since such a configuration must use a gain of less than unity for stable operation.

3-1.1 Four-Wire Terminating Set

One technique to obtain reasonable gain of up to 10 dB on two-wire trunks is to use a negative impedance repeater. Such a regenerative feedback amplifier causes an effective negative resistance to appear across the line, resulting in a net power gain. It readily passes dc signaling, maintains a signal path in case of failure, and tends to be inexpensive. Generally, not more than one can be used on a route.

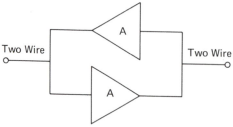

Figure 3-1 Bidirectional Amplifier

For transmission beyond the local subscriber loop, the receive and transmit paths are usually separated to give two one-way paths which result in a total of two wire pairs, or four wires. The device designed for such a splitting and recombining effort is called a four-wire terminating set, or hybrid transformer. It employs two cross-coupled transformers with identical windings, as shown in Fig. 3-2.

A signal applied to the RETURN line will generate equal voltages across each winding, but because of the antiphase in the upper windings, nothing will appear in the GO line. Equal signals will appear in the two-wire line and balance network. Similarly, any signal applied to the two-wire line will split evenly between the GO and RETURN lines. The portion appearing in the RETURN line will be absorbed in the output stages of the RETURN amplifier, whereas the signal appearing in the GO line will be passed on by the GO amplifier. This division of signals causes the signal appearing in the GO amplifier to experience a 3 dB loss. Similarly, a signal applied to the RETURN input undergoes a 3 dB loss when appearing at the two-wire line port. Because of the non-ideal nature of transformer windings, an additional 0.5 dB insertion loss is experienced in practice, resulting in an overall 3.5 dB signal loss as it traverses through a hybrid transformer. In actual practice, the hybrid is also nonsymmetrical. The loss is higher from the return to the 2 W line than from the 2 W line to GO. Figure 3-3 illustrates the losses through a hybrid transformer.

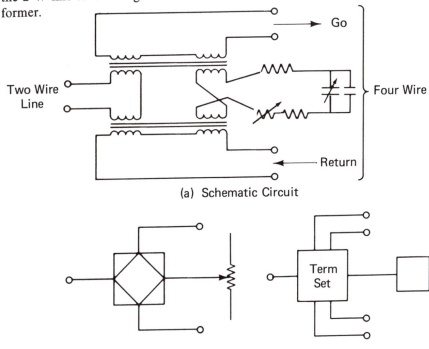

(a) Schematic Circuit

(b) Symbols

Figure 3-2   Four-Wire Terminating Set

To achieve negligible feedthrough from the RETURN to GO lines, the impedance of the balance network must be made equivalent to the characteristic impedance of the two-wire line. However, because a line characteristic impedance tends to be frequency-dependent at the lower frequencies and because of the variations between the many local loops that may be switched to the hybrid, it is impossible to achieve perfect balance with a fixed balance network. This results in some signals being coupled between the GO and RETURN paths. The loss experienced between the GO and RETURN paths is called balance-return loss and is given by:

$$B = 20 \log \left| \frac{Z_N + Z_L}{Z_N - Z_L} \right|$$

   (3-1)

where Z_N and Z_L are the balancing network and line impedances respectively. This equation ignores any imbalance in the transformers. Note that by careful matching where $Z_N = Z_L$, the balance-return loss goes to infinity. In fact 60 to 70 dB is about the highest practically available. Typically, B is in the order of 50 dB. If the line appears as an open or short circuit, then B goes to 0 dB.

When considering the stability of a network, the balance-return loss is taken across the entire frequency band from 0 to 4 kHz. The minimum value of B tends to govern the stability. When dealing with echos, however, an average value of B (B_e) over the 500 to 2500 Hz frequency range is used.

A complete long-distance telephone network is illustrated in Fig. 3-4. The dashed section can be made up of a multiplexed microwave system, a multichannel carrier on a cable, etc.

3-1.2   The Reference Equivalent

In setting up a system of international standards that will allow different national systems to be interconnected at a level of quality satisfactory to customers, it is necessary to:

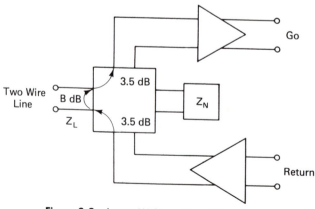

Figure 3-3   Losses Within a Hybrid Transformer

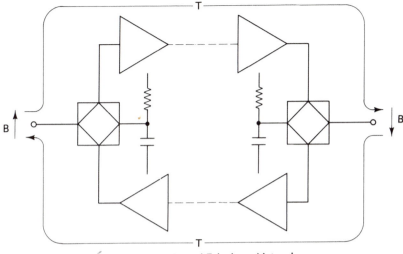

Figure 3-4   Typical Telephone Network

a. Define the sending and receiving levels.

b. Allocate maximum permissible noise contributions.

c. Allocate losses and propagation times.

To aid in achieving these objectives, the International Telecommunications Union has a reference system in Geneva called NOSFER (Nouveau Système Fundamental pour la détermination des équivalents de Reférence) against which telephone companies can compare their telephone subsets. The rating against this NOSFER standard is known as the reference equivalent. The transmitter and receiver can be rated separately and are called:

Transmit reference equivalent, tre

Receive reference equivalent, rre

A positive reference equivalent indicates that the circuit is less efficient or more objectionable to a subscriber than the laboratory standard. Adding the tre and rre produces an overall reference equivalent, ore. CCITT recommends that the maximum limits for the reference equivalent on an international connection be:

20.8 dB tre
12.2 dB rre

for 97% of the connections made.

This leads to a maximum overall reference equivalent of 33 dB. With an ore of 32 dB, the British post office has found 9.7% of the subscribers to be unsatisfied. Thus, an ore of 33 dB would prove unsatisfactory to about 10% of its subscribers. Common carriers desire to maintain as low a reference equivalence as possible for a higher percentage of subscriber satisfaction.

3-1.3 Network Stability

Consider the total loss around the telephone network of Fig. 3-4. If T represents the loss between the two-wire end points, which includes the hybrid losses (typically 7 dB), the total loop loss will be $2T + 2B$ dB for symmetrical sending and return paths. If this loss is 0 dB, the gain of the network is 1, and the circuit will oscillate or "sing." Even with some loss, the circuit may still ring, much like talking into a hollow steel barrel. To avoid such a condition, it is desirable to maintain a total loop loss of at least 6 dB. Thus:

$$2T + 2B \geq 6 \text{ dB}$$

or
$$T \geq 3 - B \text{ dB} \qquad (3\text{-}2)$$

If a shorted, or open, condition is allowed to exist on the two-wire side of the hybrid so that B is 0, Eq. (3-2) states that the T loss must be at least 3 dB. In any real system, T varies considerably due to improper alignment and the effects of temperature on amplifier gains and line losses. Thus, T must undergo a loss considerably in excess of 3 dB to assure a low probability of singing. Even with the high loss required through the network to minimize singing, the ore obtained remains below the 33 dB minimum permitted. Of more concern, however, is the echo problem on long lines.

3-1.4   Echo

When a person speaks into a telephone system, an echo or reflected signal can occur if there is an impedance mismatch in the system. Such an echo can be extremely annoying and will worsen with increased delay and loudness. As shown in Fig. 3-5, the speaker's echo experiences a delay of $2(t_1 + t_2)$ seconds and a loss of

$$(B_e + 2T) \text{ dB.} \qquad (3\text{-}3)$$

As the delay increases, more subscribers become dissatisfied with the telephone conversation. More attenuation needs to be added in the echo path in order to give improved customer satisfaction. It has been found that for a 20 ms one-way propagation delay, 50% of the customers are satisfied if 17.7 dB are added to the echo path. For a 50 ms one-way delay, 30.9 dB need to be added for 50% customer satisfaction.

As can be observed from these results, delays beyond 50 ms require even more attenuation, thus exceeding the 33 dB ore allowed. What can be done in the case of a synchronous satellite link which has about a 270 ms one-way propagation delay? Echo suppressors are usually added to the system. Generally, echo suppressors are used for near round-trip delays exceeding 45 ms. These electronic devices are inserted in the reflected path between regional centers to suppress the echo, and are activated by a speech burst from the re-

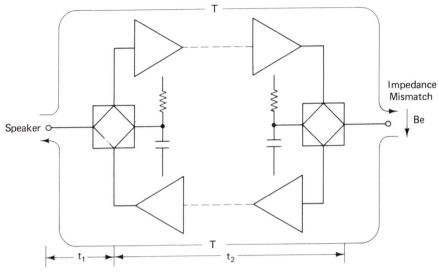

Figure 3-5   Echo Path

mote end, as shown in Fig. 3-6. When the person speaks, the echo suppressor is activated, and a 60 dB attenuator is placed in the return line. If someone begins transmitting from the oposite end, the suppressor is removed (the break-in circuit is not included), and a similar echo suppressor is inserted at the opposite end.

Because a syllable or two of speech must be detected before an echo suppressor is turned ON or OFF, significant delays and some loss of signal are experienced when echo suppressors are reversed in direction.

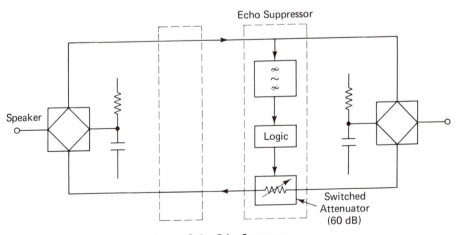

Figure 3-6   Echo Suppressor

Such loss of signal is unacceptable with data communication, and the echo suppressor must be turned OFF prior to the transmission of data. Since the turn-around time of echo suppressors is about 100 ms, when operating in half-duplex mode this turn-around time must be included in the delay between the request to send being turned on and the clear to send turning ON by the modem.

For full duplex operation, the echo suppressors must be disabled. This is accomplished by applying a tone in the 2010 to 2240 Hz range at -5dBmO for about 400 ms from either end of the circuit. No other signal should be present during this disabling period. They are again enabled when no power is present in the 300 to 3000 Hz region for a period greater than 100 ms. Echo suppressors must also be disabled when utilizing the reverse or secondary channel discussed earlier in section 1-5.5. This allows simultaneous transmission of both the main data and the oppositely directed low speed data.

Another method becoming increasingly common of reducing echo is the echo canceler. It is a device which synthesizes a replica of the echo and then subtracts it from the real echo. Unfortunately the circuitry is quite complex and expensive.

3-1.5   Via Net Loss

In designing a transmission system, a national transmission plan is normally drawn up for each country. Such a plan sets the standards regarding singing, echo levels and delays, and overall reference equivalent, so that a given percentage of subscribers will give the connection a fair or better rating. In North America a variable loss plan (VNL) is employed to meet these objectives.

One method to meet the ore and still meet the singing and echo within tolerable limits is to increase the balance-return losses. Increasing the B terms in Eqs. (3-2) and (3-3) allows a decrease of the T terms, thereby improving the ore of the connection. To achieve larger balance-return losses requires improved matching at the hybrid. This can be done by more careful adjustment of the balance network and careful control of the subscriber loop. Despite taking all reasonable precautions, reflections or echos are always present. Since echo suppressors are used for round trip delays exceeding about 45 ms, sufficient loss must be present in the loop to reduce the echo problem for delays less than this. The maximum loss in decibels at which a circuit can operate without objectionable singing or echo is known as Via Net Loss (VNL). In North America each toll trunk circuit is designed to give a loss of:

$$\text{VNL} = 0.1 \times \frac{2l}{v} + 0.4 \text{ dB} \qquad (3\text{-}4)$$

where l = circuit length − km
$\quad\quad v$ = velocity of propagation − km/ms
$\quad\quad l/v$ represents the one-way path delay in ms
$\quad\quad 0.1$ represents the number of dB loss that must be inserted per ms of

delay to maintain tolerable echo level. For the echo round trip this must be multiplied by a factor of 2.

The one-way loss, or overall connection loss, between two end-offices consists of *n* links in tandem, extrapolated from Eq. (*3-4*).

$$T = 0.2 \frac{l}{v} + 0.4n + 4 \text{ dB} \qquad (3\text{-}5)$$

where T is the overall connection loss
N = number of links
l/v = total one-way path delay in ms

The additional 4 dB come from adding an extra 2 dB between the end-office and toll center to control singing when a round trip delay approches 0 ms. This assures a minimum loss of 4.4 dB. The factor $0.2/v$ in Eqs. (*3-4*) or (*3-5*) is known as the via net loss factor (VNLF) and has typical values of

0.0015 dB/mi for a high velocity carrier system (L-carrier),

0.03 dB/mi for H-88 loading on #19AWG wire, and

0.01 dB/mi for two-wire open wire.

Consider a single trunk which has a one-way delay of 11 ms. This results in a round trip delay of 22 ms which leaves $(45 - 22)$ ms or 23 ms of delay for the other intertoll and end-office to talk links in the system. The VNL in a delay of 22 ms is

$$\text{VNL} = 0.1 \times 2 \times 11 + 0.4 = 2.6 \ dB$$

The corresponding intertoll length (one-way) for an L-carrier link can be calculated by substituting the VNLF for the $0.2/v$ terms in Eq. (3-4).

$$l = \frac{\text{VNL} - 0.4}{\text{VNLF}} = \frac{2.6 - 0.4}{0.0015} = 1467 \text{ mi}$$

Figure 3-7 shows a simplified North American transmission plan. Note that all end-office to toll center trunks operate with a minimum loss of 2 dB. To maximize the balance-return loss, these trunks also are carefully matched to the toll exchange.

3-2 Telephone System Impairments

The acceptability of the telephone system to the user is largely dependent upon information impairments suffered during transmission. Any degradation in the signal-to-noise ratio seriously affects the listener's opinion regarding the satisfactory operation of the system. Although the various types of impairment are at times grouped into two major categories, i.e. noise and distortion, it is not always easy to allocate every impairment to one category or the other. In such cases, noise is referred to as effects coming from extraneous services, and distortion is directly related to the transmission of the signal.

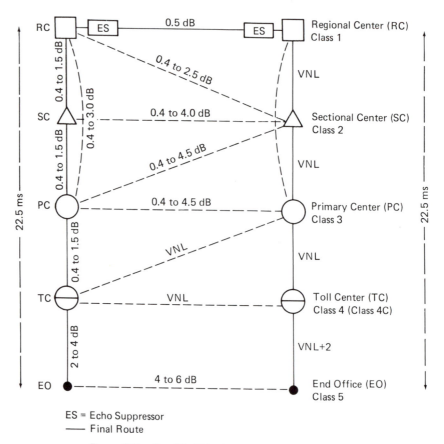

Figure 3-7 Simplified North American Transmission Plan

Noise, such as crosstalk and thermal, cannot be removed by any special operation. Intermodulation noise, which is very much dependent upon the signal amplitude, cannot be eliminated once it is present in the system. Measures can be taken, however, to minimize the introduction of such noise by reducing the signal levels or reducing the number of channels carried by the system. Distortions, such as attenuation and phase-versus-frequency distortion, can be removed by the use of equalizers, or devices which compensate out the nonlinear attenuation and phase characteristics.

In the design of a telephone network, one must make certain that the noise contributed by the various segments of the system do not exceed those set by the CCITT. For long distance transmission, the noise power is allocated on a per length basis. For example, on long distances up to 2500 km, a cable or radio-relay system can contribute up to 3 pW/km plus 2300 pW for multiplexing equipment. Radio systems operating at the very long distance of 2500 km may contribute up to 10,000 pW of noise power. These recommendations can be found in the CCITT White Book 111.

The following sections will discuss more serious noise and distortion contributors on the telephone system.

3-2.1 Crosstalk

Because of the close proximity of adjacent wire pairs in a cable, signals are easily induced into neighboring pairs by either capacitive or inductive coupling. This phenomenon, illustrated in Fig. 3-8, results in "cross-talk."

Because capacitive coupling tends to be more severe at higher signal frequencies where capacitive reactance is decreased, this capacitive coupling is of major concern when applying digital signals with their fast rise and fall times to cable installed for analog transmission.

The nature of the impedance of the coupled line determines whether the electric or the magnetic coupling is dominant. If the impedance of the coupled

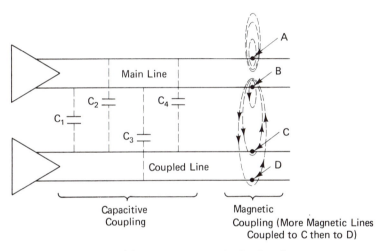

(a) Capacitive and Inductive Coupling

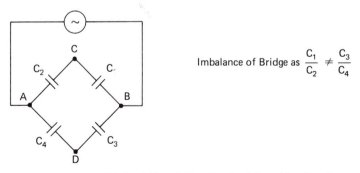

Imbalance of Bridge as $\dfrac{C_1}{C_2} \neq \dfrac{C_3}{C_4}$

(b) Equivalent Bridge Circuit of Capacitive Couplings

Figure 3-8   Illustration of Capacitive and Inductive Coupling

line is high, the capacitive pick-up is large. This is the effect observed when stray RF from fluorescent lights, etc. is coupled to a set of dangling leads attached to a high impedance input of an oscilloscope. On the other hand, if the coupled line impedance is low, the series impedance as seen by the induced voltage is small, allowing large induced currents to flow. The later effect is observed when a low frequency line runs near a coaxial cable that has each end grounded. As illustrated in Fig. 3-9, the induced voltage in the shield results in a substantial longitudinal current because of the low impedance path through the grounds.

Since this induced current is in series with the signal current, hum from nearby power lines appears in the signal.

The amount of crosstalk tolerated by the average listener establishes an upper limit on the number of wire pairs in a cable that can be used for digital transmission. Crosstalk that occurs near repeater stations is termed Near End Crosstalk (NEXT). In this region, the large amplitude signals from an amplifier output are coupled to an adjacent wire pair receiving a low-level signal, resulting in significant interference. Since NEXT is due to the pick-up of signals traveling in opposite directions, it can be reduced by physically separating the GO and RETURN wire pairs.

Far End Crosstalk (FEXT), as illustrated in Fig. 3-10, is due to coupling of signals traveling in the same direction and is of major concern for long closely spaced wire pairs.

To reduce NEXT, particularly with the digital cable, screens are inserted to separate the GO and RETURN line pairs to opposite sides of the screen. The screen acts as an electromagnetic shield, greatly reducing the crosstalk. Examples of such screens are shown in Fig. 3-11.

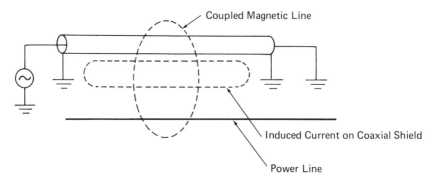

Figure 3-9   Magnetic Coupling on a Coaxial Cable

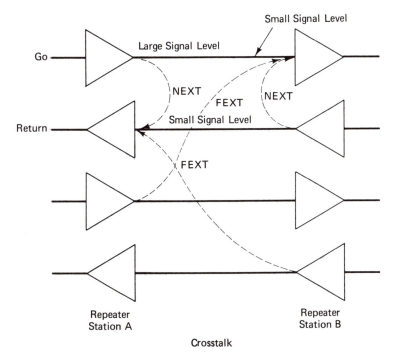

Crosstalk

Figure 3-10   Crosstalk

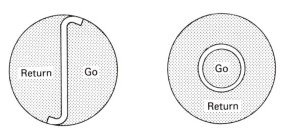

Figure 3-11   Screened Cable

3-2.2 Thermal and Shot Noise

In any long-distance communication system, the received useful signal power level is very weak. Much effort is thus expended minimizing the degradation of that signal by the receiving equipment. If the input carrier is of acceptable power level, the performance of the receiver controls the quality of the transmitted information.

The greater the distance traversed by the signal, the more stringent are the receiver sensitivity requirements. The sensitivity of the receiver is limited by the electromagnetic noise generated within the receiver. This noise can be a combination of shot noise, partition noise, thermal noise, or intermodulation noise as outlined in the next section. In evaluating the system performance quality, the quantities most frequently used are noise figure and noise temperature. The concept of these terms is applicable to the individual system components as well as the overall system. The calculation and measurement of noise power are dealt with in this and the following section.

Thermal noise from a resistor is often used as a reference noise source when determining noise figures. Thermal noise is the result of free charge carriers in a conductive material, moving about in a random motion when the material is above absolute zero in temperature. At any given instant in time a charge imbalance may exist across the material resulting in some voltage potential. The average voltage must be zero, but the rms voltage and therefore noise power available is not zero. The thermal noise spectrum has a uniform energy density distribution over the frequency spectrum to about 1000 GHz. Beyond this, the quantum effect becomes dominant, as in the case of optical lasers.

Noise in transistors stems from three major sources: thermal noise, shot noise, and partition noise.

The thermal noise is due to the resistance of the base region.

Shot noise, which sounds like a shower of lead shot striking a metallic target, is the result of recombination fluctuations of holes combining with the electrons in the base region. This shot effect is dependent upon the level of the bias current.

Partition noise is the result of the random fluctuations in the direction of carriers, when they have to divide between two or more paths. More electrodes cause an increase in partition noise. In vacuum tubes, pentodes are more noisy than triodes for this reason.

In addition to the above, at frequencies below a few kilohertz, flicker noise appears which increases as frequency decreases. It is thought to be due to surface recombination effects. It is of particular concern in low-frequency and dc amplifiers encountered in certain instrument and biomedical applications.

The maximum available thermal noise delivered by a conductor or resistor occurs when it is matched to a load. Under this condition, the available thermal noise is given by

$$P = k\,TB \tag{3-6}$$

where B is the bandwidth

T is the absolute temperature

k is Boltzmenn's constant $= 1.38 \times 10^{-23} \text{J} / °\text{K}$

The available noise power does not depend upon the resistance value. As long as the load resistance is matched to the source (generator) resistance, the available noise from the generator resistance is only dependent upon the temperature and bandwidth. This is normally the condition when connecting communication equipment in tandem. The output impedance of one unit is matched to the input impedance of the following unit.

Let us consider what occurs when a signal is processed by an element such as a repeater in a communication link. Independent of whatever other effects the element has on the incoming signal, some amount of noise will be added to the signal due to the fundamental noise sources within the element itself, as illustrated in Fig. 3-12. The signal-to-noise (S/N) power ratio at the output of the element must therefore be less than the S/N power ratio of the incoming signal. The "quality" of the signal is thus degraded.

As a measure of the amount of degradation an element will contribute, H. T. Friis defined the noise factor of an element as:

$$\text{Noise Factor} = \frac{\text{Available S/N power ratio at the input}}{\text{Available S/N power ratio at the output}}$$

In mathematical form, this can be expressed as follows:

$$F = \frac{P_{si}/P_{ni}}{P_{so}/P_{no}} = \frac{P_{si}}{P_{so}} \times \frac{P_{no}}{P_{ni}} \tag{3-7}$$

But the gain g of the element is:

$$g = \frac{P_{so}}{P_{si}} \tag{3-8}$$

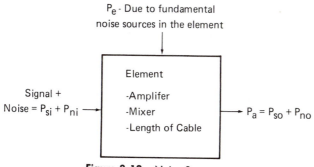

Figure 3-12   Noise Sources

Hence

$$F = \frac{P_{no}}{gP_{ni}} \qquad (3\text{-}9)$$

where

F is the element noise factor.

P_{si} is the signal power available from the generator.

P_{ni} is the signal generator available noise power.

P_{so} is the available signal output power of the element.

P_{no} is the available noise output power of the element.

g is the power gain of the element.

Equivalently, the noise factor of an element may be expressed as a noise figure, where

$$\text{Noise Figure} = F_{dB} = 10 \log (F) \qquad (3\text{-}10)$$

Note that F is always ≥ 1 and therefore $F_{dB} \geq 0$.

As part of the definition, the noise factor is determined using a simple source which has its internal impedance at the same temperature as the element (i.e. $P_{ni} = kTB$). With regard to the temperature T, it is assumed to be 290 °K (equal to 17 °C). Friis settled on this value because "it makes the value of kT a little easier to handle in computations." We will denote this reference temperature by

$$T_o = 290\,°\text{K}$$

Equation 3-9 can thus be rewritten as

$$P_{no} = F_g k T_o B \qquad (3\text{-}11)$$

where F is the repeater noise factor

g is the repeater power gain

k = Boltzmenn's constant = $1.38 \times 10^{-23} J°\text{K}$

T_o = 290 °K

B is the system bandwidth

This noise factor includes all noise due to processes within the repeater as well as room temperature thermal noise generated at repeater input. Note that the noise factor is an indication of the noise contributed by the repeater. If $F = 1$, the noise of the amplifier output is merely the noise at its input (kT_oB) multiplied by the power gain of the amplifier. The F can be measured and is usually given by the manufacturer of the equipment. It can be expressed in decibels as $F_{dB} = 10 \log F$ dB. Sometimes a noise temperature (T) is given that is related to the noise factor by:

$$F = 1 + \frac{T}{T_o} \qquad (3\text{-}12)$$

Equation ($3\text{-}11$) can also be expressed in dBm's as

$$P_{no} = F_{dB} + G - 174 + 10 \log B (\text{dBm}) \qquad (3\text{-}13)$$

where $\qquad G$ is the amplifier gain in dB's.

Repeater gains are normally designed to make up line losses, which results in a nominal 0 dB gain or loss over a complete repeater section, as shown in Fig. 3-13. In identical 0 dB gain sections, the total noise contributed will be mP_{no}. Equation (3-13) then becomes

$$P_{no} = F_{dB} + G - 174 + 10 \log B + 10 \log m (\text{dBm}) \qquad (3\text{-}14)$$

where
$\qquad F_{dB}$ = repeater noise figure (dB)
$\qquad G$ = repeater power gain in dB
$\qquad B$ = repeater bandwidth
$\qquad m$ = number of repeater sections
$\qquad P_{no}$ = output noise in dBm

Thermal and shot noise tend to be independent of signal levels, but an increase in signal levels will cause a decrease in the *effect* of thermal noise.

That is, as the signal level increases, the S/N ratio improves.

When making thermal/shot noise measurements on a voice circuit, they can be made either with the system in the idle or active state. The disadvantage of measuring noise in the absence of a signal is that the overall circuit characteristics are not necessarily the same as when the signal is present. This is chiefly due to the presence of automatic gain devices within the circuit. If a compandor is present in the circuit, the level of the applied tone should be such that no compression or expansion is occurring. In North America, it is common practice to employ C message filtering when measuring the background thermal and shot noise. Fig. 3-14 shows the test set-up for either case. The speaker is used to monitor the channel to make sure that crosstalk or a single frequency tone is not present to invalidate the measurement for the idle condition.

When the voice channel is "loaded" at the transmit end with a holding tone that represents the comparable power of an actual message, this tone must be filtered or notched out at the receiving end prior to the C weighting filter. A notch filter performs this function. Although 1000 Hz is the common frequency used in testing telephone circuits, it is harmonically related to the 8 kHz sampling frequency used in PCM carrier. This can result in significant measurement error and so the test frequency is slightly offset to 1004 Hz.

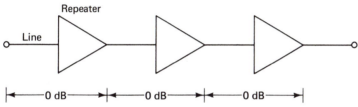

Figure 3-13   Illustration of Repeater Sections Where Line Loss = Repeater Gain

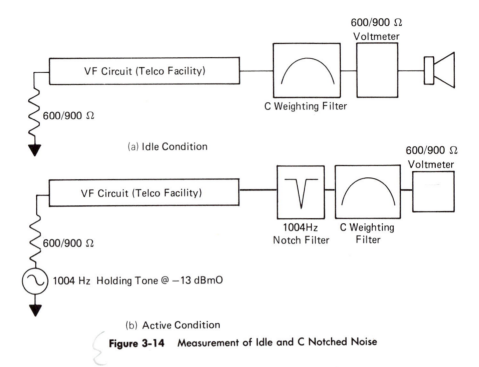

Figure 3-14 Measurement of Idle and C Notched Noise

Example 3-1

A typical 4-wire circuit has an end-to-end loss of 16 dB. With a 0 dBm 1004 Hz tone applied to this circuit, a C-notched reading of 38 dBrnc is obtained. What is the resulting signal-to-noise ratio, C message weighted?

Solution

At the receive end, the signal level will be −16dBm. The c-notched noise is at 38 − 90 = −52 dBm. Thus, the S/N = −16 − (−52) = 36 dB.

3-2.3 Harmonic Distortion

When an active circuit is driven into its nonlinear region, harmonics of the input signal begin to appear at the output. In the voice band the second and third harmonics are the most significant. Harmonic distortion is measured by applying a 704 Hz sine wave, and monitoring the fundamental (704 Hz), the second (1408 Hz), and the third (2112 Hz) harmonics with a frequency selective voltmeter. The distortion is the difference in dB between the harmonic and the fundamental component level.

Since there are several possible sources of harmonic distortion in a telephone channel, the distortion products may at some frequencies be additive and at other frequencies even cancel out because of the path length delays between these sources. This can cause a large measurement error, unless the distortion is chiefly the result of one major nonlinear source.

Total harmonic distortion is the ratio of the all-harmonic frequency products to the fundamental component, expressed as a percentage. This can be performed with a distortion meter which measures all the energy in its passband. By inserting a 704 Hz notch filter, the total remaining noise is measured. Since not only the harmonic distortion but also thermal and shot noise is present, the result must be corrected if only the harmonic distortion is to be determined.

3-2.4 Intermodulation Noise and Loading

If an amplifier is driven sufficiently hard, it is driven into a nonlinear operating region. Depending upon the nature of this nonlinearity, frequency products are generated that initially are not present. In general, the output/input voltage relationship of an amplifier may appear somewhat like that shown in Fig. 3-15, where the output-to-input voltage ratio tends to drop off at the higher voltages. The transfer function can be expressed as a MacLaurins series:

$$v_o = a_1 v_i + a_2 v_i^2 + a_3 v_i^3$$

(3-15)

where the nonlinear or higher order terms represent distortion.

When operating in the linear region, only the first term of expression 3-15 is present, and the output is an amplified replica of the input waveform. No new frequencies are generated.

Driving the circuit just a bit harder, the second order term becomes significant. As we shall soon see, this term generates a DC component—2nd harmonics of the input signals as well as the sum and difference frequency components of the input signals. Fortunately, the harmonics can frequently be rejected by filtering without compromising on the desired signal in the passband.

When driving the circuit yet harder, the third order term must be included in the output signal expression. This term produces products of the form $2f_1 - f_2$ and $2f_2 - f_1$, where f_1 and f_2 represent two applied signal frequencies. These terms can reside within the desired passband and cannot be removed without disturbing the desired signal. Terms of this nature are called intermodulation products.

Figure 3-15 Amplifier Gain Characteristic

To see how all these products are developed, consider an input signal consisting of two tones, i.e.,

$$v_i = A_1 \cos \omega_1 t + A_2 \cos \omega_2 t \tag{3-16}$$

where ω_1 and ω_2 are "in-band" information tones in the original signal.

$$\text{Let} \quad v_o = a_1 v_i + a_2 v_i^2 + a_3 v_i^3 \tag{3-17}$$

Substituting Eqs. *(3-16)* and *(3-17)* and using the identity $\cos\omega_1 t\cos\omega_2 t = \frac{1}{2}\cos(\omega_1 + \omega_2)t + \frac{1}{2}\cos(\omega_1 - \omega_2)t$, we obtain:

$$v_o = \frac{a_2}{2}(A_1^2 + A_2^2) + a_1[A_1 \cos \omega_1 t + A_2 \cos \omega_2 t]$$

$$+ a_2\left[\frac{A_1^2}{2} \cos 2\omega_1 t + \frac{A_2^2}{2} \cos 2\omega_2 t + A_1 A_2 \cos (\omega_1 + \omega_2)t\right.$$

$$+ A_1 A_2 \cos (\omega_2 - \omega_1)t\bigg] + a_3\left[\left(\frac{3A_1^3}{4} + \frac{3}{2} A_1 A_2^2\right) \cos \omega_1 t\right.$$

$$+\left(\frac{3A_2^3}{4} + \frac{3}{2}A_1^2 A_2\right) \cos \omega_2 t$$

$$+ \frac{A_1^3}{4} \cos 3\omega_1 t + \frac{A_2^3}{4} \cos 3\omega_2 t$$

$$+\frac{3}{4} A_1^2 A_2 \cos (2\omega_1 + \omega_2)t + \frac{3}{4} A_1 A_2^2 \cos (\omega_1 + 2\omega_2)t$$

$$+\frac{3}{4} A_1^2 A_2 \cos (2\omega_1 - \omega_2)t + \frac{3}{4} A_1 A_2^2 \cos (2\omega_2 - \omega_1)t\bigg] \tag{3-18}$$

Generally the two tones are adjusted to give equal amplitude sidebands, i.e., $A_1 = A_2$. Under such conditions the output spectrum appears similar to that shown in Fig. 3-16.

The two frequencies on either side of the desired fundamental tones f_1 and f_2 are the intermodulation products due to third order distortion, i.e.:

$$2f_1 - f_2 \text{ and } 2f_2 - f_1 \tag{3-19}$$

If we apply, for instance, two tones of frequencies 1k and 1.6 kHz to an SSB transmitter, the desired upper sideband components will be

$$f_1 = f_c + 1 \text{ kHz}$$
$$f_2 = f_c + 1.6 \text{ kHz}$$

where $\qquad$ f_c is the unmodulated carrier frequency.

The intermodulation products then will be:

$$2f_1 - f_2 = 2f_c + 2 \text{ kHz} - (f_c + 1.6 \text{ kHz}) = f_c + .4 \text{ kHz}$$
$$2f_2 - f_1 = 2f_c + 3.2 \text{ kHz} - (f_c + 1 \text{ kHz}) = f_c + 2.2 \text{ kHz}$$

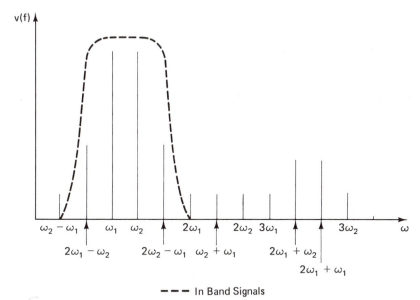

--- In Band Signals

Figure 3-16 Frequency Spectrum as a Result of Intermodulation Distortion

Note that the intermodulation products are separated from their nearest tones by the amount of the separation between the original tones. Both intermodulation products are in-band and once present cannot be removed.

Another important feature to note is the rate of increase in the amplitudes of the intermodulation terms as compared to the desired sideband components. Again, for equal amplitude side tones, the amplitudes of the third order intermodulation products are proportional to A^3 whereas the desired sidebands have amplitudes proportional to A ($a_3 << a_1$). Thus, if both fundamental components are doubled or increased by $20 \log 2 A/A = 6$ dB, the intermodulation components are increased by a factor of 8 or $20 \log (2A)^3/A^3 = 18$ dB.

For every 1 dB increase in the fundamental component, a 3 dB increase occurs in the intermodulation component. Single sideband transmitter maximum output power ratings are usually specified in terms of how far the intermodulation product must be below the power level of one of the two-frequency test signals.

To measure distortion in a telephone channel, two pairs of tones A and B centered at 860 and 1380 Hz are applied at a combined level equal to the data level. For wideband systems, white noise is applied to measure the distortion performance. This will be considered just a bit later in this section.

The in-band distortion products fall into three narrow bands centered at

$$520 \text{ Hz, 2nd order, } B - A$$
$$2240 \text{ Hz, 2nd order, } B + A$$
and
$$1900 \text{ Hz, 3rd order, } 2B - A$$

The combined power of these products are measured and compared to the received signal level. The measurement set-up is shown in Fig. 3-17. Initially S_1 is closed and S_4 open, to obtain the level of the tones. Then S_1 is opened, S_4 is closed and either S_2 or S_3 closed to determine the level of the second or third order products as so many decibels below the tone or signal level. As the power measured in the filter slots also includes shot and thermal noise, this gives a measurement of signal to distortion plus noise. To correct this, the shot and thermal noise must be measured. This is accomplished by cutting off the "B" pair and increasing the "A" pair at the transmitter by 3dB. In this way, the circuit is driven equally hard, but only the thermal and shot noise is left to pass through the filter slots. This measurement thus gives the signal-to-noise ratio. From these two measurements, i.e., $S/N + D$ and S/N, the signal-to-distortion ratio (S/D) can be determined. Let us consider an example.

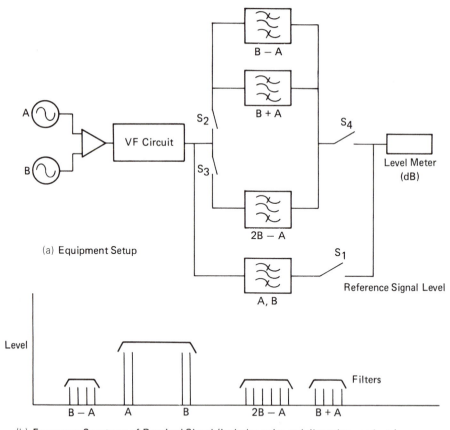

(a) Equipment Setup

(b) Frequency Spectrum of Received Signal (includes noise and distortion products)

Figure 3-17 Measurement of Nonlinear Distortion

Example 3-2

Using the measurement system of Fig. 3-17 (a), the signal-to-noise plus distortion ratios of the second and third order products are measured. This is accomplished by initially determining the signal level (S_1 closed and S_4 open) and then noting the change in dB when only the second or third order products are monitored. (S_1 open, S_4 closed, S_2 or S_3 closed). This measurement gives

	$S/(N + D)$
second order products ($B - A$, $B + A$)	26 dB (398.1)
third order products ($2B - A$)	28 dB (631.0)

Cutting off generator B and increasing A by 3 dB gives the following

	S/N
noise in second order slots	31 dB (1258.9)
noise in third order slots	32 dB (1584.9)

To determine the resulting signal-to-distortion ratio, we must solve S/D in terms of the measured $S/(N + D)$ and S/N ratios. Thus

$$\frac{S}{D} = \frac{S}{N + D - N} = \frac{1}{\dfrac{N + D}{S} - \dfrac{N}{S}} \qquad (3\text{-}20)$$

The results are tabulated as follows:

Products	$S/(N + D)$	S/N	S/D
2nd	398.1 (26dB)	1258.9 (31dB)	582.2 (27.7dB)
3rd	631.0 (28dB)	1584.9 (32dB)	1048.4 (30.2dB)

Note that the harmonics of the signal are not measured as they appear outside the second and third order slots. On a wideband system, noise-loading tests are performed to measure intermodulation noise. When several speakers are on a system, the average power is greater than it would be if there were only one speaker. With a second speaker added to a system, the average power can be expected to increase by 3 dB (10 log 2). For N talkers, each operating over a different frequency band added to a system, the average power will be increased by 10 log N dB, where N represents the number of talkers (channels). To simulate typical loads on a multichannel FDM system, CCITT recommends that white noise at the zero relative level point in the system (refer to section 1-2.5) be used. The white noise power levels are:

$$P_{avg} = -15 + 10 \log N \text{ dBm0} \quad \text{for } N \geq 240 \qquad (3\text{-}21(a))$$
$$P_{avg} = -1 + 4 \log N \text{ dBm0} \quad \text{for } 12 \leq N < 240 \qquad (3\text{-}21(b))$$

where N represents the number of channels and dBm0 is the power level in dBm at the system zero relative level point.

As more speakers or channels are added to a system, it is obvious that the average signal level increases. A good approximation of the magnitudes of these levels in a typical system is given by Eq. *(3-21)*. This formula includes the power contributed by pilot and signaling losses.

As the number of channels increases, the signal levels increase and the amplifiers can be driven into an overload, or nonlinear, state. This results in intermodulation noise. Intermodulation noise is measured by applying white noise of suitable level to the FDM system and keeping one channel-wide gap free from the applied noise. A band reject filter can be used for this purpose. At the receive end of the system, the intermodulation noise creeping into the channel gap is monitored.

Measurement of Intermodulation Noise

The measurement of intermodulation noise can be most conveniently made by applying white noise loading. Consider a noise loading set which measures the intermodulation and thermal noise in terms of a noise power ratio (NPR). The noise power ratio, expressed in decibels, is defined as the ratio of noise in the test channel with *all* channels loaded with white noise to noise in the test channel with all channels *except* the test channel fully noise loaded. To measure the NPR of a system, white noise of a suitable level (see Eq. *(3-21)* is passed through a band-stop filter to provide one "quiet" telephone channel within the baseband signal. The baseband signal with the quiet slot is applied to the transmitter. The resulting noise at the receiver output, which occurs within this quiet slot, is a measure of the thermal and intermodulation noise. Figure 3-18 illustrates the NPR measurement technique.

The channel signal-to-noise ratio or test tone level-to-noise ratio in terms of NPR is shown in Eq. *(3-22)*. The derivation of this equation is explained in Appendix B.

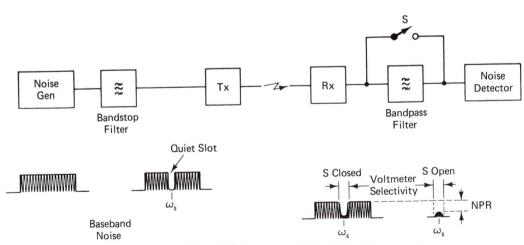

Figure 3-18 Intermodulation Noise Measurement

$$\left(\frac{S}{N}\right)_{dB} = NPR - (P_{avg})_{dBm0} + 10 \log \left(\frac{\text{original base bandwidth}}{\text{channel slot bandwidth}}\right) \qquad (3\text{-}22)$$

where   NPR is the noise power ratio
P_{avg} is given by Eq. *(3-21)*.

On first glance, it might appear that in order to improve the S/N ratio of a system, the signal level should be increased. This is certainly true if only thermal noise is considered. On the other hand, an increase in the signal level tends to increase the intermodulation noise. The goal in a communications system is to find an optimum signal level that will give a minimum amount of total noise. This is depicted in Fig. 3-19 which shows the optimum level at which all amplifiers should be operated.

Data transmission causes heavier line loading than voice communication. Data is usually transmitted at a maximum constant bit rate, operating the system at full capacity. In voice communication, the activity factor is typically down to 0.25, which represents the fraction of the system capacity being actively used at a given time. (The maximum bit rate has an activity factor of 1.) The activity factor in voice communication tends to be low because listening and talking are usually shared on a 50-50 basis and there are punctuation pauses during a conversation. In addition, a voice communications system is designed so that, on the average, less than the full complement of channels is in operation. By not fully utilizing all channels, blocking of calls is kept reasonable. If all N-lines are normally kept busy, all new calls would of necessity be turned away.

In order to cope with data transmission, data is usually permitted up to about 25% of the channels, and the rest is allocated to voice communications. Exceeding the 25% limit causes excessive intermodulation noise. In addition, data channels should not be near the group band edge where the filter group delays become highly non-linear.

3-2.5   Impulse Noise

Impulse noise, as the name suggests, is an impulsive or spikey form of interference. It is heard as a crackle and is generally not too objectionable in voice

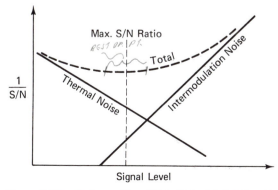

Figure 3-19   Optimum Signal(S)-to-Noise(N) Relationship

communication since it rarely causes the message to be misunderstood. It is of serious concern, however, in data communication since it can cause clusters of errors in a data bit stream.

Impulse noise appears in telephone systems chiefly because of the electro mechanical switches in the exchanges. These switches can cause vibrations to occur over entire equipment racks, inducing any contacting surfaces to create noise spikes. Impulse noise can also be the result of faulty soldering or dirty relay contacts. It can also originate from external sources, such as power lines and electrical storms.

The measurement of impulse noise requires that some present level be selected, so that a counter is actuated each time that a pulse exceeds this value. Typically this threshold level is 6 to 8 dB below the received signal level. The maximum rate of the electronic counter is also limited to about 8 counts per second.

3-2.6   Attenuation Distortion

As noted in Sec. 2-12, the attenuation of a transmission line is not flat with frequency but tends to vary as the square root of the frequency at the lower frequencies. Attenuation distortion is caused as the high frequency components in a signal experience greater attenuation than the lower frequencies as the signal travels down a transmission line.

Fortunately, this type of attenuation can be flattened out by the use of an attenuation equalizer. If the electrical network has a frequency loss that is complementary to that of the line, the net result is a flat frequency response. Inductive loading of cable also tends to flatten out the attenuation curve if the cutoff frequency is not exceeded. See Fig. 3-20.

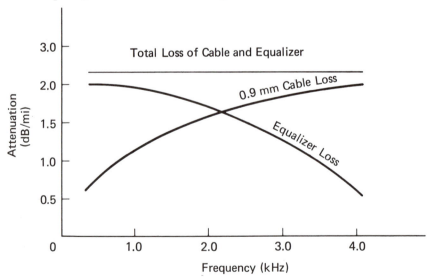

Figure 3-20   Attenuation Distortion of a 0.9 mm Cable

3-2.7 Envelope Delay Distortion

As mentioned in Sec. 2-12, the velocity of signal propagation over a transmission line tends to vary with frequency. Because of this variation, the different frequency components in a signal arrive at the receiving end at different times. This delay distortion is a problem when bandpass filtering is used since the delay at the band edges is quite different than at the midband frequency points.

Sharp filters cause greater delay distortion of frequencies, approaching the band edge than filters with a more gradual roll-off. Loaded cables also show a greater delay distortion than non-loaded cables.

To illustrate the effects of envelope delay, consider a 1000 Hz square passing through a 5 mile length of 22H88 loaded cable. Because of the low-pass filtering of the telephone facility, only the 1000 and 3000 Hz components will be present. As can be found from almost any reference on Fourier series (or as shown in Appendix A), the input signal can be expressed as

$$e_i = 3 \cos 2\pi 1000t - \cos 2\pi 3000t \qquad (3\text{-}23)$$

This waveform is plotted in Fig. 3-21.

From Table 2-7, it can be noted that the phase shift for 22H88 cable is 0.519 radians/mi at 1 kHz. At 3 kHz, this phase shift increases to 1.808 radians/mi.

For a cable circuit 5 miles long, the 1 kHz component will be shifted by 2.595 radians and the 3 kHz component will be shifted by 9.04 radians. Neglecting any attenuation effects, the output voltage can be expressed as:

$$e_o = 3 \cos (2\pi 1000t - 2.595) - \cos (2\pi 3000t - 9.04) \qquad (3\text{-}24)$$

This is plotted in Fig. 3-21(b). The distortions in the output waveform can be readily seen when compared with the input waveform.

Group or envelope delay can be measured by an amplitude-modulated signal, as shown in Fig. 3-22. The modulation frequency (ω_m) is constant and kept much lower than the carrier frequency (ω_c). As shown in Appendix C, the group delay is given by

$$T_g = \frac{db}{d\omega}\bigg|_{\omega_c} \qquad (3\text{-}25)$$

where b is the total phase shift experienced by the envelope as it travels down the line. If the modulation frequency (ω_m) is kept small enough, b will be linear (see Fig. 3-23(a)) within the $2\omega_m$ frequency band, and Eq. ($3\text{-}25$) can be approximated by:

$$T_g = \frac{\Delta b}{\Delta \omega}\bigg|_{\omega_c} = \frac{\Delta b}{\omega_m} \qquad (3\text{-}26)$$

In an actual measurement system, the modulation of frequency is 41.66 Hz as recommended by the CCITT for the 200 Hz to 200 kHz range. This is a compromise frequency since lower frequencies require an excessively high res-

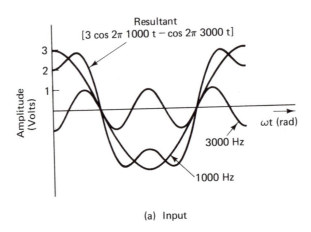

(a) Input

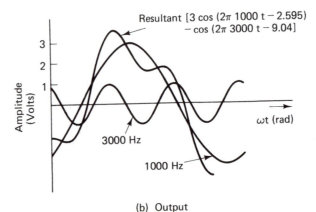

(b) Output

Figure 3-21 Input and Output Waveforms on a 5 Mile Section of 22H88 Cable. (Attenuation Neglected)

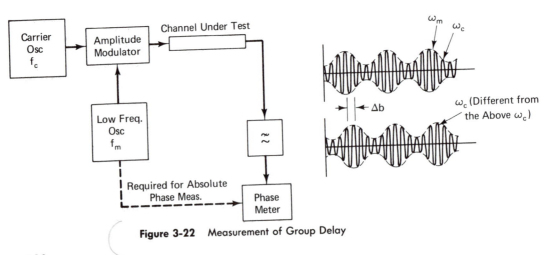

Figure 3-22 Measurement of Group Delay

olution phase meter, and higher frequencies distort the non-linearity required by the Eq. (*3-26*) relation.

After the AM signal passes through the network shown in Fig. 3-22, the output is demodulated and compared with the original modulating signal. As the systems measure the difference in phase shift between the two sidebands, the group delay can be computed as

$$T_g = \frac{\Delta b}{2\pi(f_m)}$$

By varying the carrier frequency ω_c, the entire group delay curve can be obtained across the bandwidth of the system.

Measuring the absolute delay between the sending and receiving ends is difficult to determine because of the large number of 2π radian phase shifts the signal undergoes. It is more common to measure the relative envelope delay. Relative delay uses a specific frequency as a reference, usually the one with the minimum delay, and measures the delay at the other frequencies by taking the difference between its delay and the reference frequency delay. Thus, from Fig. 3-23(b), the relative group delay at f_c is equal to

$$\Delta t_g = T_g(\omega_c) - T_g(\omega_r) \tag{3-27}$$

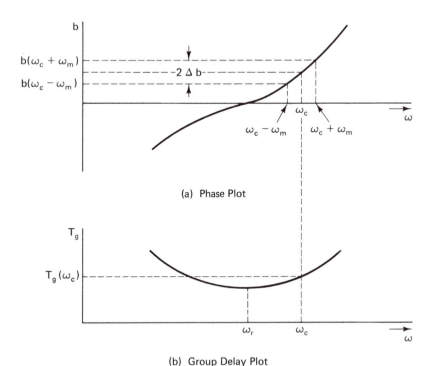

(a) Phase Plot

(b) Group Delay Plot

Figure 3-23   Group Delay Measurement Using an AM Signal with Modulation Frequency f_m.

Another technique of measuring envelope delay requires a four-wire circuit, as shown in Fig. 3-24. The test equipment transmits an AM signal whose carrier frequency, f_c, can be varied. At the receiver end, the envelope information is detected and transmitted back to the test instrument at a fixed carrier frequency. Any delay in the return loop will be constant and can be zeroed out. The test equipment compares the phase of retransmitted envelope with the original modulation signal, and the resulting delay can be monitored over the bandwidth of the channel by varying the carrier frequency.

As in the case of attenuation distortion, envelope delay distortion can be reduced by inserting delay equalizers. Delay equalizers have opposite, or complementary, phase characteristics to those of the transmission path, as shown in Fig. 3-25.

3-2.8 Line Conditioning

One method of improving the passband characteristics of a telephone channel and thereby increasing the information capacity of a telephone system is to provide special conditioning on the line. This requires the rental or leasing of a private line rather than the more common switched line or dialed-up connection since the latter are more or less randomly selected and can have no special conditioning applied. Private lines are more permanently connected and can be individually conditioned to provide improved channel characteristics. Other advantages and disadvantages of switched and private lines are listed in Table 3-1.

In order to obtain high-speed data communications on a voice-grade line, the attenuation distortion, envelope-delay distortion, the signal-to-noise ratio,

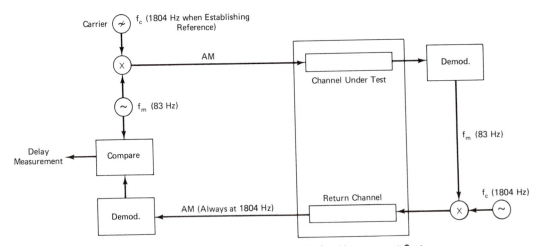

Figure 3-24 Envelope Delay Measurement System

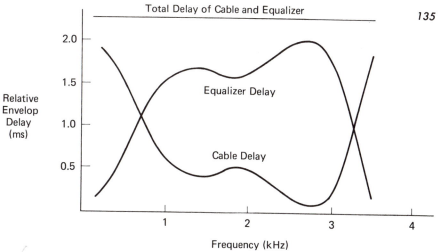

Figure 3-25 Envelope Delay Correction of a Voice Circuit Cable

Table 3-1   Switched vs. Private Lines

	Switched Lines	Private or Dedicated Lines
Advantages	Can access millions of other telephones/machines. Can access other computers if present computer is under repair or overloaded. Can use alternate line facilities in case of line failure. Low cost if usage is low. Can handle up to 9600 b/s 1600 baud.	Always available. Can be conditioned. More efficient in time as no time is spent on busy signals and handshaking. Can handle higher line speeds because of conditioning (less line noise and distortion) and no signalling frequencies to avoid. Performance quality of private lines is not as variable as for switched lines. Multiple station connections are possible.
Disadvantages	Limited to line speeds of 9600 bits/s. Lines can be busy. Subject to larger error rates due to impulse noise. Contain echo suppressors which interfere with data communications if not disabled. Time required to establish connection. Variability of performance from call to call	More costly for low usage as cost is based on distance rather than time used.

and the harmonic distortion must be controlled. The first two are controlled through C-conditioning and the latter by the more recent D-conditioning.

The commonly used 3002 unconditioned voice grade private line has certain attenuation and envelope-delay distortions as shown in Fig. 3-26(a). Attenuation distortion can be somewhat compensated for by employing circuit conditioning or equalization. This usually consists of several tandem networks near the receiving end of the line. The net result is a more flattened amplitude response, as shown in the C2-conditioned line of Fig. 3-26(b).

The delay equalizer operates in much the same fashion. The human ear tends to be insensitive to delay, and therefore, envelope delay distortion has not been of traditional concern. Delay distortion is of significance, however, for data transmission when pulses are employed.

Table 3-2 lists the attenuation and envelope delay limits as well as the signal-to-noise ratios for the most common types of channel conditioning. The attenuation distortion is defined as the level difference between the reference frequency (1004 Hz) and the actual frequency. The delay distortion indicates the maximum time differential between the arrival at the receiver of simultaneously generated signals having different frequencies.

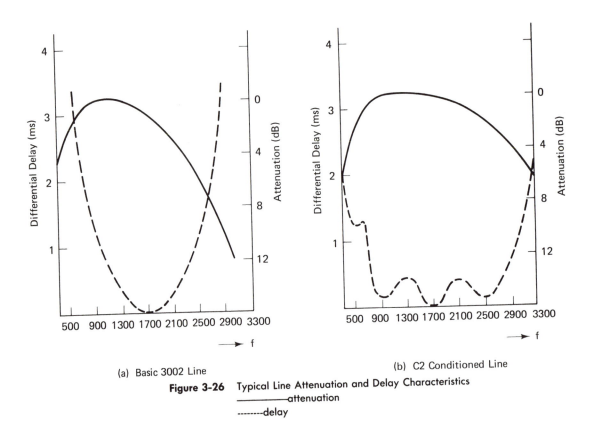

(a) Basic 3002 Line

(b) C2 Conditioned Line

Figure 3-26 Typical Line Attenuation and Delay Characteristics
——————attenuation
--------delay

Table 3-2   Conditioning Levels (Courtesy Hewlett-Packard Company)

	Non-Conditioned 3002 Channel		With C1 Conditioning		With C2 Conditioning		With C4 Conditioning		With D Conditioning
Frequency Range in Hertz (Hz)	300-3000		300-3000		300-3000		300-3200		
Attenuation Distortion (Net Loss at 1000 Hz)	Frequency Range	Decibel Variation	Frequency Response	Decibel Variation	Frequency Response	Decibel Variation	Frequency Response	Decibel Variation	
	300-3000	−3 to +12	300-2700	−2 to +6	300-3000	−2 to +6	300-3200	−2 to +6	
	500-2500	−2 to +8	1000-2400	−1 to +3	500-2800	−1 to +3	500-3000	−2 to +3	
			300-3000	−3 to +12					
Delay Distortion in Microseconds (μs)	Less than 1750 μs from 800 to 2600 Hz.		Less than 1000 μs from 1000 to 2400 Hz. Less than 1750 μs from 800 to 2600 Hz.		Less than 500 μs from 1000 to 2600 Hz. Less than 1500 μs from 600 to 2600 Hz. Less than 3000 μs from 500 to 2800 Hz.		Less than 300 μs from 1000 to 2600 Hz. Less than 500 μs from 800 to 2800 Hz. Less than 1500 μs from 600 to 3000 Hz. Less than 3000 μs from 500 to 3000 Hz.		
Signal to Noise (dB)	24		24		24		24		28
Non-Linear Distortion Signal to 2nd Harmonic (dB)	25		25		25		25		35
Signal to 3rd Harmonic (dB)	30		30		30		30		40

The cross-reference designations between the American Telephone and Telegraph Company (AT & T) and the Trans Canada Telephone System (T.C.T.S.) for the various types of conditioning are:

T.C.T.C. Schedule/Type	AT & T 3002
4/4	Basic
4/4A	C1
4/4B	C2
4/4C	C4

As an example, the recommended conditioning for the Bell 209 modem is:

3002 basic or schedule 4/type 4.

Example 3-3   When a +13 dBm 1000 Hz test tone is applied to a C2 conditioned channel, a received signal of −16 dBm is obtained. What are the received signal level limits between 300 and 3000 Hz?

Frequencies from 500 to 2800 Hz must be received at a minimum level of $-16 - 1 = -17$ dBm and at a maximum level of $-16 + 3 = -13$ dBm. Frequencies within the bands of 300 to 499 Hz and 2801 to 3000 Hz must be received at signal strengths between $-16 - 2 = -18$ dBm and $-16 + 6 = -10$ dBm. Fig. 3-27 graphically illustrates the limits on attenuation distortion for this C2 conditioned voice channel.

In carrier multiplex systems in which several channels form a group, the channels which are nearest the group band edge suffer the most severe delay distortion because of the highly nonlinear characteristics of the filter near the cut-off regions. Data transmission channels should not be allocated near these group edges but near the group center.

D-conditioning is a more recent option and not available throughout North America. It limits harmonic, or nonlinear, distortion. Nonlinear distortion is irreversible and cannot be compensated out once it is present. Once the signal has been altered by nonlinear impairments, the original signal cannot be re-created.

Attenuation delay equalizers are generally designed in a bridged-T or lattice configuration to compensate out distortions occurring in the passband. Operational amplifiers can also be employed in conjunction with suitably designed passive networks to obtain the desired passband delay characteristics. Typical circuits are illustrated in Fig. 3-28.

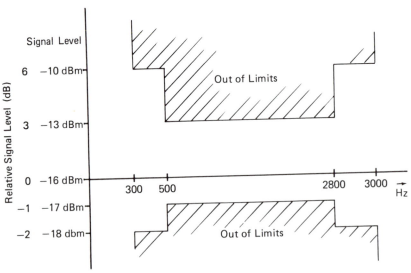

Figure 3-27 Attenuation Distortion Limits for C2 Conditioning. The Absolute Signal Levels Relate to Example 3-3.

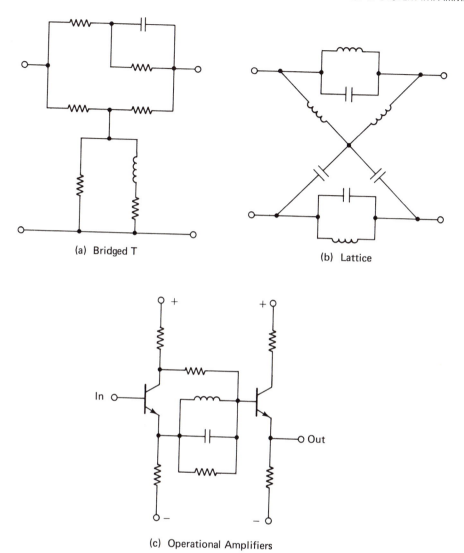

(a) Bridged T

(b) Lattice

(c) Operational Amplifiers

Figure 3-28 Attenuation and Delay Compensating Networks

3-2.9 Compandors

To reduce the dynamic range of a signal, compression is applied to the signal to slightly reduce the maximum signal amplitudes and significantly raise the minimum signal amplitudes. The intensity *range* of the input signal, normally 60 dB, is reduced to a range of 30 dB. At the receiving end, an expandor restores the signal to its original dynamic range. The joint combination of the

COMPressing-expANDING process is commonly called companding. Companding a signal results in:

    a. Improved signal-to-noise ratios, and
    b. Reduced peak powers with less likelihood of overloading.

To understand the reason for the reduction in noise contributed by the system, consider Fig. 3-29. Because the weaker signals at the receiver undergo the greatest attenuation, the low-level noise is attenuated more than even the weakest signal. This results in an improved signal-to-noise ratio as compared to that of the non-companded system.

    Compandors respond to signal level changes with a syllabic time constant, that is in the order of 20 ms. Instantaneous companding is not normally employed since it requires excessive bandwidths. A simplified diagram of a compressor and expandor circuit is shown in Fig. 3-30.

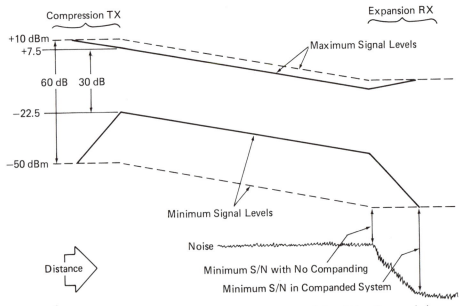

Figure 3-29 Signal-to-Noise (S/N) Ratios in Companded and Non-Companded Systems. The dotted lines refer to non-companded system.

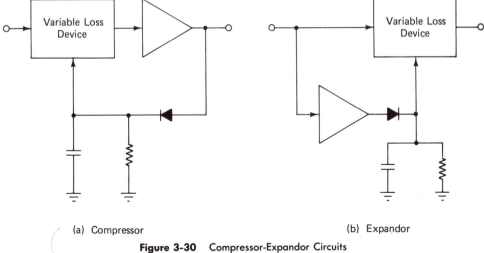

(a) Compressor   (b) Expandor

Figure 3-30   Compressor-Expandor Circuits

Problems

1. a. Why is it so difficult to match the balance network in a fourwire terminating set to the line characteristic impedance?
   b. What is the characteristic impedance of #22 CSA NL line at 1000 Hz? (See Table 2-7.)
   c. What is the balance-return loss of a 1 kHz signal applied to the return arm of a four-wire terminating set when the balance network impedance is 600 Ω?
   d. If a signal experiences 3.5 dB loss as it traverses through a hybrid transformer (refer to Fig. 3-3), what residual power appears in the GO arm when a 0.1 mW signal appears in the RETURN arm under the conditions given in (c)?

2. What does the reference equivalent of a telephone circuit represent?

3. a. What is a singing circuit, and how can it be prevented?
   b. In the telephone network of Fig. 3-4, the balance-return loss of the east circuit is 40 dB and that of the west circuit is 50 dB. If the two hybrid coils each have a transmission loss of 3.5 dB in either direction, calculate the total repeater gain if a 6 dB gain margin against singing is provided.

4. a. What is an echo suppressor?
   b. How long a loop of #19 DNB NL can be used before an echo suppressor must be included? With the same line loaded with H88, how long may it be before a suppressor must be added? (Refer to Table 2.7 for velocity data.)

5. a. What does the Via Net Loss of a circuit represent?

   b. Find the Via Net Loss required for a 1000 mi L-carrier trunk circuit.

   c. If the 1000 mi L-carrier is the only link between end offices, what is the desired overall connection loss?

6. What causes crosstalk, and how can it be reduced?

7. a. What causes intermodulation noise?

   b. In Eq. (*3-18*), what is the second harmonic term?

   c. When monitoring the harmonic products and intermodulation products on a spectrum analyzer, how can one determine which product is a harmonic and which is an intermodulation term? (Note: Varying the amplitudes of the test tones results in different amplitude charges of the unknown products.)

   d. Why are intermodulation products often of more concern than harmonic, sum and difference products, etc.?

8. The NPR in a 12-channel CCITT group is measured at 25 dB following the CCIR procedure of measuring signal-to-intermodulation noise under busy conditions. (See Eq. *3-22*). What is the intermodulation S/N ratio?

9. a. What causes thermal and shot noise?

   b. What thermal noise power is available from a 100 kΩ resistor at 20 °C over a 1 MHz bandwidth?

10. a. A 2000 km system consists of 100 repeater sections. If the repeater gain at the highest 300 to 3400 Hz bandwidth channel frequency is 50 dB and the noise figure is 10 dB at this frequency, what is the total noise from the system in dBm?

   b. If the relative output level of the repeater is -10 dBr, what is the noise referred to the 0 TLP in dBm0's?

   c. If this white noise over the 4 kHz bandwidth is C-message weighted, what is the C-message-weighted noise at the 0TLP in dBm's and in picowatts?

11. Define activity factor, and give reasons why it is less than 100%.

12. To achieve maximum S/N ratios, can signal levels be continuously increased? Why or why not?

13. State the nature and origin of impulse noise in the telephone network.

14. a. State the chief causes of attenuation and envelope delay distortion in a cable pair.

   b. What name is given to the devices that partially compensate out the attenuation and envelope delay distortion?

15. The phase delay representing the transmission time of a single frequency signal component through a transmission line is given by

$$T = \frac{b}{\omega}$$

where b is total phase shift in radians
                               $\omega = 2\pi f$
                               T = phase delay in unit of time

The envelope or group delay is given by:

$$T_g = \frac{db}{d\omega} \text{ (see Eq. } (3\text{-}25))$$

This represents the *slope* of the phase curve at the frequency in question.

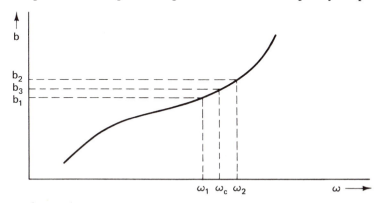

a. Under what conditions are the phase delay and envelope delays identical?
b. Use the following graph to illustrate how T and T_g can be determined at ω_c.
c. Use Eq. (*3-26*) to explain why the modulation frequency in the group delay measurement must be kept small.
d. What is the difference between absolute and relative envelope delay?

16. Find the difference in phase delay in units of time of the two frequency components illustrated in Fig. 3-21. Refer to Eq. (*3-24*).

17. a. State the advantages and disadvantages of switched and private telephone lines.
b. What is a conditioned line?
c. What is the difference between C- and D-conditioning?

18. What is the purpose of a compandor?
What is its principle of operation?

19. Using the measurement system of Fig. 3-17, the following signal-to-noise plus distortion and signal-to-noise ratios are measured.

	$\dfrac{S}{N\&D}$	$\dfrac{S}{N}$
Second order product	25 dB	30 dB
Third order product	27 dB	31 dB

Determine the resultant signal-to-distortion ratios.

4

Physical Layer Protocols

4-1   Introduction

In order to interconnect the various types of digital equipment supplied by different manufacturers or to interconnect the devices in a distributed data processing system, international standards are developed and adhered to. The International Organization for Standardization (ISO), a voluntary association of national standards bodies, and the Consultative Committee for International Telephony and Telegraphy (CCITT) are currently defining a hierarchy of coordination procedures for interconnecting computers, computers-to-terminals and terminals-to-terminals. Such a standard defines a set of conventions governing the format and relative timing of messages (protocol) to make communication possible between terminals and networks.

In these and other standards the electric and physical connection procedures are defined at the location between the Data Terminal Equipment (DTE) and the Data Communication Equipment (DCE). The DTE comprises the data source, the data sink, or both. It usually provides for control logic and buffer store. Examples are terminals, computers, etc.

The DCE provides the functions to establish, maintain and terminate a connection, and to code/decode the signals between the DTE and the data channel. Examples are modems or data sets, line drivers, etc.

Fig. 4-1 is a simplified diagram of the seven-layer Open Systems Interconnections (OSI) model developed by the ISO to standardize procedures for exchanging information among computers, terminals and networks. Although much development work still needs to be done on the higher level layers, this model is gaining increasing acceptance. It should be noted that the internal method of generating the OSI protocol is not standardized, only how each system follows the communications protocol. The essence of the layer concept is that a layer uses services provided by the adjacent lower layers and adds function to provide its own services. These layers are defined as:

Level 1. The physical layer concerns itself with the link's physical interfaces; e.g., EIA RS-232-C, CCITT, V. 24 and V. 28.

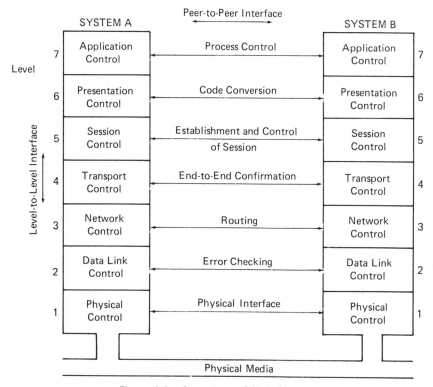

Figure 4-1   Seven-Layer OSI Architecture

Level 2. The Data Link layer sets up and disconnects a link, controls the flow of data between generator and receiver, detects errors in bit stream transmission; e.g., High Level Data Link Control (HDLC), X. 25 Link Access Procedure.

Level 3. The network control includes the procedures for establishing and disconnecting a vitual circuit and for controlling the flow of data packets in a packet network. In a packet network the physical circuit is dynamically allocated. Digital data is sent in short packets and routed through a system via modes and links. At present, most data communication systems still use a physical circuit which is assigned either permanently (private or leased line) or for the duration of the call (dial-up line); e.g., X. 25 Packet Level.

Higher Levels. Manages the exchange of data under user control.

(In this chapter, frequent references are made to several of the Electronic Industries Association standards. Permission has been granted by the Electronic Industries Association, 2001 Eye Street, N.W., Washington, D.C. 20006 to refer to and to reprint certain portions of the RS-232-C, RS-422-A, RS-423-A, RS-449 standards and the IEB-12 bulletin.)

4-1 EIA RS-232-C Interface Between DTE and DCE

The first level protocol standard commonly in use throughout North America to connect DTE such as computers, teletypewriters, business machines, and printers to DCE or more specifically Modems or Data Sets is the EIA RS-232-C standard developed by the Electronic Industries Association. This standard is defined for serial binary data communications, and conforms to the internationally recognized CCITT V. 24 standard. The reader may refer to the document RS-232-C "Interface between Data Terminal Equipment and Data Communication Equipment Employing Serial Binary Data Interchange," Aug. 1969, Electronic Industries Association, for the complete description of this standard. This set of standards ensures:

a. Compatible voltage and signal levels

b. Common pin wiring configurations

c. A minimum amount of control information between the DTE and DCE as illustrated in Fig. 4-2.

The RS-232-C standard not only conforms to the V. 24 standard but it also specifies the mechanical connector type and pin assignments for each signal. The DB-25 (D-type 25 pin connector shown in Fig. 4.3) is the most common plug and socket associated with the RS-232-C standard, although this standard makes no mention of it.

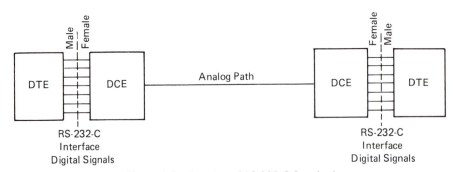

Figure 4-2 Location of RS-232-C Standard

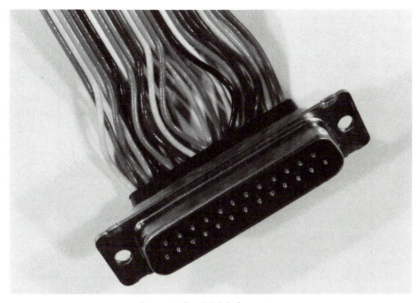

Figure 4-3 DB-25 Connector

In addition to the protective and signal grounds, all circuits carry bipolar voltage signals. Voltages at the connector pins with respect to the signal ground (circuit AB) may not exceed ±25V. Any pin must be able to withstand a short circuit to any other pin without sustaining damage to itself or any associated equipment. With a 3 k to 7 k ohm load, the driven output represents a logic "0" for voltages between +5 to +15 volts and a logic "1" for voltages between −5 and −15 volts. A voltage at the receiver from +3 to +15 volts represents a "0" while a voltage between −3 and −15 volts represents a "1." Voltages between ±3 V lie in the transition region and are not defined.

Fig. 4-4 graphically represents the relationship between voltage level and signal condition. As can be seen, a 2 volt departure in the output voltage level can be sustained without causing any error at the receiver. The interchange circuit can be used for control, data, or timing and its drives can be located in the DTE or DCE. Its equivalent circuit (see Fig. 4-5) must always meet the following criteria.

$$|\text{open-circuit voltage, } V_o| < 25$$
$$3000\Omega < R_L < 7000$$
$$C_L < 2500 \, pF$$
$$\text{voltage slew rate} = \frac{dv}{dt} < \frac{30 \, V}{\mu S}$$

Maximum time for signal to pass through transition region: the lesser of 1 ms or 4% of signal element duration.

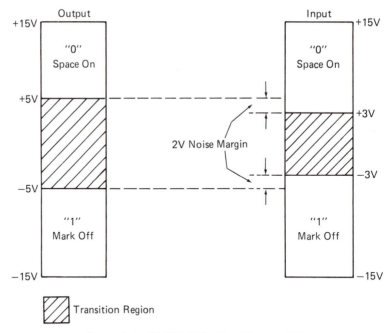

Figure 4-4 RS-232-C Electrical Characteristics

The limitation on slew rate is concerned with the problem of cross talk between conductors in a multiconductor cable. The faster the transition the greater the cross talk. Fig. 4-6 illustrates the slew rate restriction on a voltage waveform. Both the driver and receiver circuits use a common signal ground for all the circuits. Such single-ended operation severely restricts the maximum signaling rates because of the noise introduced by ground current.

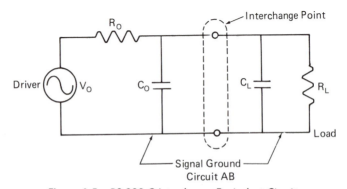

Figure 4-5 RS-232-C Interchange Equivalent Circuit

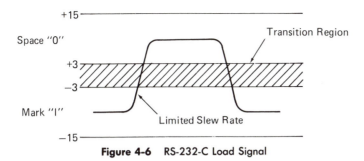

Figure 4-6 RS-232-C Load Signal

The RS-232-C standard is applicable to signaling rates up to 20,000 b/s. The female side of the 25-pin connector is associated with the DCE. Since cables must not present more than 2500 pF at the interface point, the distances between the DCE and DTE are usually kept to below 100 m. Short cables of less than 50 feet (15 m) are recommended.

Just because a device adheres to the RS-232-C specifications, this does not mean that it will operate with other equipment following the same EIA standards. The EIA standard, for instance, does not control clock timing, idling characteristics, or control characteristics. These may be specified by the user or manufacturer.

The data and control lead terms can be summarized as follows:

Data Circuit Interface Terms (BA and BB Circuits)		
Binary state	"1"	"0"
Signal condition	Mark	Space
Voltage	Negative	Positive
Control Lead Terms		
Control function	OFF	ON
Voltage	Negative	Positive

Table 4-1 identifies and briefly describes each interface circuit, gives its pin assignment, and the direction of the signal flow.

Although the two-letter circuit designated is not too enlightening in its abbreviation, the first letter indicates into which category the circuit falls:

A—Ground or Common Return

B—Data Circuits

C—Controls Circuits

D—Timing Circuits

S—Secondary Channel Circuits

The function of each of these circuits is described in the following.

Table 4-1   RS-232-C Interchange Circuits

Interchange Circuit	Pin Assignment	Description	Mnemonic	Data		Control		Timing	
				From DCE	To DCE	From DCE	To DCE	From DCE	To DCE
AA	1	Protective Ground	PG						
BB	7	Signal Ground/Common Return	SG						
BA	2	Transmitted Data	TD		X				
BB	3	Received Data	RD	X					
CA	4	Request to Send	RTS				X		
CB	5	Clear to Send	CTS			X			
CC	6	Data Set Ready	DSR			X			
CD	20	Data Terminal Ready	DTR				X		
CE	22	Ring Indicator	RI			X			
CF	8	Received Line Signal Detector	DCD			X			
CG	21	Signal Quality Detector	SQ			X			
CH	23	Data Signal Rate Selector (DTE)					X		
CI		Data Signal Rate Selector (DCE)				X			
DA	24	Transmitter Signal Element Timing (DTE)	TSET						X
DB	15	Transmitter Signal Element Timing (DCE)	TSET					X	
DD	17	Receiver Signal Element Timing (DCE)	RSET					X	
SBA	14	Secondary Transmitted Data	(S)TD		X				
SBB	16	Secondary Received Data	(S)RD	X					
SCA	19	Secondary Request to Send	(S) RTS				X		
SCB	13	Secondary Clear to Send	(S)CTS			X			
SCF	12	Secondary Rec'd Line Signal Detector	(S)DCD			X			

Circuit AA (pin 1): Protective Ground

This conductor is electrically bonded to the equipment frame. It is usually connected to external grounds, allowing any sizable static charge to be bled away without affecting any signal lines.

Circuit AB (pin 7): Signal Ground or Common Return

This conductor is the signal ground line which provides the common ground reference level for all the interchange circuits. Except for the protective ground, all signals flow through this one common return line.

Circuit BA (pin 2): Transmitted Data

The signals on this circuit are transmitted to the DCE from the DTE. It transfers serial data and is held in a MARK state when no data is being transmitted.

If the RS-232-C is fully implemented, the following four circuits must be in the ON state for data to be transmitted:

DSR
DTR
RTS
CTS

Circuit BB (pin 3): Received Data

The signals on this circuit are transferred in bit serial form from the DCE to the DTE when receiving. It must be held in the MARK condition when the received line signal detector is in the OFF condition. On half-duplex operation, circuit BB must be in the mark condition anytime RTS is ON.

Circuit CA (pin 4): Request to Send

This circuit is turned ON by the DTE to indicate that it is ready to transmit and that the DCE must begin to prepare to receive data from the local DTE and enter the transmit mode. In half duplex operation, it also inhibits the receive mode. Turning this lead ON will cause the DCE to turn on its carrier.

In some instances, when the RTS lead goes ON, a complete connect sequence to set up the communications is performed—usually by an autodialing unit. After some delay, the DCE turns ON the Clear to Send (CTS) circuit. This informs the local DTE that it can begin to transmit data.

When the load DTE has no more data to transmit, it turns the RTS line from ON to OFF. After a brief time delay to ensure all the data that has crossed the interface point has been transmitted, the local DCE turns the CTS line OFF.

Once the RTS circuit is turned OFF, it may not be turned ON again until the CTS circuit has been turned OFF by the DCE. Fig. 4-7 illustrates the RTS and CTS handshaking procedure.

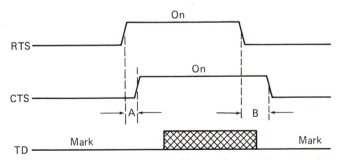

Figure 4-7 RTS and CTS Handshaking, Showing Set-up Delay (A) and Turn-off Delay (B) of the CTS Circuit

Circuit CB (pin 5): Clear to Send

Signals on this circuit activated by the DCE inform the DTE that it is ready to receive and to retransmit data from the DTE.

CTS is turned on in response to simultaneous ON conditions of the RTS, DSR and, where implemented, DTR circuits. Where RTS is not implemented, RTS is assumed to be always ON.

Circuit CC (pin 6): Data Set Ready

This circuit is turned ON by the DCE to indicate to the DTE that it is connected to the line. In automatic calling situations, this means that the number has been dialed by the DCE, the call establishment completed and in data mode.

Circuit CD (pin 20): Data Terminal Ready

This line is turned ON by the DTE, indicating that it is ready to transmit or receive data. DTR must be ON before the DCE can turn on DSR, indicating that the Data Set has been connected to the communications channel. DTR and DSR indicate equipment readiness. RTS and CTS indicate channel readiness.

When DTS goes OFF, the DCE is removed from the communication channel following the completion of any "in process" transmission. In switched network applications, when DTR is turned OFF, it cannot be turned ON again until DSR is turned OFF.

Circuit CE (pin 22): Ring Indicator

The ring indicator is turned ON by the DCE while ringing is being received. It is not disabled by the OFF condition of DTR.

Circuit CF (pin 8): Received Line Signal Detector

The ON condition on this circuit sent by the DCE informs the DTE that it is receiving a carrier signal which meets its suitability criteria from the remote DCE. This circuit is commonly called "data carrier detect" or "carrier detect" (DCD or CD).

In modems, this circuit is held on as long as it is receiving a signal that can be recognized as a carrier. On half duplex channels, DCD is held OFF when RTS is in the ON condition.

Circuit CG (pin 21): Signal Quality Detector

Signals on this circuit from the DCE indicate whether or not there is a high probability of an error in the received data. An OFF condition indicates there is a high probability of an error. An ON condition suggests that no errors have occurred.

Circuit CH (pin 23): Data Signal Rate Selector

Signals on this circuit control the feature of some data sets which allows them to transmit at two different signaling rates. When turned ON by the DTE, the higher signaling rate is selected.

Circuit CI (C.C.I.T.T., circuit 112)

This is similar to circuit CH; however, rate selected is from the DCE.

Circuit DA (pin 24): Transmitter Signal Element Timing (DTE Source)

This infrequently used option clocks the modem transmitter from an external clock. The clocking signal is developed by the DTE. The serial bit stream to be transmitted is applied to pin 2 with each bit occurring at the positive-going transition of the clock. The first bit must appear at the first positive-going transition after CTS goes ON.

Circuit DB (pin 15): Transmitter Signal Element Timing (DCE Source)

This "transmit clock" is used to provide the DTE with signal element timing to clock the bit serial data from the DTE to the modem for transmission. This clocking signal is developed by synchronous modems only.

Circuit DD (pin 17): Receiver Signal Element Timing

This "receive clock" is used to provide the DTE with received signal element timing to clock the bit serial received data from the modem to the DTE. This clocking signal is developed by synchronous modems only.

Circuit SBA (pin 14): Secondary Transmitted Data
SBB (pin 16): Secondary Received Data
SCA (pin 19): Secondary Request to Send
SCB (pin 13): Secondary Clear to Send
SCF (pin 12): Secondary Received Line Signal Detector

The secondary channel circuits control the secondary channel in much the same manner that the primary channel is controlled. Reverse or secondary channel transmission flows in a direction opposite to the primary channel transmission. The secondary channel is also under the control of DTR and DSR. These circuits function only with modems that have the "reverse channel" feature or the ability to operate full-duplex over a two-wire channel. As an example, a "high" on the secondary received line signal detector (S) DCD indicates that a reverse channel is being received.

Secondary channels are often used for circuit assurance; e.g., the receiving modem is operable or inoperable or as a supervisory channel for error detection and control. The aim may be to retransmit the signal that was received in error or to interrupt the flow of data in the primary channel. No actual data is transmitted in the secondary channel, only the presence or absence of the secondary channel carrier.

In this application, the STD, SRD, and SCTS circuits are not provided. Circuit (S) DCD recognizes the presence or absence of the second channel carrier, switched ON or OFF by the SRTS circuit.

4-3   Bauds and Bits

A baud (Bd) is a term used in data communications to indicate the rate at which changes occur in a signal over a given period of time. It is the unit of signaling speed, or modulation rate, and is found by taking the reciprocal of the shortest signaling element. For the ASCII code illustrated in Fig. 1-19, if we assume that the shortest signaling element is 208.3 μs in duration, the signaling rate is $1/T = 1/208.3 \times 10^{-6} = 4800$ Bd.

The baud rate should not be confused with the information rate, bits per second, the rate at which the actual data is transmitted. For the same ASCII code word just mentioned, if each 7-bit character is surrounded by one-unit stop and start bits and one parity bit, the byte becomes 10 elements long. The actual information rate is then $7/10 \times 4800 = 3360$ b/s. Thus 1440 b/s are lost due to error detection and synchronization.

A signal which can take on n-signaling conditions, such as a multifrequency or multilevel signal having n unique frequencies or levels, can be represented in its various conditions by $\log_2 n$ bits. For 4 signaling conditions, $\log_2 4 = 2$ bits; i.e., 00, 01, 10 and 11.

To express the rate at which the information is transmitted in b/s, if each signaling element is T seconds in duration, and if each signaling element represents $\log_2 n$ bits, the information rate is given by:

$$\frac{1}{T} \log_2 n \quad \text{b/s} \qquad (4\text{-}1)$$

i.e., bit rate = baud rate $\times$ no. of bits/baud. This expression assumes no start/stop or parity bits.

If we have a two-level system, i.e., the binary system, then $\log_2 2 = 1$ and we say that there is one bit per baud. For a four-level system, there are two bits per baud; for eight levels, three bits per baud; and so forth. In an eight-state signaling system with 10 ms unit signaling elements, the data information rate would be

$$\frac{1}{0.01} \log_2 8 = 300 \text{ b/s}$$

The signaling rate is 1/0.01 or 100 baud.

To obtain the information rate for a parallel system, the information rate for each channel is calculated and the results summed.

Take the case of a modem that transmits a constant amplitude signal and takes on one of four different phases. The four phases can be represented by two bits as 00, 01, 10, and 11. In this case, each baud represents two bits of information. For a 1200 baud system, the information rate will be 2400 b/s.

The Data Transfer Rate is another term used to describe data transfer rates on an operating system. It is defined by the CCITT as "the average number of bits, characters or blocks per unit time passing between corresponding equipment in a data transmission system." Since systems normally introduce errors, the information bits accepted by a terminal are less than the number that are transmitted. To distinguish between data transfer rate and information rate, it should be noted that data transfer rate equals information rate less error rate.

4-4   Balanced and Unbalanced Electrical Circuits

Two of the real drawbacks of the RS-232-C standard are the limited data transmission rate of 20 k b/s and the 15 m cable length. The unbalanced nature of the circuitry employed is the main cause for this. Although unbalanced lines (see Fig. 4-8(a)) in which one line is kept at ground are still the most common, the trend is towards a balanced system (see Fig. 4-8(b)). This is clearly illustrated by the more recent EIA RS-422 and RS-449 standards, recommending the use of a balanced interface.

To more clearly understand the advantages of balanced lines, consider two typical applications: a single-ended operation using an unbalanced line and a differential operation using a balanced line. In a single-ended operation, as illustrated in Fig. 4-9, induced noise is added to the signal voltage as a result of magnetic or electric coupling of adjacent signal lines and as a result of any ground voltage difference due to the presence of 60 Hz. There is no way of removing such noise once it is present in the system.

In addition, the logic ground at the receiver may differ from the logic ground at the driver because of the ground currents produced by other signals in the ground wire or because these devices are on different electrical distribution systems. This can cause severe voltage sensing problems. If, for instance, the receiver ground (circuit AB) is 4 volts higher than the driver ground, with a +6 V asserted by the driver on the transmit data line (BA), the receiver senses only 2 V. With a space transmitted under RS-232-C standards, the received signal falls into the transition region or no man's land.

The differential operation can ignore any ground voltages since they are

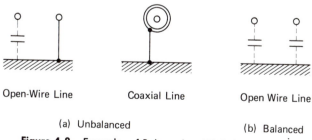

Open-Wire Line	Coaxial Line	Open Wire Line
(a) Unbalanced		(b) Balanced

Figure 4-8   Examples of Balanced and Unbalanced Lines

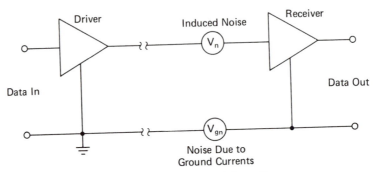

Figure 4-9 Single-ended Operation

common to both inputs of the differential receiver. (Refer to Fig. 4-10.) In addition, the differential operation is insensitive to any inductively or capacitively coupled noise which is commonly found in both conductors of the balanced line. Net noise resulting from any unbalance in coupling can be reduced by employing twisted lines which cancel out any transverse magnetic field by alternating the polarity of the induced voltage.

Capacitively coupled noise can be minimized by assuming longitudinal balance so that both lines are equally affected by noise. Shielding and keeping all line lengths short also aids in minimizing noise.

An illustration of a properly terminated balanced transmission line, in which each conductor has identical impedances with respect to ground, is shown in Fig. 4-11. This system employs a differential line driver and receiver. Each leg of the line is terminated with an $R_o/2$ resistor, resulting in no reflections if a balanced line or characteristic impedance R_o is used. Data is inserted on the balanced line by unbalancing the line voltage with driver current. The strobe feature will allow line-sharing or party-line operation, thus conserving twisted-pair line costs.

When the transmitting distances go beyond a few thousand meters, modems or data sets must be used to convert the digital or two-state signals to analog signals at the transmitting end and to perform the reverse function at the receiving end. The resulting analog line signal has less distortion than if

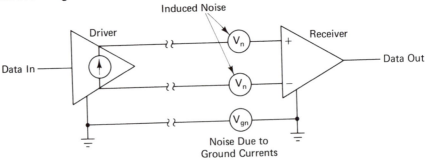

Figure 4-10 Differential Circuit Operation

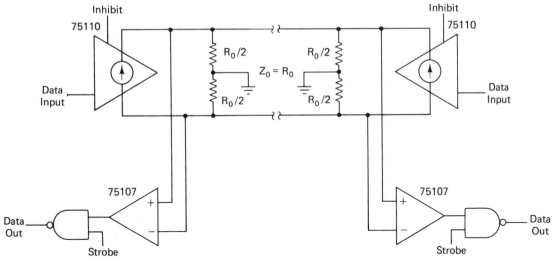

Figure 4-11   Differential Transmission Employing Fairchild Devices

the original baseband had been employed and therefore can be transmitted over much longer distances.

4-5   EIA RS-422-A, 423-A and 449 Interface Circuits

In order to improve upon the signaling rate and distance characteristics of the EIA RS232-C standard for binary data interchange, the EIA has introduced two standards, the RS-422-A and the RS-423-A, that employ differential inputs at the receiver. (EIA Standard RS-422-A, "Electrical Characteristics of Balanced Voltage Digital Interface Circuits," Dec. 1978. EIA Standard RS-423-A, "Electrical Characteristics of Unbalanced Voltage Digital Interface Circuits," Sept. 1978.) Fig. 4-12 shows the schematic diagram for the circuits associated with these standards.

Although the RS-423-A still uses an unbalanced line, only one end is ground, preventing low frequency ground current loops. Also, each signal has its own ground return line, preventing the accumulation of large current flow in the return path as in the case of the RS-232-C standard. By using lower resistance or capacitance cable, the cable lengths can be increased.

The relationship between signaling rate, signal risetime, and cable length is specified in RS-423-A, and given in Fig. 4-12(b). These curves relate to #24 AWG twisted wire pair having a shunt capacitance of 52.5 pF/m, connected to a 50 ohm generator delivering 12 V peak-to-peak. The signal risetime, specified from 10% to 90% of the differences of the steady-stage voltages, limits the crosstalk to 1 V peak. This is generally much slower than the RS-232-C rise-

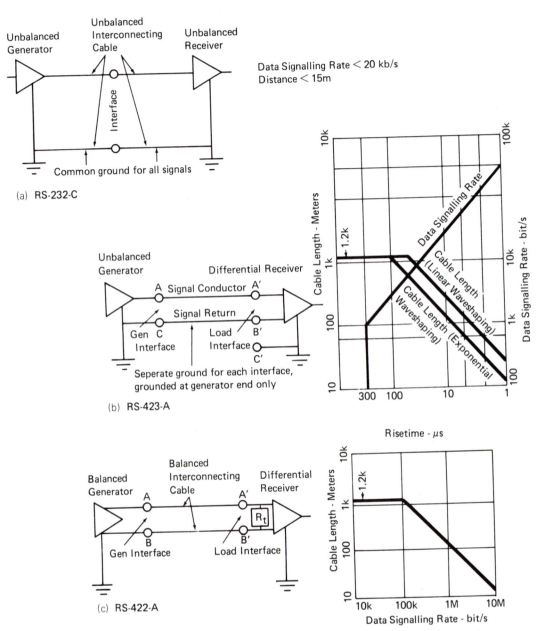

(a) RS-232-C

(b) RS-423-A

(c) RS-422-A

Figure 4-12 EIA Electrical Interface Circuits with Corresponding Data Signaling Rates versus Cable Length Using #24 AWG Twisted Pair Cable, 52.5 pF/m

time, which specifies that the time for the signal to pass through the ±3 volt transition region shall not exceed 4% of the signal element duration.

For comparison, at 20 kHz the maximum time allowable in the transition region is $\dfrac{.04}{20 \times 10^3}$ or 2 μs for the RS-232-C specification. From Fig. 4-12(b),

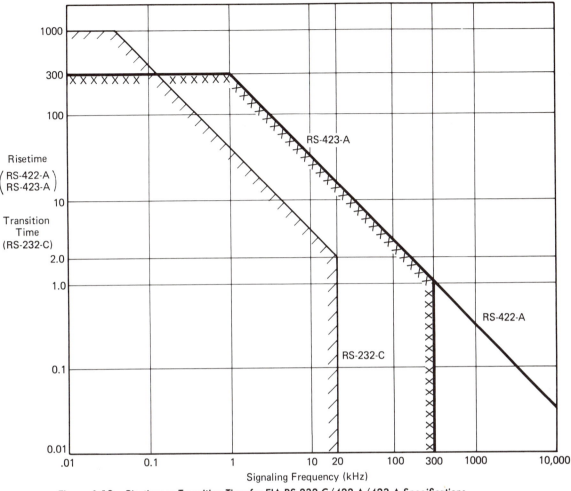

Figure 4-13   Risetime or Transition Time for EIA RS-232-C/422-A/423-A Specifications

the maximum allowable risetime within the RS-423-A specification is 15 μs. RS-422-A specification has the same risetime as RS-423-A at 20 kHz. Fig. 4-13 summarizes the rise/transition time for the three specifications.

The slower risetime result in reduced cross-talk over similar distances, and thus longer distances are achievable before the cross-talk becomes excessive. The curves given in Fig. 4-12(b) relate to either linear or exponential risetimes, with the linear expected to be the more typical implementation.

Given a cable length, the minimum signal risetime using the RS-423-A interface can be obtained by moving towards the right to the appropriate cable length waveshaping curve of Fig. 4-12 and reading the bottom abscissa scale. The associated maximum data signaling rate can be determined by moving from the risetime scale upwards until the data signaling rate curve is intercepted. The righthand scale ordinate gives the maximum data rate.

As an example, for a 100 m cable length, the minimum linear risetime permitted is 3.5 μs. The maximum data signaling rate would be 80 kb/s to ensure less than 1 V peak near end crosstalk. For 60 m operation, the maximum signaling rate is 60 kb/s for exponential waveshaping and 138 kb/s for linear wave shaping.

The relationship between the data signaling rate and the cable length is also specified for the RS-422-A standard. This curve is shown in Fig. 4-12(c). For signaling rates beyond 90 kb/s, the maximum cable length is inversely proportional to the data rate. For operation on a 60 m cable, the signaling rate is limited to 2 Mb/s. The signal rise time at the load end of the cable should be kept four times greater than the one-way propagation delay time on the cable.

When using cable with different electrical characteristics, the curves can still be used, but a correction based on the RC time constant must be made. Obtaining the cable l for a particular data signaling rate on the #24AWG cable, the length of l' of the actual cable can be obtained from the relation

$$Rl\, Cl = R'\, l'\, c'\, l'$$

or

$$l' = \sqrt{\frac{RC}{R'C'}}\, l \qquad (4\text{-}2)$$

where R, R', C, C' are the loop resistances and capacitances; unprimed values relate to #24 AWG cable and primed values relate to the cable actually used.

RS 449 Interface Standard

In order to take advantage of developments in IC technology and to increase the number of interchange circuits used in the RS-232 standard, the EIA has issued the EIA RS-449 standard. (EIA Standard RS-449, "General Purpose 37-Position and 9-Position Interface for Data Terminal Equipment and Data Circuit-Terminating Equipment Employing Serial Binary Data Interchange," Nov. 1977.)

This standard in conjunction with the RS-422-A and RS-423-A standards is intended to gradually replace the RS-232-C standard. This standard offers greater immunity to noise, and permits greater cable lengths and higher data signaling rates. In addition, it defines 10 new interchange circuits, chiefly for control and testing purposes. It also accommodates 10 balanced interchange circuits. The EIA has elected to use two connectors: a 37-pin and a 9-pin connector. The separate 9-pin connector accommodates secondary channel interchange circuits. These connectors come from the same family as the DB-25 connector that is usually associated with the RS-232-C interface. In addition to specifying dimensions, it also specifies a clip-on latch, so that no special tool will be required to couple or decouple the connectors. The chief drawback of this dual connector is the sizeable space it takes up.

The RS-449 specification divides the interchange circuits into two categories: Category I circuits provide interconnection of either balanced (RS-422) or unbalanced (RS-423) generators via a pair of wires with a differential receiver. Category II circuits provide interconnection of only unbalanced generators (RS-423) via a single wire to an unbalanced receiver which uses a common signal return circuit. Category I circuits are applied on the data interchange circuits, timing circuits, and five selected control circuits while Category II applies to all other circuits.

The following circuits are classified as Category I circuits:

SD (Send Data)   RS (Request to Send)
RC (Receive Data)   CS (Clear to Send)
TT (Terminal Timing)   RR (Receiver Ready)
ST (Send Timing)   TR (Terminal Ready)
RT (Receive Timing)   DM (Data Mode)

This standard limits the data signaling rate on a balanced interchange circuit to a nominal upper limit of 2 Mb/s for operation up to 60 m of interconnecting cable. For unbalanced interchange circuit, the data rate is that for the RS 423 interchange, 60 kb/s or 138 kb/s for 60 m of cable, depending upon whether exponential or linear waveshaping is used. The maximum cable length permitted is a function of the data transmission rate, and can be extrapolated from the curves given in Fig. 4-12(b) or (c).

For signaling rates under 20 kb/s, the circuits may use either the unbalanced RS-423 or balanced RS-422 category I circuits that interconnect a balance or an unbalanced generator and a differential receiver. For signaling rates exceeding 20 kb/s, Category I (RS-422) circuits must be used. Figure 4-14 shows the proper generator/receiver connections for each category. Table 4-2 gives a complete listing of the RS-449 circuits and the equivalent RS-232-C circuit where applicable. Note that the naming conventions of the RS-449 circuit closely reflect the circuit function.

Pin 1 is normally connected to the shield of the interconnecting cable at the DTE end. This is to suppress electromagnetic interference (EMI).

A example of RS 449 pin connections for either a balanced or unbalanced interchange circuit, complete with grounding is shown in Fig. 4-15. The shield of the interconnecting cable used to suppress electromagnetic interference (EMI) is normally connected only to the DTE frame ground. The frame ground is a point that is electrically bonded to the equipment frame. The signal ground interchange circuit (SG) provides a path between the DTE circuit ground and the DCE circuit ground.

Fig. 4-15 shows a 100 ohm resistance between the frame and circuit grounds. The circuit ground may also be directly coupled to the frame ground, although caution should be used to prevent excessively large ground loop circuits.

Table 4-2   RS-449 and RS-232-C Equivalent Interface Circuits.

RS-449		Pin Assign-ment	AA	RS-232C
			AA	PROTECTIVE GROUND
SG	SIGNAL GROUND	19	AB	SIGNAL GROUND
SC	SEND COMMON	37		
RC	RECEIVE COMMON	20		
IS	TERMINAL IN SERVICE	28		
IC	INCOMING CALL	15	CE	RING INDICATOR
TR*	TERMINAL READY	12, 30	CD	DATA TERMINAL READY
DM*	DATA MODE	11, 29	CC	DATA SET READY
SD*	SEND DATA	4, 22	BA	TRANSMITTED DATA
RD*	RECEIVE DATA	6, 24	BB	RECEIVED DATA
TT*	TERMINAL TIMING	17, 35	DA	TRANSMITTER SIGNAL ELEMENT TIMING (DTE SOURCE)
ST*	SEND TIMING	5, 23	DB	TRANSMITTER SIGNAL ELEMENT TIMING (DCE SOURCE)
RT*	RECEIVE TIMING	8, 26	DD	RECEIVER SIGNAL ELEMENT TIMING
RS*	REQUEST TO SEND	7, 25	CA	REQUEST TO SEND
CS*	CLEAR TO SEND	9, 27	CB	CLEAR TO SEND
RR*	RECEIVER READY	13, 31	CF	RECEIVED LINE SIGNAL DETECTOR
SQ	SIGNAL QUALITY	33	CG	SIGNAL QUALITY DETECTOR
NS	NEW SIGNAL	34		
SF	SELECT FREQUENCY	16		
SR	SIGNALING RATE SELECTOR	16	CH	DATA SIGNAL RATE SELECTOR (DTE SOURCE)
SI	SIGNALING RATE INDICATOR	2	CI	DATA SIGNAL RATE SELECTOR (DCE SOURCE)
SSD	SECONDARY SEND DATA	3	SBA	SECONDARY TRANSMITTED DATA
SRD	SECONDARY RECEIVE DATA	4	SBB	SECONDARY RECEIVED DATA
SRS	SECONDARY REQUEST TO SEND	7	SCA	SECONDARY REQUEST TO SEND
SCS	SECONDARY CLEAR TO SEND	8	SCB	SECONDARY CLEAR TO SEND
SRR	SECONDARY RECEIVER READY	6	SCF	SECONDARY RECEIVED LINE SIGNAL DETECTOR
LL	LOCAL LOOPBACK	10		
RL	REMOTE LOOPBACK	14		
TM	TEST MODE	18		
				PINS 9 & 10 TEST FUNCTION
SS	SELECT STANDBY	32		
SB	STANDBY INDICATOR	36		

* Category I Circuits

162

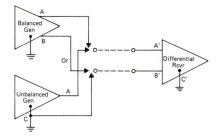

(a) Category I Circuits - Data Signaling Rate<20k Bits/s

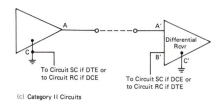

* Optional Cable Termination Resistance

(b) Category I Circuits - Data Signaling Rate>20k Bits/s.

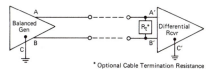

(c) Category II Circuits

Figure 4-14 **Generator and Receiver Connections at RS-449 Interface.**

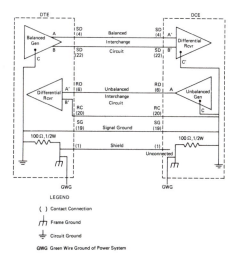

LEGEND

( ) Contact Connection

⊓ Frame Ground

⊥ Circuit Ground

GWG Green Wire Ground of Power System

Figure 4-15 **Grounding Arrangements for RS-449 Interface**

4-6 Mating Interfaces

In order to interconnect equipment using different families of digital receivers or transmitters, one has to assure: level compatibility, loading and sinking capabilities, switching rates, etc. Although it is occasionally possible to interconnect different units directly, it is more normal to use a level translation stage. This is particularly true if different supply levels are used.

Fig. 4-16 shows the voltages required to represent the logic "1" and "0" for the TTL and CMOS logic families. A supply voltage of 5 V is assumed, but the CMOS family can operate over the 3 to 18 V range.

From this figure we can see that a TTL driver cannot directly connect to a CMOS receiver, for if the driver in the logic "1" state outputs a voltage in the 2.4 to 3.5 volts range, it will not necessarily be interpreted as a "high." It appears, however, that the CMOS could drive a TTL circuit from a level point of view.

The input to a TTL gate consists of a multiple emitter transistor whose base is connected via a resistor to V_{cc}, as shown in Fig. 4-17. If its input is driven low, a sink current flows from the input and must be handled by the transmitter output. This sink current is 1.6 mA for TTL. The output of a CMOS driving gate can handle a sink current if only 0.4 mA. Thus it is not possible to connect CMOS directly to TTL, even if their voltage levels are compatible.

To interconnect these different families, a discrete transistor level shifter such as that shown in Fig. 4-18 can be used, but it is more common to employ IC packages which contain several buffers. The logic levels for the RS-422-A and RS-423-A are shown in Fig. 4-19. The RS-232-C levels were illustrated earlier in Fig. 4-4. Since the differential receivers of the RS-422-A and RS-423-A are electrically identical, it is possible to interconnect these interfaces.

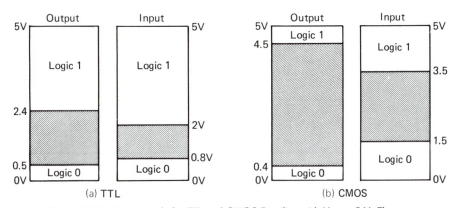

Figure 4-16 Logic Levels for TTL and CMOS Families with $V_{CC} = 5$ V. The area between the logic 1 and 0 is a "no man's land."

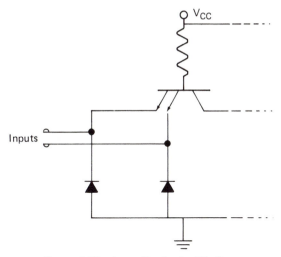

Figure 4-17   Input Circuit of a TTL Gate

The balanced RS-422-A cannot, however, be connected to the RS-232-C interface. The RS-423-A generator can usually operate satisfactorily with an RS-232-C receiver, provided that the risetimes are compatible. Fig. 4-20 shows the data rate permissible when operating an RS-423 generator with an RS-232-C receiver for a given signal risetime. The two sets of curves apply to signals with an exponential risetime and a linear (ramp) risetime. The area of interoperability for RS-423-A and RS-232-C is defined by the three parameters—risetime, generator output voltage V_o, and the data signaling rate. As an example, for a linear risetime of 10 μs, and a generator output of 4 volts, interoperation is possible up to about 4200 b/s. If the generator voltage is increased to 6 V, interoperation up to around 6000 b/s is possible.

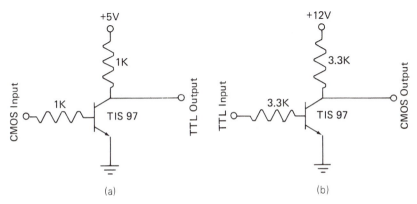

(a)   (b)

Figure 4-18   CMOS to TTL (a) and TTL to CMOS (b) Level Shifters

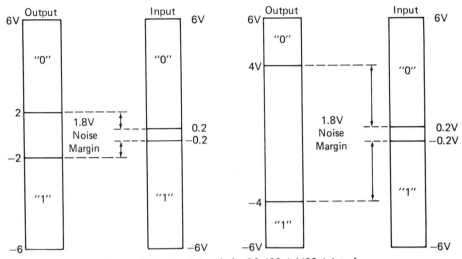

Figure 4-19   Logic Levels for RS-422-A/423-A Interfaces

RS-423-A specifies two separate common return paths, one for each direction of transmission, in addition to the signal ground circuit SG. RS-232-C specifies only the signal ground circuit AB. When connecting RS-449 equipment to RS-232-C equipment, the circuits SG, RC, and SC as well as the "return or B leads" of the TR, SD, TT, RS for DCE's and the DM, RD, ST, RT, CS, and RR circuits for DTE's must be connected to the AB circuit in the RS-232-C equipment.

Since an RS-423 receiver can only withstand 12 V, when an RS-232 generator is connected to an RS-423 receiver an inverted L pad consisting of a 2 k ohm series resistance and a 3.3 k ohm shunt resistance should be placed within 3 meters from the receiver input in RS-449 equipment. The pad appears as a high impedance and the distance restriction ensures that the near-end crosstalk will not exceed 1 volt peak. Some manufacturers design RS-423 receivers that can withstand the higher RS-232-C voltages.

On RS-449 equipment, RS-423 generators must be implemented on all Category I circuits. The total load resistance must not be less than 3 k ohm on each interchange circuit. The cable length must not exceed 15 meters when interconnecting RS-449 equipment with RS-232-C equipment.

4-6.1   Converting RS-232-C Levels

When interfacing DCE to a microcomputer or Terminal UART, it is necessary to convert from RS-232-C voltages to TTL voltages and vice versa. In addition, there is the requirement to invert the logic levels because a "1" is a positive voltage in TTL and a negative voltage in RS-232-C. Manufacturers have developed special level shifter I.C.'s that perform these functions.

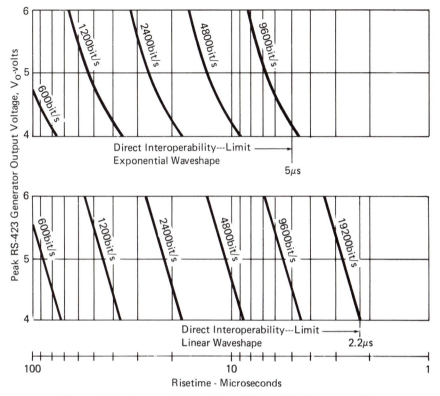

Figure 4-20 Risetime Relationship to RS-449/232C Interoperability

The transmitter MC1488 shown in Fig. 4-21 translates the TTL levels to the RS-232-C standard level. It also acts as a line driver and contains three NAND gates and a single inverter. The receiver MC1488 converts the RS-232-C standard level back to the TTL level. The additional external response control pins 2, 5, 9, and 12 of the MC1489 IC allow for variation in the input threshold voltage levels. This is accomplished by tying a resistor between this pin and the power supply voltage. The pin is left open for TTL to RS-232-C level shifting. This pin is often bypassed to ground by a capacitor for noise filtering.

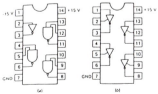

Figure 4-21 IC Converters (a) MC1488 and (b) MC1489

4-7 20-Milliampere Current Loop Interfaces

A 20 mA loop circuit is often used to connect computer or data terminal equipment to a teletypewriter, such as the Model 33 Teletype. It transfers data in a bit serial manner by turning the current ON and OFF, depending upon the logic level. Several devices can be connected serially, but the drive must be sufficient to deliver a minimum of 20 mA. One major difficulty with the 20 mA loop is that there is a lack of any mechanical or electrical standard. Thus one must take care when attaching two devices together, that the voltage from a generator can be handled by the receiver.

A typical 20 mA interface is shown in Fig. 4-22, where the current source is assumed to be in the transmitter (active transmitter). In practice, it can also be in the receiver (active receiver). If both the receiver and transmitter are active, a special interface is required. This is commonly achieved by an optical coupler as described a bit later.

If both the transmitter and the receiver are passive, one must be equipped with a voltage supply.

An example of a 20 mA loop is the interconnection between an Intel 8080 microcomputer and a teletypewriter (TTY) via an 8251 Universal Synchronous Asynchronous Receiver/Transmitter (USART). The USART accepts data from the microprocessor in parallel format and converts it into a serial data stream for transmission. It can also simultaneously receive a serial data stream and convert it into parallel data characters.

Pin Number 19 of the USART, shown in Fig. 4-23, is the Transmitter Data Connection (TxD), while Pin Number 3 is the Receiver Data Connection (RxD). With TTY strapping, a LO state on the TxD lead causes Q_2 and Q_3 to remain in cut-off preventing any current to flow to the TTY. On the other

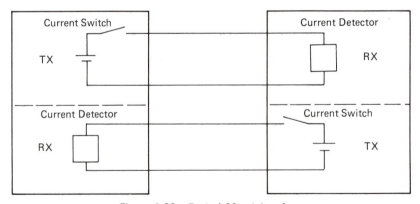

Figure 4-22 Typical 20 mA Interface

hand, a HI state on the TxD lead causes Q_2 and thereby Q_3 to go into saturation. Current can now flow from the +12 volt supply via Q_3 and through R_{13}. The TTY and its associated line flow through R_{20} to the negative 10 V supply line.

By selecting the smaller R_{13} value, longer line lengths can be used to the TTY. Applying Ohm's Law:

$$20 \times 10^{-3} = \frac{12 + 10}{430 + R_{13} + R_{coil} + R_{line}}$$

where R_{coil} represents the resistance of the relay coil in the teletype, normally in the 100 to 200 ohm range, and R_{line} represents the resistance of the connecting cable.

If a minimum of 20 mA is to be maintained, and R_{13} is selected at 390 ohms, with a relay coil resistance of 150 ohms, the line length may be such that it may contribute a maximum total resistance of 130 ohms.

When receiving data from the TTY, no current can flow between pins 2 and 7 of J3 with the brush contacts open. The -10 volt supply causes Q_1 to remain OFF, resulting in a HI state at its output because of the 5.1 k pull-up resistor. The 7406 inverter causes the RxD pin to see a LO. If the TTY brushes are closed, the +12 V supply causes Q_1 to turn ON, lowering the input voltage to the 7406 inverter. This results in a HI state at the RxD terminal.

Fig. 4-23 also shows the standard TTL (Transistor-Transistor Logic) levels strapped to the input/output terminals. In this case 0.0 to 0.8 V represents a logical LO, and 2 to 5 V represents a logical HI.

For data transmission rates up to 9600 b/s, the 20 mA interface can achieve cable lengths of up to 600 m, much greater than the RS-232-C interface. In some applications the line lengths allowed are considerably less to keep the pulse distortion to a reasonable level. For longer liner lengths, line drivers can be employed. The line driver system reduces the line capacitance problem by translating the signal into a low impedance output so that the time constant is lowered to an acceptable level. In this manner the pulse rise and fall times are greatly reduced.

When two active devices are interfaced, the optical coupler shown in Fig. 4-24 can be used. The light-emitting diode (LED) is electrically isolated from the phototransistor, and so the supply voltages at either side are decoupled. When a current flows through the LED, it emits light, which the phototransistor senses. When the phototransistor detects light, it begins to conduct current from the active receiver.

One manufacturer adds a full wave rectifier at the optical coupler receiver input and at the transmitter output of another optical coupler, along with a +12 V/ground strapping option to create a current source (active) or alternatively to act as a current sink (passive) in the case of an external voltage source.

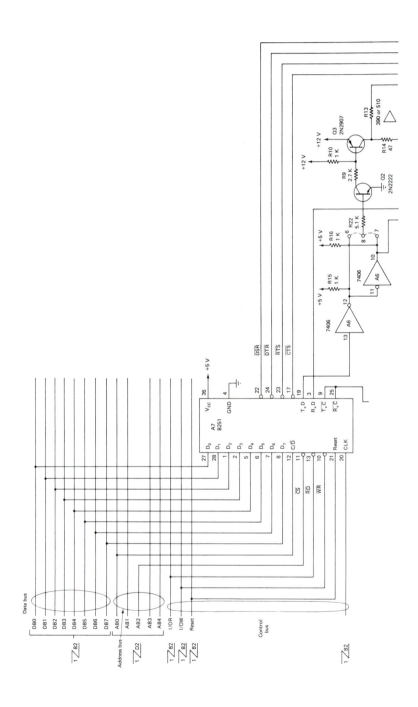

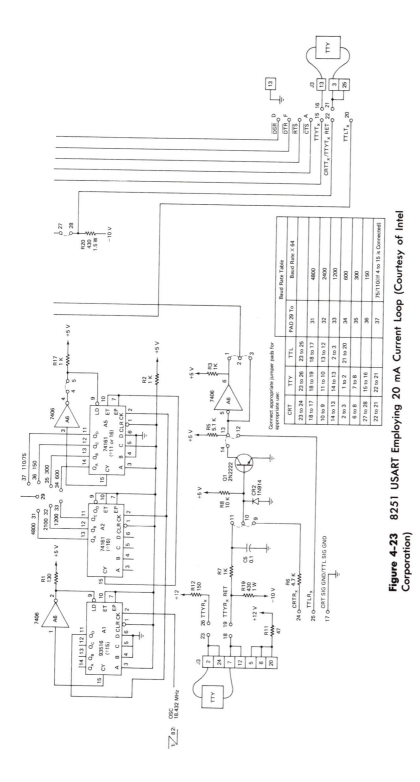

Figure 4-23 8251 USART Employing 20 mA Current Loop (Courtesy of Intel Corporation)

171

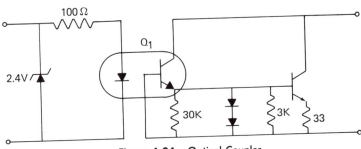

Figure 4-24   Optical Coupler

Problems

1. Define the terms
   a. Protocol
   b. DCE
   c. DTE
   d. OSI architecture

2. RS-232-C is a specification for
   a. Parallel ASCII data transmission
   b. Series binary data transmission
   c. Data communication cable
   d. IEEE interface (GPIB)

3. RS-232-C specifies
   a. Signal levels and pin wiring configurations
   b. ASCII code
   c. The use of balanced lines
   d. All of the above

4. The most common type of mechanical connector used with RS-232-C is the
   a. Cannon plug/socket CA3106B14S-2P/S
   b. SMA connector
   c. DB-25 connector
   d. BNC connector

5. RS-232-C signals are
   a. Polar RZ
   b. Unipolar NRZ
   c. Duobinary
   d. Polar NRZ

6. In the RS-232-C specification a mark condition at the receiver is represented by a voltage that lies between
   a. +5 to +15 V
   b. +3 to +15 V
   c. −3 to −15 V
   d. −3 to +3 V

7. The voltage slew rate in the RS-232-C specification is limited because of
   a. Crosstalk between conductors
   b. Transmitter circuit switching time limitation
   c. Receiver circuit switching time limitation
   d. Cable resistance
8. The RS-232-C standard recommends
   a. Maximum signaling rates of 20 k b/s and 15 m cable length
   b. Maximum cable length of 50 ft and an idling condition in the SPACE state
   c. The bonding of the signal ground to the equipment frame
   d. The physical dimensions of the connector
9. Equipment readiness is indicated by the ON state of the following circuits:
   a. RTS and CTS
   b. RI and DTR
   c. DTR and DSR
   d. DCD and DTR
10. Explain why on HDX the RD circuit must be in the mark condition anytime RTS is ON.
11. The delay between the CTS and RTS state allows for
   a. The removal of the MARK HOLD clamp from the Receive Data Interchange Circuit of the remote data set to a remote DTE
   b. The ringing indicator to turn ON at the remote DTE
   c. The turning ON of the data carrier from the remote DTE
   d. All of the above
12. Data channel readiness is indicated by the ON state of the following circuits:
   a. RTS and CTS
   b. RI and DTR
   c. DTR and DSR
   d. DCD and DTR
13. Indication of readiness to receive is given by circuit
   a. RI
   b. CD
   c. TD
   d. DTR
14. The RI circuit remains ON
   a. After RD turns ON
   b. After ringing has been received
   c. Only while ringing is being received
   d. During the transmission and reception of data
15. The OFF condition on the DSR circuit shall not impair the operations of circuits
   a. RI and CTS
   b. RI and RD

c. RD and TD

d. RI and DTR

16. An ON condition on the signal quality detector circuit suggests that
    a. No errors have occurred in the received data
    b. There is a high probability of error in the received data
    c. The higher data signaling rate has been selected
    d. The clock resides within the DCE

17. An ON condition on the transmitter element timing circuit DB indicates
    a. That the DTE provides the timing information on circuit DB to the DCE
    b. That the DCE provides the timing information on circuit DB to the DTE

18. When the secondary channel is used for circuit assurance, only circuit
    a. STD
    b. SRD
    c. SRTS
    d. SCTS
    is usually provided.

19. Where a synchronous DCE is used, the ON condition on circuit CTS
    a. Guarantees that the remote receiver is listening
    b. Ensures circuit assurance
    c. Implies that the remote clock of the remote DCE is synchronized

20. In the baudot code, the element J is represented by the following waveform. If alphanumeric characters are sent sequentially, with zero intercharacter space, and an error rate of 1×10^{-2} is experienced, obtain the:
    a. Baud rate
    b. Information rate, assuming no errors
    c. Data transfer rate

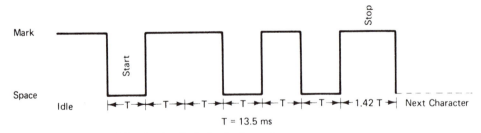

Figure P4-20 Baudot Code for the Letter J

21. a. Explain the difference in electrical characteristics of a balanced and an unbalanced line.
    b. Explain why differential circuit operation can ignore ground voltages.
    c. State the electrical advantage of twisting wire pairs.

22. Give the electrical interface circuits of the
    RS-232-C
    RS-423-A
    RS-422-A
    standards.

23. a. Given a maximum signaling data rate of 10 k b/s on #24 AWG twisted
       wire pair, determine the maximum cable length allowed and the maxi-
       mum linear waveshaping risetime permitted when using the RS-423-A
       specification. Sketch a few pulses, indicating risetime and pulse dura-
       tion.
    b. Can the signaling rate be decreased on this length of cable determined
       in part (a) and still meet the RS-423-A standard? Explain.
    c. For the signaling rate determined in part (a), can the cable length be re-
       duced and still meet the RS-423-A standard? Explain.
    d. For the signaling rate and cable length of part (a), can the risetime be
       increased?

24. For a 100 m length of cable, what may be maximum signaling rate using
    RS-422-A standards?

25. a. What standard is the EIA RS-449 intending to replace?
    b. Explain the difference between category 1 and category 11 circuits.
    c. Which category of circuit is used on the TD (SD) and RD lines? Why?
    d. What is the chief disadvantage in implementing the RS-449 standard?
    e. What is connected to pin 1 of the RS-449 connector?

26. a. What precautions must be observed when attempting to interconnect
       different interfaces?
    b. Why may an RS-422-A receiver not be connected to an RS-232-C
       transmitter?

27. Given a baud rate of 300 baud on AWG #24 twisted wire cable, determine
    the maximum cable length permitted and the maximum linear risetime,
    using the RS-423-A specification.

28. For a generator output of 4 volts and a linear risetime of 10 μs, what sig-
    naling rate is possible when interconnecting an RS-423-A generator to an
    RS-232-C receiver?

29. a. How far from the USART of Fig. 4-23 can a teleprinter be located if
       #26 AWG copper was used for the interconnecting line and R13 is se-
       lected at 390 Ω? Assume an R_{coil} of 200 Ω and a minimum loop current
       of 18 mA.
    b. What alterations can be made to the circuit in (a) to increase the dis-
       tance to the teleprinter?

30. Explain how a mark (current flow) in the current loop is converted to a
    mark suitable for RS-232-C equipment, in the current loop to RS-232-C
    converter.

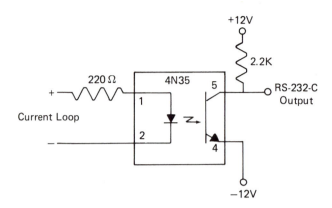

31. Explain the functions of the diodes and the two inverters of the RS-232-C to current loop converter. Explain the circuit operation.

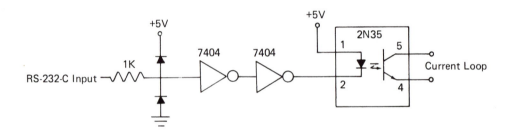

32. If both the receiver and transmitter in a 20 mA interface are active, how can they be coupled?

33. a. Using 3-cycle by 6-cycle log-log paper, sketch a graph of cable length vs. data rate for the RS-423-A linear waveshaping and RS-422-A standards. Use Fig. 4-12 as your source of data. The abscissa should show data signaling rate from 10^2 b/s to 10^8 b/s. The ordinate axis should show cable lengths from 10 m to 10^4 m.

b. Use your graph to determine the maximum data rate possible for a cable length of 200 m, using each standard.

c. If the capacitance of the cable is halved, what cable length can be used for the data rates of part (b)?

34. An ON condition of a control circuit in the RS-232-C specification at the receiver is represented by a voltage that lies between

a. +5 to +15 V

b. +3 to +15 V

c. −3 to −15 V

d. −5 to −15 V

35. The RS-232-C standard is applicable to signaling rates up to
    a. 9600 b/s
    b. 20 k b/s
    c. 2 M b/s
    d. 10 M B/s

36. In the RS-232-C specification, the Request to Send control signal is originated by the
    a. DTE
    b. DCE
    c. Modem or data set
    d. Central telephone office

37. In the RS-232-C specification
    a. The received data lead must be held at ground anytime RTS is ON
    b. Each signal has its own common return line
    c. The return line is electrically bonded to the equipment frame
    d. All signals flow through one common return line

38. A baud is related to
    a. The shortest signaling element
    b. The number of independent states that can appear
    c. The type of code used
    d. Whether synchronous or asynchronous transmission is used

39. After transmission, when the RTS lead is turned OFF, the local modem
    a. Forces the received data line to be in the MARK condition
    b. Immediately lowers the CTS line
    c. Turns off the ringing indicator signal
    d. Turns off its carrier

40. In the RS-232-C specification, on "ON" condition of the Received Line Signal Detector (carrier detect) indicates that
    a. The CTS signal has been activated
    b. A suitable carrier signal is being received
    c. No errors have occurred in the received data
    d. The RTS signal is in the ON condition

41. Assuming no errors the baud and bit rates are identical when
    a. 4ϕ PSK is used
    b. A two-level system is used
    c. QAM is used
    d. Four frequency FSK is used

42. For a cable length of 50m, what maximum signaling rate can be used with the RS-232-C, the RS-423-A, and RS-422-A specifications?

43. The extra pins of the RS-449-A specification, when compared to the RS-232-C specification, are used
    a. To provide a balanced circuit for each signal
    b. To permit the possibility of automatic calling

   c. For secondary channel signals

   d. To provide some extra control circuits and some high signaling rate data circuits

44. Line drivers require

   a. High impedance circuits

   b. DC continuity

   c. Unbalanced lines

   d. Polar RZ signal

45. Simple line drivers which put high voltage signals on a line may not be used on telephone lines because

   a. It may interfere with signals on adjacent lines

   b. The bandwidth of the line is limited to voice band

   c. It may cause the line to become hot and become an open circuit

   d. It may cause ringing at the local exchange

46. A private line metallic circuit

   a. Has dc continuity

   b. Has the limited 300-3400 Hz voice bandwidth

   c. Can only be obtained on dial-up lines

   d. Has a copper line end-to-end but not necessarily dc continuity

47. Balanced lines and differential drivers and receivers

   a. Reduce line capacitance

   b. Allow for higher baud rates

   c. Improve immunity to electromagnetic interference noise

   d. Allow for longer lines

   e. None of the above

48. What end result(s) do the RS-422 and 423 specifications aim at?

   a. Less line capacitance

   b. Balanced, twisted and shielded line

   c. Longer distances and higher data speeds

   d. Differential drivers and receivers

   e. None of the above

49. How can crosstalk be reduced?

   a. Twist the wires

   b. Reduce the baud rate

   c. Use balanced lines and differential drivers and receivers

   d. Reduce the slew rate (i.e. make the rise time longer)

   e. All of the above

50. The following circuit:

RS-232-C Rxd Data Line

TTL Out

a. Is perfectly satisfactory for receiving bipolar (e.g. ± 12 V) data signals from an RS-232-C line
b. Is not satisfactory, since RS-232-C does not specify a 0 volt reference decision level to decide between a "0" (positive) or a "1" (negative) signal
c. Is perfectly satisfactory for most practical purposes
d. Is not satisfactory, since it does not guarantee good noise immunity as specified by the RS-232-C voltage levels
e. Is not satisfactory, since it is not short-circuit-proof, which is required by RS-232-C

51. A DTE's desire to transmit must first be signaled to the DCE via the ___ line, and acknowledged by the ___ line(s) before it can actually go ahead and transmit.
a. RTS, DCE's CTS
b. DTR, DCE's CTS
c. RTS, other end DTE's CTS
d. DTR, DCE's DSR, CTS and CD

52. At a baud rate of 2400, using QAM modulation (16 phase-amplitude levels), one start bit, one stop bit, 7-bit ASCII plus one even parity bit, calculate:
a. The baud rate
b. The information rate assuming start, stop, and parity bits are information
c. The DATA TRANSFER RATE (start, stop, and parity bits are not information)

53. Describe the nature and direction of the signal associated with the following interchange circuits: BA, CA, CB, and CE.

54. Given that we want to send data at up to 10 k bits/sec over an RS-423-A line, what restrictions must be observed regarding cable length and rise-time T_R (with linear waveshaping)?
a. Length < 1 km, and $T_R > 40$ μsec
b. Length < 700 m, and $T_R < 40$ μsec
c. Length < 800 m, and $T_R > 30$ μsec
d. Length < 1 km, and $T_R < 30$ μsec
e. Any of the above is correct, since the RS-423-A only specifies balanced lines and differential drivers and receivers

55. The RS-232-C interface implements
a. The bottom-most physical protocol layer of the OSI architecture
b. The second level protocol layer, since the modem carrier tones actually carry the data at the bottom-most physical level
c. A level-to-level interface between the physical level modem DCE, and the second level DTE
d. The second-level data link control protocol including parity error checking

56. For asynchronous communication, the RS-232-C specifies
   a. At least one start and one stop bit
   b. That a parity bit must be provided for each data byte
   c. An optional parity bit to check on data and start and stop bits
   d. Nothing
   e. Certain things, but none of the above

57. a. An RS-423-A interface is to be 800 m long. For linear waveshaping find (a) the maximum signaling rate and (b) the maximum risetime.

5

Modems

When moving signals such as data from a terminal to a computer, digital pulses are often encountered. To understand some of the major difficulties in transmitting pulses, consider the case of transmitting a voltage pulse down a wire pair. Such pulses, which usually contain a dc component, cannot be transmitted between telephone exchanges through the switching devices, but can be brought into the local exchange. If a dc path can be maintained, the pulses can be transmitted to another subscriber connected to the same exchange, provided the circuit does not pass through the switching machine. This necessitates the use of a private line. If the terminal is within relatively close proximity to the computer, say less than 100 m distant, wires of relatively small gauge can directly interconnect the machines.

The distance limitation on pulse transmission is due to distortion introduced by resistance, shunt capacitance, and series inductance associated with any real transmission line. The dielectric or shunt losses due to leakage through the insulation material are usually insignificant and therefore ignored. Line resistance is very much a function of frequency which causes the various frequency components making up a voltage pulse to experience differences in attenuation and propagation velocity. Resistance variation due to frequency is chiefly the result of skin effect, proximity effect, and radiation losses.

Skin Effect

Because the center of a conductor experiences more self-flux linkages than the outer region of the conductor (see Fig. 5-1), self-inductance increases upon approaching the center region of a conductor. Thus, the asssociated impedance is less near the conductor surface which causes more of the current to flow near the conductor surface than in the central region. This effect, known as skin effect, becomes more and more pronounced as the frequency is increased. Typical current distributions for an isolated conductor are shown in Fig. 5-2.

Because current flows through a smaller net area at higher frequencies, there is an increase in the effective resistance of the conductor. This in turn causes the signal to experience greater attenuation at the higher frequencies.

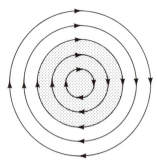

Figure 5-1   Flux Around and In Current-Carrying Conductor

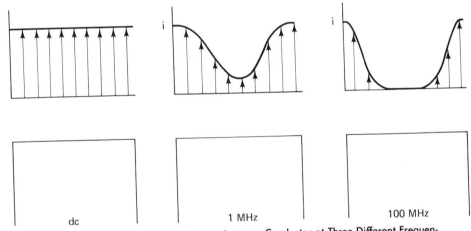

Figure 5-2   Current Distribution Across a Conductor at Three Different Frequencies

Proximity Effect

Effective resistance is not only frequency dependent but is also dependent upon other conductors that are located in its proximity. Because of mutual repulsion or attraction of the charges flowing in nearby conductors, the current density undergoes a redistribution which tends to further increase the effective resistance of the conductor. This proximity effect is a function of conductor diameter, spacing, and frequency. It contributes significantly to the loss in twisted wire cable and parallel wire lines. Because of symmetry and shielding, it is of no consequence in coaxial lines.

Radiation

Radiation losses can contribute to an increased effective resistance, particularly at the higher frequencies. For closely spaced conductors, the loss is not significant unless the spacing is some appreciable portion of the electrical wavelength in size ($\lambda = C / f$). Coaxial lines experience no radiation loss because of shielding.

Because both the line attenuation and signal propagation velocity along the line are resistance dependent (see Sec. 2-12), both the line attenuation and signal velocity are frequency dependent. Typical plots are shown in Figs. 5-3 and 5-4 for a #22 twisted-wire pair. Both attenuation and signal velocity increase with frequency. The variation of signal velocity with frequency is termed dispersion.

Dispersion and attenuation applied to a transmission line both delay and distort the pulse, so that the detected pulse at the receiving end is more difficult to interpret. The leading or lagging edges of a pulse containing high frequency components experience minimum delay, but the increased attenuation masks this effect. The pulse top with low frequency content tends to be subject to extreme delays, but the reduced attenuation causes a greater influence on the resultant signal. This causes the various parts of the pulse to arrive at the receiver at different times, producing distortion of the original signal. If a pulse train is present, as is normal, intersymbol interference, in which portions of neighboring pulses spill over into each pulse, results. Intersymbol interference (ISI) forces a reduction in the allowable permitted pulse rate for a given line length in order to maintain adequate distinction between adjacent pulses. As the line

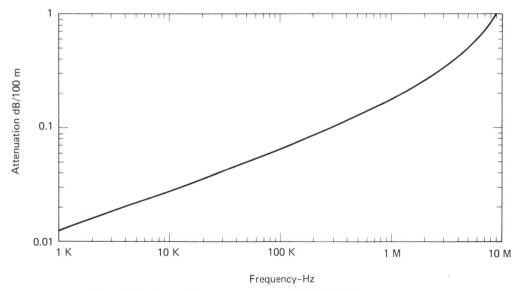

Figure 5-3 Attenuation vs. Frequency on #22 AWG Line

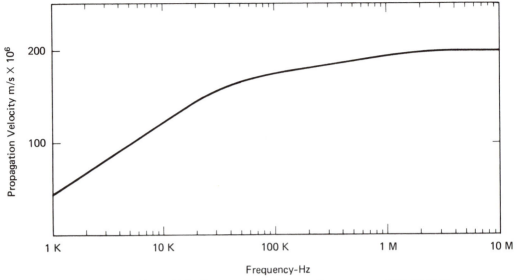

Figure 5-4 Propagation Velocity vs. Frequency on #22 AWG Line

length is increased, dispersion increases, and the pulse rate must be further re-duced. Delay equalizers must reduce this type of interference if high data rates are to be used where the pulses are more closely spaced.

Figure 5-5 shows an example of baseband data transmitted down a line. For each pulse transition applied to the line, the received pulse is transformed to a rounded-out and delayed transition as a result of line attenuation and dis-persion. The line detector has a decision threshold, or slicing level, midway be-tween the voltage extremes. When the received voltage exceeds this threshold, a one logic state is produced. A zero logic state is produced when the voltage falls below the threshold level.

The threshold can be considered the current drive necessary to change the state of a set of relay contacts. Exceeding this threshold current may cause the relay contacts to close, which would result in a sharp data output transition. As a first approximation, the pulses at the received end can contain gradually increasing or decreasing transitions due to the resistive-capacitive nature of the line.

This is illustrated by the arrival transition sketch of Fig. 5-5. Observe the severe distortion of the received bit pattern because of the line dispersion.

If the pulse rate increases much more, even by adjusting the threshold level, the received bit pattern would still be in error. Varying the setting of the threshold level bias can cause time expansion or elongation of either the high state or low state, depending upon the direction of threshold movement. If the threshold level is raised, the low-state times are increased at the expense of the higher state time. An incorrectly adjusted threshold level can severely degrade

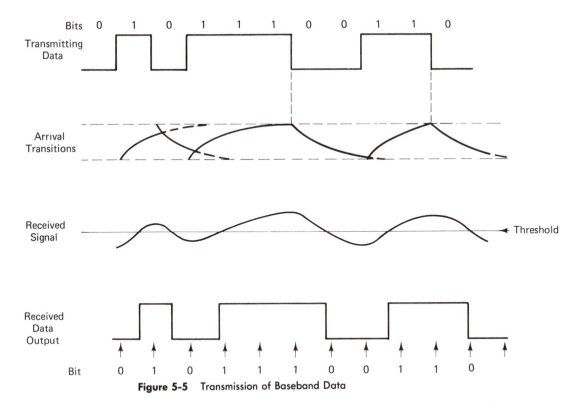

Figure 5-5   Transmission of Baseband Data

a system, particularly when several links are operated in tandem. This elongation of the high-state bits (or the low-state bits, if that is the case) is known as bias distortion.

The output bit stream is obtained by sampling the received data at the instant noted by the vertical arrow. This is merely a delayed version of the input clock pulse at the transmitting end.

As the public-switched telephone network has been specifically designed to carry voice, it is not particularly well-equipped to handle data transmission. If data is to be transmitted over such a network, the signals must be converted to reside within the audio frequency spectrum range from 300 to 3400 Hz. Modems have been designated to provide this function. The word modem is a contraction of the terms *mod*ulation and *dem*odulation. The modem converts digital data signals to suitable audio signals at the transmitter, a process referred to as modulation; while it converts the audio signals back into digital data signals at the receiver end, commonly referred to as demodulation.

The modem can be directly or indirectly coupled to the communications line. With indirect, or acoustic coupling, the data signals are converted into audible sounds with a speaker—sounds readily picked up by the standard tele-

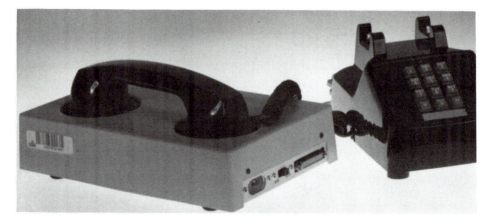

Figure 5-6   An Acoustic Coupler

phone handset microphone or transmitter. The audible signals are then converted back into electrical signals in the handset and transmitted over the telephone system. When receiving, the audible tones from the handset receiver are coupled into the acoustic coupler where they are converted back into electrical signals.

The acoustic coupler and speaker are permanently mounted into the ends of two cylindrical rubber grommets in such a manner as to allow the telephone handle to fit into the opposite and open ends of the grommet. This results in tight acoustical coupling between the handset and the modem speaker and microphone.

Fig. 5-6 illustrates such an arrangement. Because of the extra conversions involved, acoustical coupling tends to introduce more noise and distortion than direct coupling and hence is limited to data rates of less than 1200 b/s.

If modems are approved and registered by the FCC in the United States or the Department of Communications in Canada, they can be connected directly to the dial-up telephone network.

In determining which method must be used in transmitting data-type signals, distance and data rates constitute the key criteria. The five most common methods of transmitting data, in order of transmission distance, are as follows:

1. Direct connection by copper wire
2. Direct connection by special cable
3. Line drivers and cable
4. Limited distance modems and cable
5. Standard telephone modems

As a general rule, the greater the distance the greater the cost. We will initially consider the first four on the list as short-haul techniques, where terminal equipment must lie within a few meters to several kilometers from each other. Then we will consider the long distance modem.

5-2 Short-Haul Communications

5-2.1 Direct Connection

In a direct RS-232-C connection, the EIA specification limits the interface cable length to 15 m (50 ft.) for data rates up to 19.2 k b/s. Longer distances can practically be achieved, particularly if the data rates are less than the maximum in the specification. It usually is possible to run cable lengths of 60 m for data rates up to 4800 b/s, provided that the magnetic or electric noise level is not too severe. Special low-capacitance cable (less than 30 pF/m) that is also shielded can extend this distance up to 200 m at 9600 b/s.

When interconnecting a terminal to an asynchronous or synchronous modem, Data Terminal Equipment is being interconnected to Data Communication Equipment. If an extension cable is needed, a female connector is required at the cable, to be attached to the terminal, and a male connector at the opposite end for connection to the modem.

Some machines, particularly computers, can be DCE or DTE. If a terminal and a computer are both DTE's (or DCE's) how is one to interconnect them? A simple extension cable will not suffice as the Transmit Data and Receive Data signals will collide; that is, both machines will attempt to transmit on the line connected to pin 2 of the RS-232-connector and receive on the line connected to pin 3 (see Fig. 5-7).

To resolve this difficulty, one of the machines must be fooled into thinking that it is connected to a modem (DCE). This is done by using a null-modem or no-modem cable adapter. It merely transposes the leads to pins 2 and 3 of the cable. Fig. 5-8 illustrates how a null modem can interface two DTE's or two DCE's. The sexes of the connectors must also be suitably matched.

5-2.2 Line Drivers

For asynchronous transmission over distances from 30 m to 8 km, a line driver should be considered. Line drivers employ low impedance balanced lines in conjunction with differential receivers to minimize the problem of noise and crosstalk. Line drivers can typically operate at speeds up to 19.2 k b/s. The longer this distance, however, the lower the maximum data rate. Transmission rates of 1200 b/s may be feasible up to distances of 8 km, reducing to 2 km for speeds of 9600 b/s.

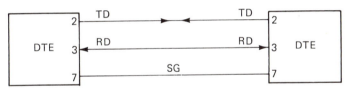

Figure 5-7 Improper Connection of DTE to DTE

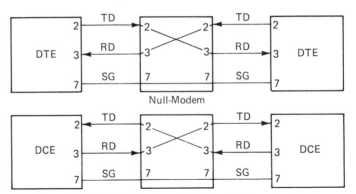

Figure 5-8   Null Modem Interfacing Two DTE's or two DCE's

Line drivers require dc continuity as no modulation is involved. Although 2 W transmission is possible, normally 4 W is used for the send and receive data. They are primarily used in point-to-point applications, as shown in Fig. 5-9.

If a line driver places a large current or voltage on a line, it should not operate over a leased line from the telephone company, as the energy level may exceed that permitted by the company, resulting in interference on adjacent local lines. DC continuity is required, limiting transmission to private lines connected to the local exchange. The transmission lines should consist of non-loaded cable if the data speeds exceed 4800 b/s. Line drivers are not normally conditioned and the Tx levels are set in volts, e.g., 6 V, rather than in dBms.

5-2.3   Limited Distance Modems

An upgrade on the line driver is the limited distance modem (LDM). The data stream contains no dc as in the case of the line driver, and yet it is not as complex and expensive as the conventional modem. The LDM requires a "metallic" circuit, a circuit which is a copper line end to end and that contains no multiplexing or carrier equipment. Such a circuit usually does not have dc continuity, because of amplifiers or isolation transformers in the line. The metallic circuit has a bandwidth which is larger than the normal telephone channel. LDM's are restricted for use on unloaded cables connected to the local ex-

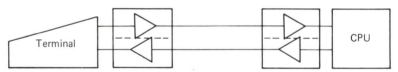

Figure 5-9   Pair of Line Drivers

change. LDM's cannot be used on the normal voice-grade channels that employ carrier equipment. Reliable data transmission can be achieved with power levels in the 40 to 50 mW range.

LDMs are available in both synchronous and asynchronous models. A typical synchronous model operates at speeds of 2400 to 19,200 b/s on private line service, and can be used in point-to-point or multidrop applications. Many use a filtered encoded waveform that conforms to AT & T Pub. 43401 "Transmission Specifications for Private Line Metallic Circuits."

Full-duplex and half-duplex models are available, and they usually operate over nonconditioned lines. The communication range of LDM's is dependent upon the cable gauge of the metallic circuit. For example, at 2400 b/s on #AWG 19, the range may be 30 km; this is reduced to 18 km for #AWG 24. At 19,200 b/s the range may be 16 km for #AWG 19, reduced to only 7 km at 19,200.

5-2.4   Modem Eliminator

In interconnecting two terminals or a terminal and a computer which are separated by more than 15 m, two data sets operating back to back are required. Synchronous terminals can also be interconnected by using a single modem eliminator that provides clock and signal regeneration, which effectively doubles the distance between terminals to 30 m. If the machines are both DTE's, then it must also contain a null-modem. Strappable RTS delay and CTS delay are also normally available.

5-3   Modems

In order to furnish computer communications over large distances, it has been proven convenient to use the widespread network of voice-bandwidth channels provided by the common carriers. To transmit digital signals over these analog channels, which have a nominal bandwidth from 300 to 3400 Hz, it is necessary for a data transmitter to MOdulate a voice frequency carrier signal and for a receiver to DEModulate this signal back to a digital format. These units are consequently called MODEMs. In addition to translating data signals, a number of control functions are also provided to coordinate the flow of data between the data terminal equipment.

Most modems are designed to accept and transmit a serial stream of binary data at the data terminal interface.

5-3.1   Types of Modulation

The voice frequency or carrier signal transmitted between modems is modulated into intervals which vary in amplitude, frequency, phase, or a combination of amplitude and phase. Each variation is used to represent the binary digital information presented to the transmitter.

The type of modulation used depends largely upon the data rates. For the narrow band reverse or secondary channel in HDX applications, AM is used, where ON-OFF conditions are employed and speeds are in the 0 to 5 b/s range. AM is not used in the main channel because of line noise problems and the variations in line loss which would require elaborate automatic gain control circuits.

In asynchronous modems, *frequency shift keying* (FSK) is used. One frequency represents a mark and the other frequency represents a space. The FSK waveform contains no DC component.

In some asynchronous modems, *phase shift keying* (PSK) is used. Here the carrier frequency is not altered, only the phase is shifted. Usually, modems employing PSK encode two bits, called *dibits,* of information at a time. Since a "dibit" encodes two bits at a time, the bit rate is twice the baud rate, i.e., there are two bits per baud. By encoding even more bits at a time, the data rate can be further increased. If, for instance, 8-phase PSK is employed, three bits of data at a time can be encoded, resulting in three bits per baud.

A variant of PSK is the differential phase shift keying, or DPSK technique. Here the phase shift, rather than the absolute phase, is used to represent the signal element being transmitted. It is used in the 1200 b/s to 4800 b/s modems and has the advantage of not requiring an absolute phase reference signal. Fig. 5-10 illustrates the four basic types of modulation. In the DPSK example, a "0" data bit is represented by a zero phase shift and a "1" by a 180° phase shift from the previous signal element.

5-3.2 Types of Modems

To illustrate the more common types of modems in use, consider some of the members in the Bell family of modems.

Bell 103/113

The Bell 103/113 modem is capable of full duplex operation over a 2 W channel. The voice bandwidth is divided into two, to provide for simultaneous transmission in both directions. FSK is the modulation technique employed, using a 200 Hz shift in frequency to distinguish a mark from a space. Operation is limited to 300 b/s asynchronous.

When the modem is in the originate mode, it transmits on 1070/1270 Hz. When it is in the answer mode, it transmits on the 2025/2225 Hz frequencies. The lower frequency in a subchannel represents a space, whereas the higher frequencies in a subchannel represent a mark. The modem goes into the originate mode when a call is originated through a modem. It goes into the answer mode if it initially answers a call. Fig. 5-11 illustrates the spectrum utilization of a typical full-duplex modem using a two-wire line.

The Bell 103/113 can be employed on either leased or dial-up lines, i.e., 3002 basic or schedule 4/type 4 line, and can be strapped to operate as an originate 113 A/D or as an answer 113 B/C modem only. When a leased line or manual dialing/answering is involved, only one type of each is used.

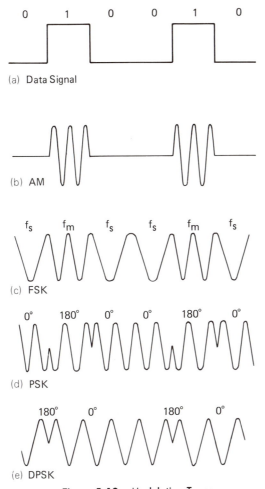

Figure 5-10   Modulation Types

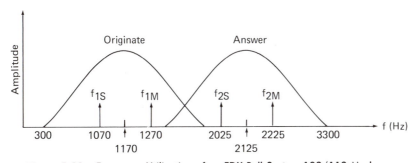

Figure 5-11   Frequency Utilization of an FDX Bell System 103/113 Modem

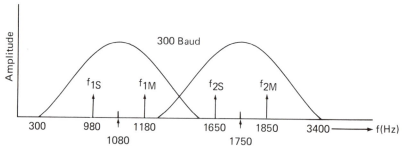

Figure 5-12   V.21 Channel Assignment

The V.21 international channel assignment shown in Fig. 5-12 is similar to the 103. The mark and space tones are reversed, the mark being either 980 or 1650 Hz, and the space being 1180 or 1850 Hz.

Bell 202

The Bell 202 is a half duplex modem, designed to operate at 1200 b/s over the switched network, from 1200–1400 b/s on leased lines with C1 or schedule 4/type 4A conditioning, and from 1400–1800 b/s on leased lines with C2 or schedule 4/type 4B conditioning. It employs FSK, with a tone of 1200 Hz representing a mark and a tone of 2200 Hz representing a space. By employing RTS and CTS signals, the modem is informed when to transmit and when to receive. A carrier should not be turned off abruptly by the dropping of the RTS line as the carrier turn-off at the remote receiver may produce transmit signals that may affect the reception of the last bits transmitted. Soft carrier turnoff is used, whereby the carrier frequency is shifted to 900 Hz to prevent transmit on the line for about 10 ms to 30 ms prior to being turned off. This informs the remote receiver to inhibit the receiving data line before the carrier is completely turned off.

When switching from transmit to the receive mode, a sufficient amount of time must be allowed for the line to turn around. On long distance calls the echo suppressors must have time to turn around. The Telephone Company (Telco) specifies around 100 ms for the turnaround of a 2 W circuit. In addition, it takes time for the carrier detect lead to turn on after the carrier has entered the modem receiver. This delay varies from 5 to 50 ms. Then net result is that the clear-to-send delay must be somewhere in the 150 ms to 200 ms range on a 2 W operation after the RTS line has been enabled.

When the RTS lead is turned off, the local receiver should be disabled until the start of the Carrier Detect signal. This procedure, called receiver squelch, is to prevent the detection of noise during the arrival waiting period of the remote carrier. The duration of squelch may be selectable but should be applied as long as possible, but must be less than the CTS delay of the remote modem.

On 4 W operation where the direction of transmission over a pair of wires never changes, only the carrier "detect" time needs to be taken into account by the CTS delay.

An optional 5 b/s AM reverse channel is also available on the Bell 202 modem. This slow speed channel is often used to request retransmission of a block of data which was received in error. If, for example, a modem at station A is transmitting information to a modem at station B, station B modem transmits a reverse channel carrier frequency to indicate that error-free signals are being received. Station B controls this information by maintaining the (S)RTS lead on the RS-232-C interface. As long as the reverse channel carrier frequency is being received by station A modem, its secondary data carrier detect lead (S)DCD will be indicating error-free transmission.

If the receiving station B detects errors, it will turn off the (S)RTS, thereby causing loss of reverse channel carrier frequency and a drop in the (S)DCD at modem A. This alerts the transmitting station of a problem.

The reverse channel can also be used to transmit a tone whenever the main channel RTS line is off, assuring that a tone is always present on the line and thus maintaining the echo suppressors in the disabled position.

Figure 5-13 shows the frequency utilization of an HDX BELL 202 modem c/w reverse channel. The CCITT V.23 international standard shown in Fig. 5-14 is the closest one to the Bell 202 modem. It recommends a reverse channel with a modulation rate up to 75 baud, mark at 390 Hz. and space at 450 Hz. Note that V.23 uses real FSK rather than the ON/OFF keying of the Bell 202.

Bell 212A

The Bell 212A is a two-speed modem in that it can operate asynchronously from 0 to 300 b/s using the identical FSK modulation scheme of the Bell 103 modem, or it can operate at 1200 b/s using a differential-phase-shift-keyed (DPSK) carrier. In the low speed mode, frequencies of 1070/1270 Hz are transmitted by the originating modem and frequencies of 2025/2225 Hz by the answering modem. In the high speed mode, the originate modem phase shifts a

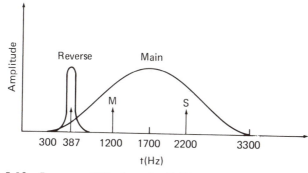

Figure 5-13 Frequency Utilization of a Half-Duplex Bell System 202 Modem

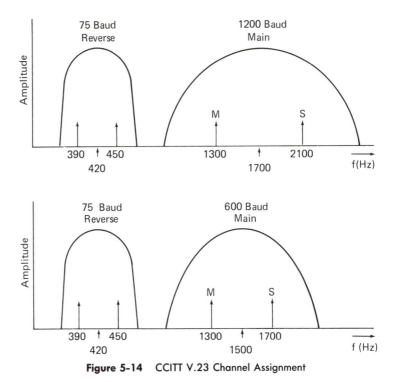

Figure 5-14   CCITT V.23 Channel Assignment

1200 Hz tone, whereas the answer modem phase shifts a 2400 Hz tone. The Bell 212A operates full duplex over the 2 W switched network or two-wire (point-to-point) private line.

In the high speed mode, it can operate in a bit-synchronous format or in a character-asynchronous format. The asynchronous character length can be selected to 8, 9, 10, or 11 bits.

In the high speed mode, two bits, called dibits, are taken at a time and each dibit causes the carrier to phase shift by a specific amount. Each dibit is assigned a specific phase shift such that if the receiver picks up a single error, only a one bit error is produced. In particular, the Bell 212A shifts phase 90° for dibit 00, 0° for 01, 180° for 10 and 270° for 11. This is shown in Fig. 5-15.

The phase shift is differential, indicating that the phase of the carrier is shifted by these phase shifts in reference to the phase of the carrier during the previous dibit interval. The advantage of this technique over PSK is that no phase reference tone needs to be transmitted or synthesized. These modems have a high speed control switch (HS), which when depressed on the originating modem forces it into the high speed mode. The speed mode must be selected before the data operating mode is entered, and cannot be changed once the modem has entered the data mode. When the HS switch is released, the modem originates the call in the low speed mode. The answering modem mode

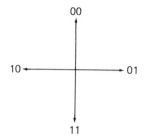

Figure 5-15 Phase Shift for Bell 212A Modem

speed is always determined by the originating modem, regardless of its HS switch position.

An interesting variation of the Bell 201 modem is the GDC 212A/ED modem, manufactured by General Data Comm. It is designed to operate on the 2 W switched network and has a built-in auto-dial unit. The number can be dialed from a terminal or from the modem's front panel buttons. These dialed numbers are stored in the modem.

Bell 201 B/C Modem

The Bell 201 B/C is a synchronous modem designed to operate at 2400 b/s. The 201 B operates over conditioned private lines, a 3002 channel with C2 or schedule 4/type 4B conditioning. The 201 C operates on the switched network or an unconditioned 3002 channel basic or schedule 4/type 4 channel. The modem can be used on 4 W lines in FDX, but usually operates two-way alternate. On 2 W lines it runs HDX. It is a good choice for multidrop 2 W private line applications as a new sync feature can be optioned. This option assures rapid re-sync of the modem receiver clock when the interval between messages is short, i.e., less than 100 ms. The RTS/CTS time delay is reduced to about 7.5 ms. When used on the switched 2 W line, the modem's turn-around time must be increased to about 150 ms.

The 201 uses DPSK, with the phase shift corresponding to the dibits as shown in Fig. 5-16.

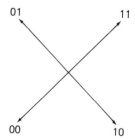

Figure 5-16 Phase Shift for Bell 201 B/C Modems

Bell 208 Modems

The Bell 208 is a synchronous DPSK modem, but uses eight-phase differential phase shift on a carrier of 1800 Hz to increase the data rate to 4800 b/s. In this case three bits are encoded at a time and the signaling rate is 1600 baud. The Bell 208A is designed to operate over a 4 W private or leased line. The Bell 208B is designed to operate over the switched network.

High Speed Modems

Theoretically it is possible to increase data rates up, increasing the number of phase shifts. If 4 bits at a time are encoded, 16 carrier phase are obtained or 4 bits per baud are obtained. With a signaling rate of 1600 baud, a data rate of 9600 b/s is obtained. The difficulty experienced with this technique is that a phase jitter on the line of only 11.25° would cause an error as the difference between adjacent symbols is 360°/16 or 22.5°. On a long haul it is extremely difficult to maintain the phase jitter down to this level.

By employing both phase and amplitude modulation, called Quadrature Amplitude Modulation (QAM), supported with amplitude and delay equalizers, and Automatic Gain Control circuits, these higher data rates can be obtained. At present 9600 b/s FDX can be transmitted over private 4 W voice grade lines using these techniques and designs.

The Bell 209 A is a synchronous modem designed for operation over a 4 W private line with D1 conditioning and no C type conditioning. The modem provides for Time Division Multiplexing of up to four 2400 b/s channels as shown in Fig. 5–17.

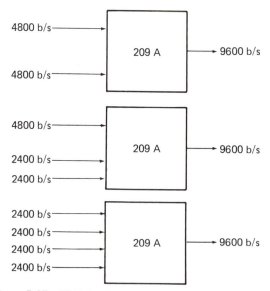

Figure 5-17 TDM Configurations on the Bell 209 A Modem

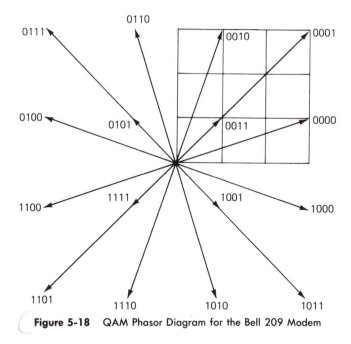

Figure 5-18   QAM Phasor Diagram for the Bell 209 Modem

The Bell 209 modem uses the QAM phasor diagram shown in Fig. 5-18. It consists of 12 different phases and 3 different amplitudes. The choice of the QAM points is such as to minimize errors in the presence of line impairment. The spacing between amplitudes determines how susceptible the modem will be to amplitude jitter, hits, and dropouts. The phase angle spacing will determine susceptibility to phase jitter. Many 9600 b/s modems provide for an eye diagram where the phase pattern can be viewed on an oscilloscope while the modem is operating. If there is amplitude jitter, the points will move radially in and out. If only phase jitter is present, the points will move concentrically about a radius from the origin. If both amplitude and phase jitter are present, the points will tend to smear. Hits or impulse noise will cause random stray points.

5-4   Equalization

In order to reduce the distortion that a signal encounters on unconditional lines, modems are often equipped with line equalizers. The telephone companies use "amplitude" and "delay" equalizers to compensate for high frequency attenuation and delay distortion on private line circuits, but this may be insufficient to reduce the error rate to a desired number. In addition, this conditioning increases the cost of the channel. Modem equalizers can be used to compensate for the effects of the remaining distortion or can be more economically feasible than having the TELCO condition a private line.

The modem equalizers can be fixed, manually adjusted, or automatic. When a modem is supplied with a *fixed equalizer,* its compensation is based on the typical amplitude and phase characteristics of the standard 3002 unconditioned line. Usually such an equalizer can be switched in or out, the choice being determined upon whether the error rate is improved. A *manual equalizer* allows for several adjustable controls, the settings being dependent upon a minimum meter reading. *Automatic equalizers* are used in the higher speed modems, 4800 b/s and greater. These equalizers are typically digital, in which the signal to be equalized is encoded and then delayed by shift registers. The tap gains are varied by digital multiplication. Some automatic equalizers are adaptive, in that they cause the equalizer to follow and track the channel variations and constantly update the equalization process. These adaptive equalizers in the initial startup equalize the line in about 120 ms and it then continually adjusts itself to any channel variations.

Usually the equalizers are placed in the receiver circuits, but they can also be located at the transmitter. In the latter case, the equalizer predistorts the signal and the line acts as the equalizer. This feature is used to advantage when transmitting from several transmitters, as in the case of a multidrop polled system. This eliminates the need to re-adjust the receiver in the master station each time a signal is received from a different polled station.

5-5   Modem Modulation Techniques

As the basic types of modulation schemes used to convert digital data into suitable analog signals for telephone lines have been described earlier in Section 5-3.1, little more needs to be discussed in this regard.

Because FSK is relatively simple in its operation and thus inexpensive to implement, it is used extensively in low speed modems for data rates below 1200 b/s. FSK is not efficient, however, in its use of spectrum space.

For higher data rates DPSK synchronous data transmission is generally used. Typically, four-phase DPSK or 4∅ DPSK is used at 2400 b/s and 8∅ DPSK is used for 4800 b/s. A typical phase diagram for a 4∅ DPSK modem is shown in Fig. 5-15. In this case, the Gray code is used in which adjacent phase shifts undergo only a one-bit change in the two bits (dibits). This scheme reduces the effects of noise on a data signal. In the straight binary code, going from a 1 (01) to a 2 (10) represents a two-bit error, whereas for the Gray code only a single bit error is produced.

For data transmission rates of 9600 b/s, QAM is finding acceptance (see Fig. 5-18). Because 16 different Gray-encoded signal states are possible, $\log_2 16$ or 4 bits can be transmitted per signaling pulse. Thus, for a 2400 baud signal, a 4×2400 or 9600 b/s information rate is obtainable.

Table 5-1 Bell System Modem Types*

	103J	113D	202S	202T	212A	201C	208A	208B	209A	LADS	500A Data Service Unit
Data rates (bits/s)	Up to 300	Up to 300	Up to 1200	Up to 1800	Up to 300 and 1200	2400	4800	4800	9600	2400/4800/ 7200/9600/ 19,200	2400/4800/ 9600/ 56,000
Channel types — Line interface	2-wire switched	2-wire switched	2-wire switched	2/4-wire private	2-wire switched	2/4-wire private 2-wire switched	4-wire private	2-wire switched	4-wire private	4-wire[1]	4-wire[2]
Line conditioning	None	None	None	C2 over 1200 bits/s	None	None	None	None	D1	None	N/A
Transmission mode[3]	Full duplex	Full duplex	Half duplex[4]	Half-duplex (2-wire) Full-duplex (4-wire)[4]	Full duplex	Half-duplex (2-wire) Full-duplex (4-wire) Half duplex	Full duplex	Half duplex	Full duplex	Full duplex	Full duplex
Modulation scheme	FSK	FSK	FSK	FSK	FSK	PSK 4-Phase	PSK 8-phase	PSK 8-phase	QAM	PSK 2-phase	None—digital regeneration
Diagnostics	Remote & local loopback, Self-test	Remote & local loopback, Self-test	Remote & local loopback, Self-test	Remote & local loopback, Self-test	Remote & local loopback, Self-test	Remote & local loopback, Self-test	Remote & local loopback, Self-test	Remote & local loopback, Self-test	Remote & local loopback, Self-test	Remote & local loopback	Remote & local loopback
Equalization	Fixed	Fixed	Fixed	Fixed	Fixed	Fixed	Automatic	Automatic	Automatic	Fixed	N/A
Line turn-around	None	None	—	—	None	—	—	—	—	None	None
Synchronization	Asynchronous	Asynchronous	Asynchronous	Asynchronous	Asynchronous or Synchronous[5]	Synchronous	Synchronous	Synchronous	Synchronous	Synchronous	Synchronous
Business-machine interface	RS232C	RS232C	RS232C	RS232C	RS232C	RS232C	RS232C	RS232C	RS232C	RS232	RS232C CCITT V.35[6]

1. Local Area Data Set (LADS) operates over Type-83 Local Area Data channel only.
2. 500A Data Service Unit operates over private DATAPHONE (service mark of AT&T) Digital Service (DDS) channel only.
3. Transmission mode indicated is that of the transmission facility; full-duplex lines support full and half-duplex protocols; half-duplex lines support half-duplex protocols only.
4. 202S and 202T are optionally available with a 5-bit/s, 387-Hz reverse channel.
5. Asynchronous operation at 0 to 300 bits/s or 1200 bits/s; synchronous operation at 1200 bits/s only.
6. EIA RS232C interface provided at 2400, 4800, 9600 bits/s; CCITT V.35 interface provided at 56 kbits/s.
* Reprinted with permission from Electronic Design, vol. 27, no. 16, © Hayden Publishing Co., Inc., 1979.

Table 5-1 offers a quick glimpse of the variety of Bell system modems presently in use. Note that for the higher bit rates, a separate pair of wires is used for each direction (4-W line).

The signal shown in Fig. 5-19(a) illustrates an FSK signal for alternating Marks and Spaces.

5-5.1 FSK Frequency Response

By using the switching function outlined in Appendix A, the FSK waveform can be constructed by multiplying an f_2 sinusoid by $S(t)$ and adding the result to an f_1 sinusoid multiplied by $1 - S(t)$. This is illustrated in Fig. 5-19(b) and (c).

Before obtaining the frequency content of the FSK signal, the frequency content of the waveforms, shown in Fig. 5-19(b) and (c), must be found. Since the switching function is a square wave with a period of 2τ, it contains all the odd harmonics of $1/2\tau$ Hz which roll-off at the $\sin f/f$ rate (see Appendix A).

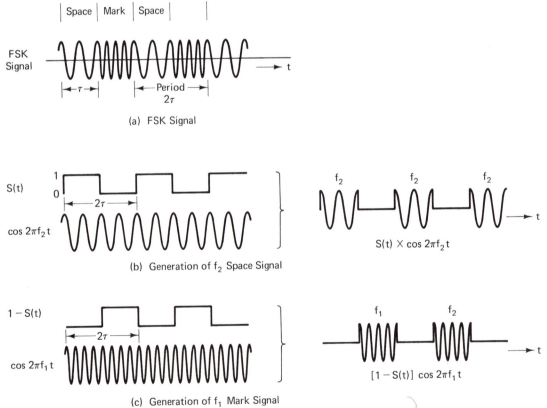

(a) FSK Signal

(b) Generation of f_2 Space Signal

(c) Generation of f_1 Mark Signal

Figure 5-19 Generation of FSK Signal

Since multiplying two sinusoids results in the sum and difference frequencies (see Eq. (5-2)), when the sinusoid $\cos 2\pi f_2 t$ is multiplied by the switching function containing odd harmonic sinusoids, the following sum and difference frequencies are obtained:

$$f_2 + \frac{1}{2\tau}, f_2 - \frac{1}{2\tau}, f_2 + \frac{3}{2\tau}, f_2 - \frac{3}{2\tau} \cdots$$

Similarly, when $\cos 2\pi f_1 t$ is multiplied by $1 - S(t)$, the following sum and difference frequencies are obtained:

$$f_1 + \frac{1}{2\tau}, f_1 - \frac{1}{2\tau}, f_1 + \frac{3}{2\tau}, f_1 - \frac{3}{2\tau} \cdots$$

The negative sign on $S(t)$ indicates that the resultant signal undergoes an extra 180° phase shift.

Figure 5-20 illustrates the resulting frequency spectrum of the FSK signal. The bandwidth required to pass an FSK signal usually includes the components that reside within the first $\sin f/f$ zero. For the signal under consideration, this would be $f_1 - f_2 + 2/\tau$ Hz.

5-5.2 Theory of Phase Shift Keying

PSK is a commonly used digital modulation technique which is very easy to handle analytically. This section will discuss a two-phase or binary phase shift keying system and then briefly discuss the implementation of a four-phase PSK system.

Binary phase shift keying (BPSK) is obtained through the system shown in Fig. 5-21. The data signal is translated into a non-return-to-zero (NRZ) code having no dc component. The resulting pulses of amplitudes $+A$ or $-A$ represent the binary 1's and 0's. The multiplier output is given by:

$$v'_o = \pm A \cos \omega_c t \qquad (5\text{-}1)$$

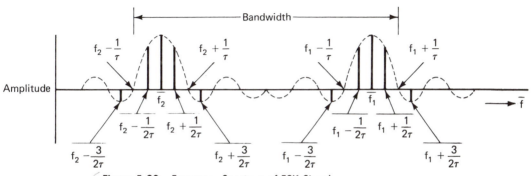

Figure 5-20 Frequency Spectrum of FSK Signal

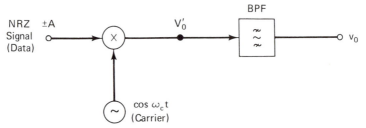

Figure 5-21 The BPSK Modulator

where it is assumed that the gain of the multiplier is unity and where $f_c = \omega_c/2\pi$ is the unmodulated carrier frequency.

Since $-\cos \omega_c t = \cos (\omega_c t + 180°)$, the output voltage can be expressed by:

$$v'_o \nearrow \begin{array}{c} A \cos \omega_c t \\ \text{or} \\ A \cos (\omega_c t + 180°) \end{array}$$

This modulated signal has a constant maximum amplitude of A and can alternate only at $0°$ and $180°$ phase.

To obtain a general impression of the v'_o spectrum, consider the modulating NRZ signal to consist of a train of square wave pulses without a dc bias. If the pulse rate is f_m, the spectrum of this pulse train would have the typical $(\sin x)/x$ amplitude response for the odd harmonics of f_m as shown in Fig. 5-22.

When one sinusoid is multiplied by another sinusoid, the following sum and difference frequencies are noted:

$$\cos \omega_1 t \cos \omega_2 t = \frac{1}{2} \cos (\omega_1 + \omega_2)t + \frac{1}{2} \cos (\omega_1 - \omega_2)t \qquad (5\text{-}2)$$

Thus when the square wave, which contains only odd harmonics of f_m, is multiplied by the sinusoid of frequency f_c, as dictated by Eq. (5-1), the harmonics of f_m are obtained as sidebands around the frequency f_c where $f_c \gg f_m$. This is shown in Fig. 5-23. Since no dc is present in the NRZ signal, the carrier at f_c is suppressed.

The spectrum indicates that the bandwidth taken up by the modulated signal is twice the baseband or modulation bandwidth. The bandpass filter is inserted to limit the transmitted spectrum. A possible negative effect of adding such a filter might be a non-uniform time delay of the various frequency components in the signal, particularly for those lying near the filter skirts if the filter is not correctly designed. This results in the system being more susceptible to noise. The block diagram of a corresponding BPSK demodulator or detector is shown in Fig. 5-24.

Ignoring all filtering effects of the filter and transmission medium aside from the inherent delay of θ radians between the transmitter and receiver, the

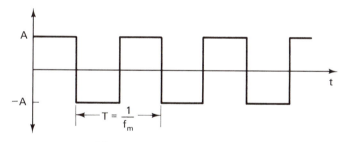

(a) Time Domain Representation of
the NRZ Signal

Wave
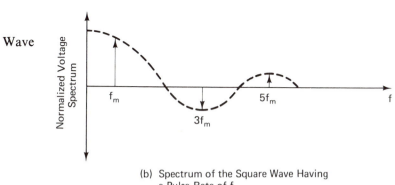

(b) Spectrum of the Square Wave Having
a Pulse Rate of f_m

Figure 5-22 Frequency and Time Domain Representation of a NRZ Square Wave

received signal, v_s, will be:

$$v_s = \pm A \cos (\omega_c t + \theta) \qquad (5\text{-}3)$$

In order to demodulate this signal, the carrier at the correct phase must be extracted. This is accomplished in the carrier recovery section by squaring the received signal and then filtering out the desired second harmonic with a bandpass filter (or a phase-lock loop). The squared signal can be expressed as:

$$v_s^2 = A^2 \cos^2 (\omega_c t + \theta) = \frac{A^2}{2} + \frac{A^2}{2} \cos 2 (\omega_c t + \theta) \qquad (5\text{-}4)$$

where the latter term represents a frequency component at twice the carrier frequency. The divide-by-two circuit yields the desired $\cos (\omega_c t + \theta)$ carrier.

Mixing or multiplying the extracted carrier with the received modulated signal results in:

$$\pm A \cos (\omega_c t + \theta) \cos (\omega_c t + \theta) = \pm \frac{A}{2} \pm \frac{A}{2} \cos 2 (\omega_c + \theta) \qquad (5\text{-}5)$$

$$\underbrace{\phantom{\pm \frac{A}{2}}}_{\substack{\text{desired}\\\text{signal}}} \quad \underbrace{\phantom{\pm \frac{A}{2} \cos 2 (\omega_c + \theta)}}_{\substack{\text{removed by}\\\text{LP filter}}}$$

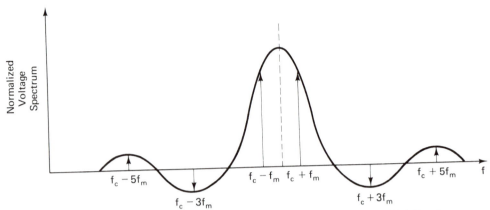

Figure 5-23   Spectrum of a Square Wave Modulated Sinusoidal Signal

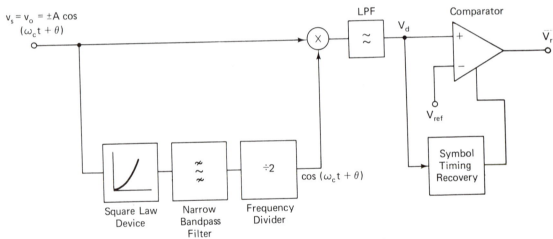

Figure 5-24   A BPSK Demodulator

The desired signal, which contains the original information, is extracted by the low-pass filter. The comparator circuit, which is clocked at the symbol rate, finally determines whether a 1 or a 0 logic state is detected. Typical waveforms at the indicated locations of Figs. 5-21 and 5-24 are illustrated in Fig. 5-25.

To achieve higher bit rates, four- or eight-phase modulation is used in which each interval or symbol carries two or three bits of information. At the transmitter, to serial-to-parallel conversion is performed to group the incoming data stream into groups of two or three bits, depending upon whether four- or eight-phase modulation is used. The receiver performs the complementary parallel-to-serial conversion.

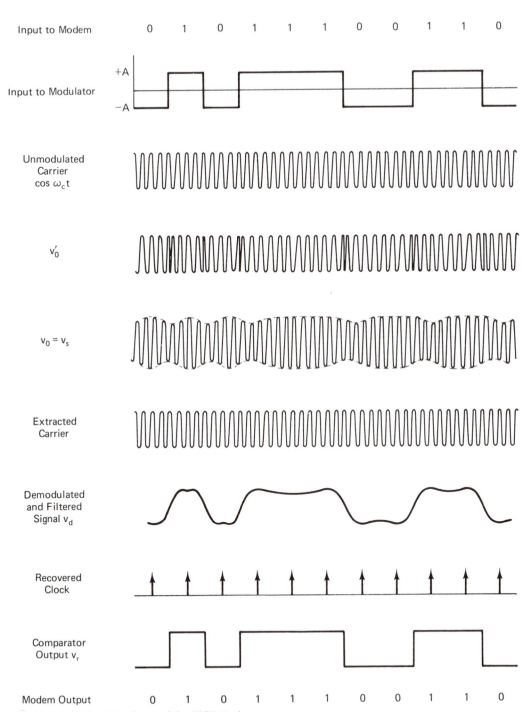

Figure 5-25 Waveforms of the BPSK Modem

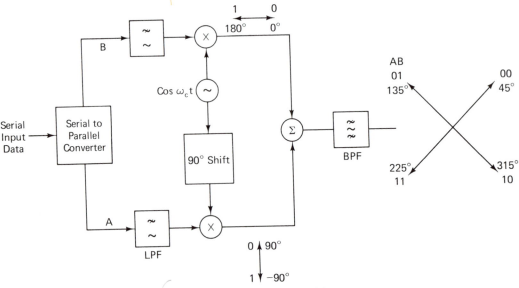

Figure 5-26 Four-Phase Modulator

Figure 5-26 shows the block diagram of a four-phase modulator. The serial-to-parallel encoder pairs up the binary digits, called dibits, which results in a transmission speed of the A and B channels that is half that of the incoming serial data stream. Figure 5-28 shows the channel waveform corresponding to the signal shown in Fig. 5-27.

The lower modulator in Fig. 5-26 is fed with a 90° carrier (cos $\omega_c t + \pi/2$) and the upper modulator with a 0° carrier (cos $\omega_c t$). When the lower modulator obtains an input A digit of 0, the output consists of a 90° carrier. When the input A digit is a 1, the output consists of a −90° carrier. Similarly, the output from the upper modulator is at 0° or 180°, depending upon whether the input B digit is a 0 or a 1. For instance, if the binary values of A and B represent the voltage signals listed in Table 5-2, the corresponding output voltage from the modulator can be represented by the phasor shown in Fig. 5-26. By summing the two output signals from the modulators, the output phasor diagram is obtained. (See Fig. 5-27.)

The four-phase signal can be readily detected if a fixed reference phase signal is available at the receiver. Figure 5-29 shows such a receiver.

Using Eq. (5-2) with a representative received signal of cos ($\omega_c t + \theta$), the output from the multipliers or demodulators will be:

$$\cos (\omega_c t + \theta) \cos (\omega_c t + \phi) =$$
$$\frac{1}{2} \cos (\theta - \phi) + \frac{1}{2} \cos (2 \omega_c t + \theta + \phi) \qquad (5\text{-}6)$$

where ϕ takes on a value of 0 or 90°, depending upon which modulator is being considered.

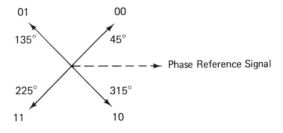

(a) Phaser Diagram

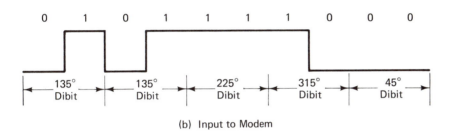

(b) Input to Modem

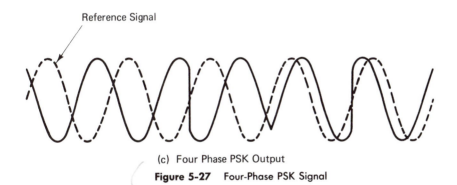

(c) Four Phase PSK Output

Figure 5-27   Four-Phase PSK Signal

   The low-pass filter will remove the second harmonic components of $2\omega_c$, leaving the term of the form $\frac{1}{2} \cos (\theta - \phi)$. Since the incoming signal phase must take on one of the θ values of $\pm 45°$ or $\pm 135°$, when this signal is multiplied by the reference oscillator of phase $\phi = 0°$ or $90°$, the signal obtained at the LPF output is one tabulated in Table 5-3. The comparator circuits readily detect the dc signal level from the LP filters to note the digital assignment. With the comparators shown, the comparator output goes LO when the input exceeds 0 volts, indicating a 0. The parallel-to-serial converter performs the process of converting the two-bit symbol back to two bits of serial data.

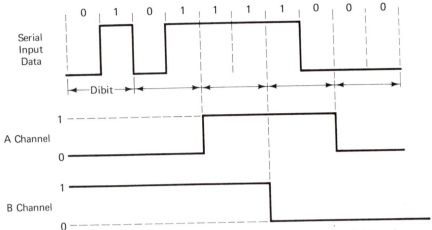

Figure 5-28 Channel Waveforms from a Four-Phase Serial-to-Parallel Encoder

Table 5-2 Four-Phase Modulator Signal

	Binary Value	Voltage	Modulator Output
B	0	V	$V \cos \omega_c t$
	1	$-V$	$-V \cos \omega_c t = V \cos (\omega_c t + \pi)$
A	0	V	$V \cos (\omega_c t + \pi/2)$
	1	$-V$	$-V \cos (\omega_c t + \pi/2) = V \cos (\omega_c t - \pi/2)$

Table 5-3 Four-Phase Demodulator Signal

Received Signal Phase	A Output	B Output	Data Output A	B
45°	$\tfrac{1}{2} \cos 45° = .35$	$\tfrac{1}{2} \cos 45° = .35$	0	0
135°	$\tfrac{1}{2} \cos 45° = .35$	$\tfrac{1}{2} \cos 135° = -.35$	0	1
225°	$\tfrac{1}{2} \cos 135° = -.35$	$\tfrac{1}{2} \cos 225° = -.35$	1	1
315°	$\tfrac{1}{2} \cos 225° = -.35$	$\tfrac{1}{2} \cos 315° = .35$	1	0

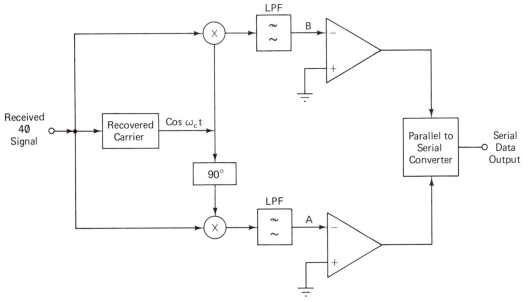

Figure 5-29   Four-Phase Demodulator

Although it is quite feasible to obtain a fixed phase recovered carrier, it is difficult to obtain the required *absolute* phase. The receiver must be given some indication of the proper phase reference since it is relatively easy for the carrier recovery phase to rest at 90° or 180° from the desired phase reference. Because of this difficulty, differential phase modulation and detection is preferred. In this modulation scheme, the information is encoded in terms of phase *changes* between adjacent symbols, rather than as an absolute phase for each symbol.

In the two-phase, coherent PSK system previously examined, a 1 corresponded to a 0° phase signal and a 0 corresponded to a 180° phase signal. In the modified differential phase modulation system, a 1 may correspond to a 0° *change* between successive signals and a 0 correspond to a 180° *change* between successive signals. At the receiver, each succeeding signal is compared with the phase of the previously received signal. Since noise interferes with *both* the current signal and the reference signal, the signal-to-noise ratio must be slightly greater in order to achieve similar error rates.

The encoding of a differential phase shift keyed (DPSK) signal is shown in the block diagram of Fig. 5-30.

The exclusive-NOR gate has a truth table, which is given in Table 5-4.

The encoding of a data bit stream is illustrated in Table 5-5. To begin the encoded signal, a reference binary digit of 1 is chosen. Because of the delay in the feedback path, the present message bit (data input) is compared with the previously encoded signal bit to arrive at the following digit for the encoded signal. In the example given, the first digit in the message, namely a 0, is com-

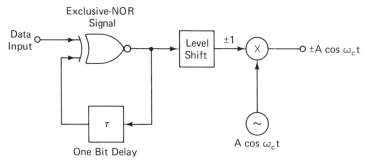

Figure 5-30   DPSK Modulator

pared with the initialization bit of 1, resulting in a 0 appearing as the next encoded bit. This latter 0 bit is the reference for the next digit to be encoded. Comparing this 0 with the next message digit of 1 results in another 0 as the next encoded digit. The resulting encoded message is then level shifted to a + or −1, to be multiplied by the unmodulated carrier $A \cos \omega_c t$, which results in a phase-shifted carrier of 0 or π radian.

The block diagram of Fig. 5-31 can be used for demodulation. The received signal is compared bit by bit, or, in other words, adjacent bits are com-

Table 5-4   Exclusive-NOR Truth Table

	Inputs	Output
A	B	C
0	0	1
0	1	0
1	0	0
1	1	1

Table 5-5   DPSK Example

Data Input		0	1	0	1	1	1	0	0	1	1	0
Encoded Message	1	0	0	1	1	1	1	0	1	1	1	0
reference digit ⟶												
Transmitted Phase	0	π	π	0	0	0	0	π	0	0	0	π

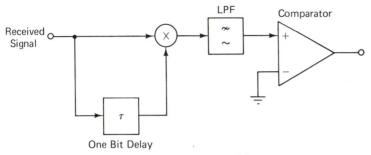

Figure 5-31 DPSK Demodulator

pared. If they are of the same phase, the multiplier output will be positive, i.e.:

$$(A \cos \omega_c t)^2 = [A \cos (\omega_c t + \pi)]^2 = \underbrace{\frac{A^2}{2}}_{\text{desired component}} + \frac{A^2}{2} \cos 2 \omega_c t$$

desired
component

If they are of different phases, the multiplier output will be negative, i.e.:

$$(A \cos \omega_c t) \, A \cos (\omega_c t + \pi) = \frac{A^2}{2} \cos \pi + \frac{A^2}{2} \cos (2 \omega_c t + \pi)$$

$$= \underbrace{\frac{-A^2}{2}}_{\text{desired component}} + \frac{A^2}{2} \cos (2 \omega_c t + \pi)$$

desired
component

The dc term is allowed to go through the low-pass filter so it can be applied to a comparator. The comparator translates the dc levels of $+A^2/2$ and $-A^2/2$ to 1's and 0's. When the threshold voltage of 0 volts is exceeded, a 1 results. If the input voltage is below the 0 threshold voltage, a 0 results. The detected message of Table 5-5 is illustrated in Table 5-6.

Table 5-6   Detection of DPSK Signal Given in Table 5-5

Received Message	0	π	π	0	0	0	0	π	0	0	0	π
Comparator Input		$\frac{-A^2}{2}$	$\frac{A^2}{2}$	$\frac{-A^2}{2}$	$\frac{A^2}{2}$	$\frac{A^2}{2}$	$\frac{A^2}{2}$	$\frac{-A^2}{2}$	$\frac{-A^2}{2}$	$\frac{A^2}{2}$	$\frac{A^2}{2}$	$\frac{-A^2}{2}$
Output Bit Stream		0	1	0	1	1	1	0	0	1	1	0

5-6   Interface Control

Prior to transmitting data, a valid circuit must be established between the DCE's. Each DCE must communicate with the other end of the telecommunication channel to make certain that the circuit is operational and that data can be transmitted and received. This procedure is called handshaking. The handshaking procedure for several modes of modem-TELCO line operation are described in the following five sections.

5-6.1   Procedures on a Switched Network

The handshaking procedure and the application of the RS-232-C specifications can be seen in more detail by examining the commonly used T103A data set, or modem, which achieves bit rates of 300 b/s when it is used in full-duplex operations over the direct distance dialing network (DDD). This examination will specifically concentrate on the RIXON T103A Data Set, the simplified block diagram of which is shown in Fig. 5-32.

The system block diagram of Fig. 5-32, which uses the RIXON data set, illustrates the interconnections of the DTE, the data set, and the control unit as well as the tie-in to the DDD network.

Full-duplex operation is achieved by operating the data sets over two frequency bands, f_1 and f_2 (see Fig. 5-11). Within each band, there is a MARK and a SPACE frequency. One data set operates as the originating station, while the other acts as the answering station. Both data sets receive and transmit, but the station which initiates a call becomes the originating station. The originating station transmits in the f_1 band and receives in the f_2 band. The answering station transmits in the f_2 band and receives in the f_1 band. Table 5-7 shows the frequency allocations.

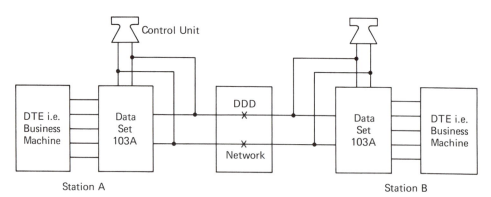

Figure 5-32   Data Phone System

Table 5-7   T103A Data Set Frequency Allocation

Data Set Mode	Transmit Band	Designation	Frequency
Originating	f_1	f_{1M} (MARK)	1270
		f_{1S} (SPACE)	1070
Answering	f_2	f_{2M} (MARK)	2225
		f_{2S} (SPACE)	2025

Depending upon which key is depressed on the control unit, the customer can go into one of the several available modes of operation. Figure 5-33 shows the control unit with its keys, and Table 5-8 gives the options available to each key position. To initiate a call, the customer dials the location he wishes to transmit data to or from via the DDD network. At the receive end, the attendant may answer the call and go into the data mode by pushing the data key on his telephone set. If the telephone set is placed on automatic answer, the data set will receive the call and control the line and business machine.

In general, the data set accepts serial binary data from the DTE and converts the binary data into a FSK signal. The FSK signal, in turn, is transmitted over the voice band telephone system. At the receiving end, the FSK signal is demodulated into a binary data stream and sent to the adjoining DTE. Either

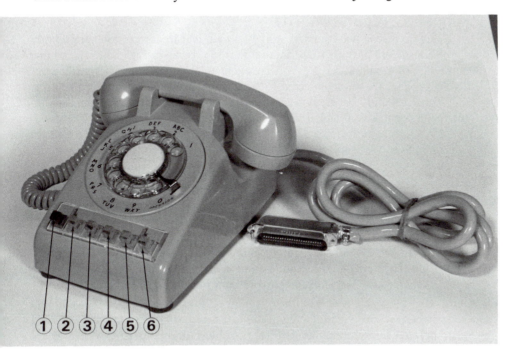

Figure 5-33   567PB-61 Telephone/Control Set (Courtesy of Rixon Inc.)

Table 5-8   Key Designations 567PB-61 Telephone Control Set (Courtesy of Rixon Inc.)

Index Number	Key Designation	Function
1	DATA	Nonlocking key used to shift from talk to data mode (also takes data set out of test mode if connected in error).
2	TALK	Locking key used to obtain normal telephone operation.
3	TEST 1	Nonlocking key used for remote testing of answer mode.
4	TEST 2	Nonlocking key used for remote testing of originate mode.
5	LOCAL	Locking key used to allow the telephone or business machine to operate to half duplex to make tapes or check their accuracy.
6	AUTO	Locking key used to provide automatic answer of incoming calls.

station may be the originating or answering station. The calling station becomes the originating station. If no carrier is present, the data set signals a MARK condition to the local DTE.

Before discussing the processing of a call, it is important to examine the significant blocks of the T103A data set that are shown in Fig. 5-34. The triangular shaped termination points correspond to the pin numbers specified in Table 4-1 and are interconnected to the local DTE. Note that the protective and signal grounds are connected together (Pins 1 and 7). These provide the reference point for all signal voltages. The following discussion will assume that the Data Terminal Read lead (CD) is ON, which causes the RR relay to operate.

Data Set Block Diagram

1. **Transmitter:** When transmitting data, the multivibrator (MV) operates at one frequency for a MARK input on the data lead and at another frequency for a SPACE input on the data lead. The time constant required to bring the OFF transistor to an ON state in the multivibrator circuit is altered by selecting a different charging resistance. Figure 5-35 gives a simplified version of the modulator. In order to remove undesirable harmonics and reduce sideband levels, bandpass filter Z_1 bandlimits the signal from the modulator.

2. **Receiver:** The Tip and Ring (T-R) leads from the telephone line are connected across the ringing detector (a voltage doubler circuit) which, when activated, causes the R relay to operate. Figure 5-36 illustrates the ringing relay circuit.

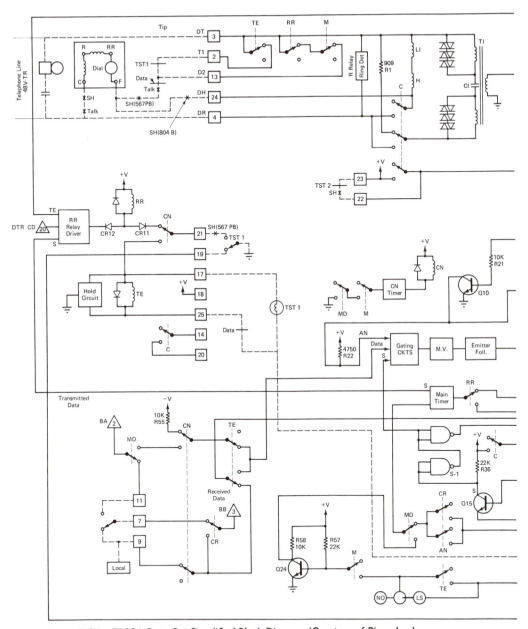

Figure 5-34 T103A Data Set Simplified Block Diagram (Courtesy of Rixon Inc.)

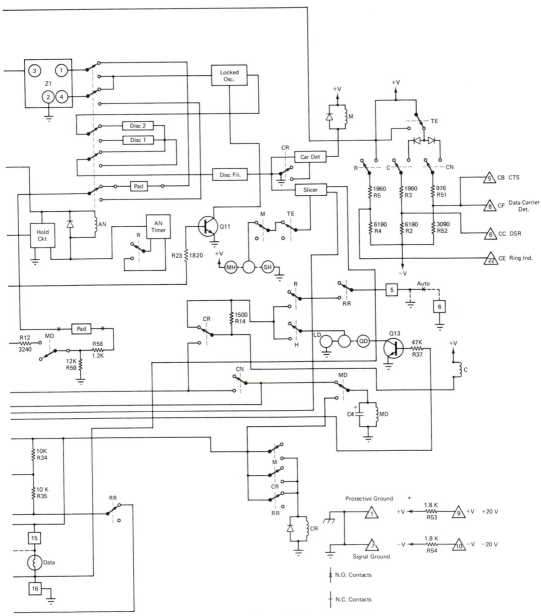

Figure 5-34 Concluded

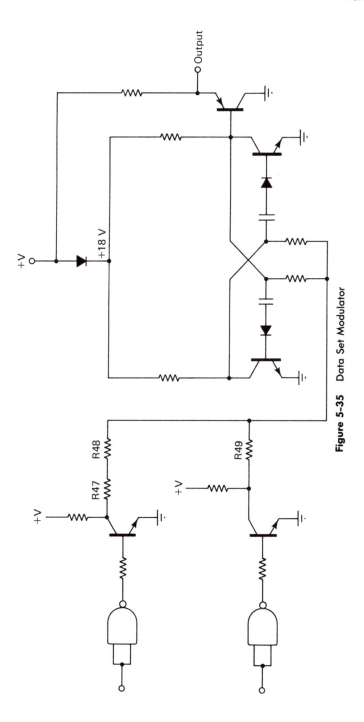

Figure 5-35 Data Set Modulator

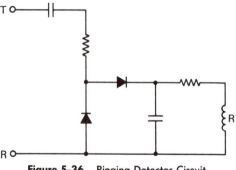

Figure 5-36 Ringing Detector Circuit

With the R contacts switched, ring indicator Pin 22 (CE) goes high and the line transformer, T_1, is connected to the T-R leads provided that the CD lead (Pin 20) is ON. (Relay C is operated via contacts R and RR.) The secondary of the line transformer is connected to a filter which, in turn, directs the signal to the locked oscillator. When no signal is present, the oscillator free-wheels at a frequency centered midway between the MARK and SPACE frequency: 1170 Hz for f_1 and 2125 for f_2. The discriminator is tuned negatively for 1270 Hz and positively for 1070 Hz when in the f_1 range. It is tuned negatively for 2225 Hz and positively for 2025 Hz when in the f_2 range. Thus, the discriminator output will be at 0 when the oscillator is free-wheeling, i.e., running midway between the MARK and SPACE frequencies. When a signal of sufficient magnitude appears, the oscillator locks to the frequency of the applied signal, causing the discriminator output to go positive or negative. The f_2 band discriminator is shown in Fig. 5-37.

If the discriminator output is significantly large and stable, the carrier detector circuit activates the M relay. High-level noise tends to be rejected.

The slicer circuit is a zero crossing detector that reshapes the discriminator output into a digital data wave. Depending upon the strapping used, the slicer is either clamped in the marking (MARK HOLD) or the spacing (SPACE HOLD) condition in the absence of carriers. The output from the slicer appears at the received data BB leads (Pin 3).

3. **Timer Circuits:**

Main timer: The main timer generates a number of delays while a call is being set up. These delays include a guard interval of about 150 ms, a 1.5 s delay for disabling the echo suppressors, and disconnect time intervals.

AN timer: AN is a memory device which locks up after ringing has occurred.

CN timer: The CN timer delays informing the local DTE that it may transmit data in order to ensure that the distant station has detected the carrier and is prepared to receive data.

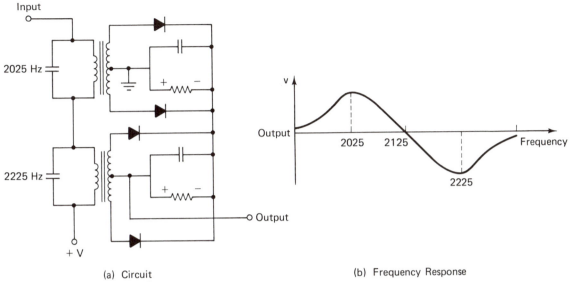

(a) Circuit

(b) Frequency Response

Figure 5-37   Discriminator Circuit and Action

Now that the workings of the T103A data set have been described, call processing will be more easily understood.

The discussion on processing a call, or of establishing a connection from a calling station to a called station, is outlined in tabular form in Table 5-9. It is assumed in the table that the CD terminal is ON at both ends; that is, that RR relay is activated, and that time advances with each step. Notes written alongside the steps explain what is happening within the modem (Fig. 5-34) when the AUTOMATIC ANSWER is in operation.

While following the channel establishment sequence, it may be helpful to observe the graphical counterpart illustrated in Fig. 5-38. Figure 5-39 graphically illustrates the disconnect sequence when the called station is in AUTO mode. A summary of the events is also included.

In the local mode, the BA (transmitted data) and BB (received data) leads are interconnected via MD, LOCAL and CR.

Table 5-9 Processing of a Call via a T103A Data Set

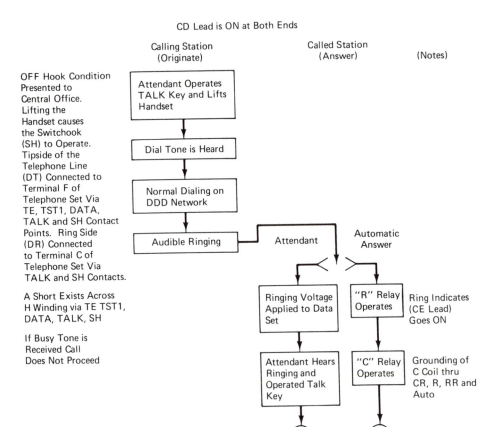

Table 5-9 (Continued)

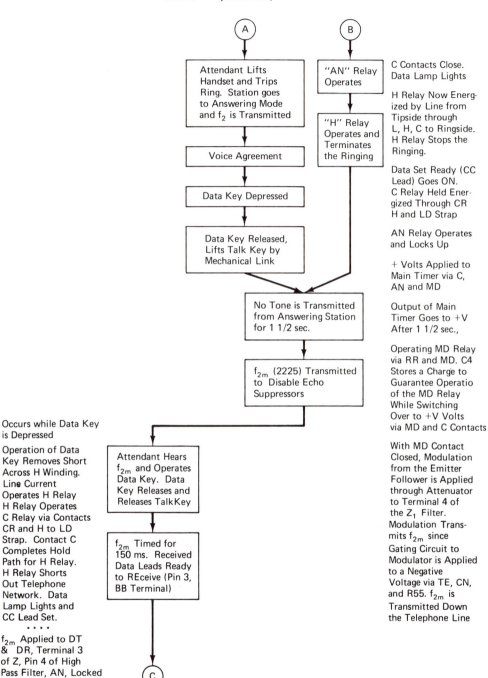

C Contacts Close.
Data Lamp Lights

H Relay Now Energized by Line from Tipside through L, H, C to Ringside. H Relay Stops the Ringing.

Data Set Ready (CC Lead) Goes ON. C Relay Held Energized Through CR H and LD Strap

AN Relay Operates and Locks Up

+ Volts Applied to Main Timer via C, AN and MD

Output of Main Timer Goes to +V After 1 1/2 sec.,

Operating MD Relay via RR and MD. C4 Stores a Charge to Guarantee Operatio of the MD Relay While Switching Over to +V Volts via MD and C Contacts

With MD Contact Closed, Modulation from the Emitter Follower is Applied through Attenuator to Terminal 4 of the Z_1 Filter. Modulation Transmits f_{2m} since Gating Circuit to Modulator is Applied to a Negative Voltage via TE, CN, and R55. f_{2m} is Transmitted Down the Telephone Line

Occurs while Data Key is Depressed

Operation of Data Key Removes Short Across H Winding. Line Current Operates H Relay H Relay Operates C Relay via Contacts CR and H to LD Strap. Contact C Completes Hold Path for H Relay. H Relay Shorts Out Telephone Network. Data Lamp Lights and CC Lead Set.
• • • •
f_{2m} Applied to DT & DR, Terminal 3 of Z, Pin 4 of High Pass Filter, AN, Locked Oscillator

Table 5-9 (Continued)

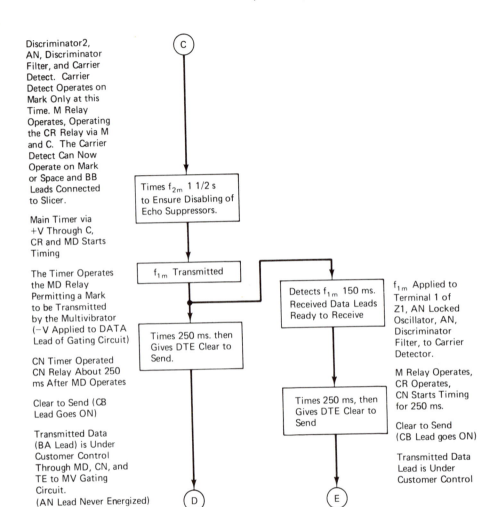

Discriminator2, AN, Discriminator Filter, and Carrier Detect. Carrier Detect Operates on Mark Only at this Time. M Relay Operates, Operating the CR Relay via M and C. The Carrier Detect Can Now Operate on Mark or Space and BB Leads Connected to Slicer.

Main Timer via +V Through C, CR and MD Starts Timing

The Timer Operates the MD Relay Permitting a Mark to be Transmitted by the Multivibrator (−V Applied to DATA Lead of Gating Circuit)

CN Timer Operated CN Relay About 250 ms After MD Operates

Clear to Send (CB Lead Goes ON)

Transmitted Data (BA Lead) is Under Customer Control Through MD, CN, and TE to MV Gating Circuit. (AN Lead Never Energized)

C

Times f_{2m} 1 1/2 s to Ensure Disabling of Echo Suppressors.

f_{1m} Transmitted

Times 250 ms. then Gives DTE Clear to Send.

D

Detects f_{1m} 150 ms. Received Data Leads Ready to Receive

Times 250 ms, then Gives DTE Clear to Send

E

f_{1m} Applied to Terminal 1 of Z1, AN Locked Oscillator, AN, Discriminator Filter, to Carrier Detector.

M Relay Operates, CR Operates, CN Starts Timing for 250 ms.

Clear to Send (CB Lead goes ON)

Transmitted Data Lead is Under Customer Control

Table 5-9   (Concluded)

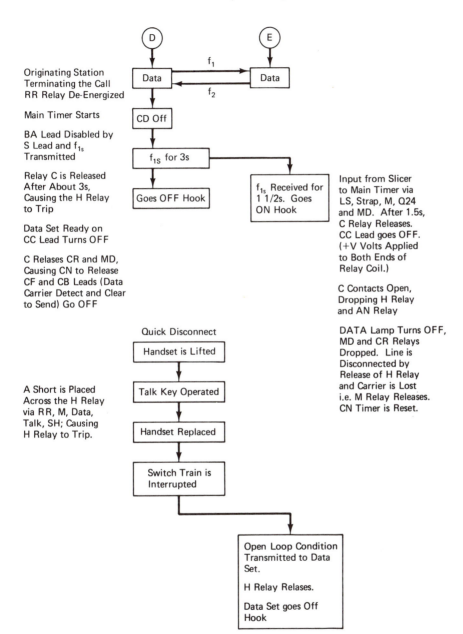

Originating Station
Terminating the Call
RR Relay De-Energized

Main Timer Starts

BA Lead Disabled by
S Lead and f$_{1s}$
Transmitted

Relay C is Released
After About 3s,
Causing the H Relay
to Trip

Data Set Ready on
CC Lead Turns OFF

C Relases CR and MD,
Causing CN to Release
CF and CB Leads (Data
Carrier Detect and Clear
to Send) Go OFF

D → E

Data —f$_1$→ Data
      ←f$_2$—

CD Off

f$_{1s}$ for 3s

Goes OFF Hook

f$_{1s}$ Received for
1 1/2s. Goes
ON Hook

Input from Slicer
to Main Timer via
LS, Strap, M, Q24
and MD. After 1.5s,
C Relay Releases.
CC Lead goes OFF.
(+V Volts Applied
to Both Ends of
Relay Coil.)

C Contacts Open,
Dropping H Relay
and AN Relay

DATA Lamp Turns OFF,
MD and CR Relays
Dropped. Line is
Disconnected by
Release of H Relay
and Carrier is Lost
i.e. M Relay Releases.
CN Timer is Reset.

Quick Disconnect

Handset is Lifted

Talk Key Operated

Handset Replaced

Switch Train is
Interrupted

A Short is Placed
Across the H Relay
via RR, M, Data,
Talk, SH; Causing
H Relay to Trip.

Open Loop Condition
Transmitted to Data
Set.

H Relay Relases.

Data Set goes Off
Hook

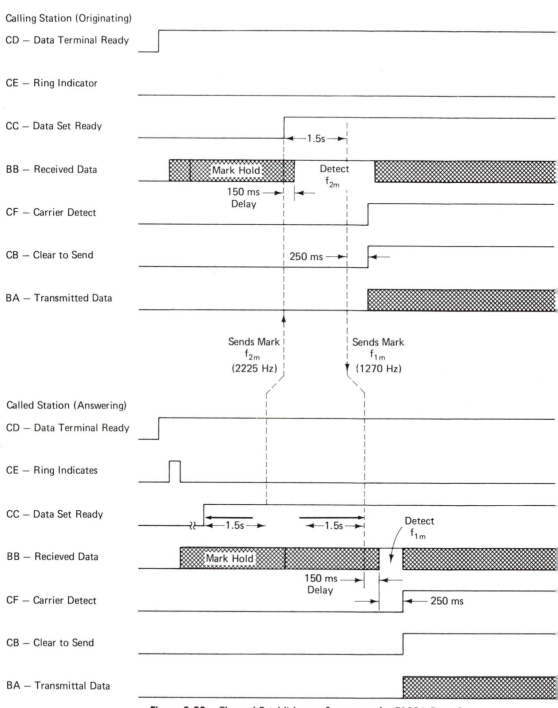

Figure 5-38 Channel Establishment Sequence of a T103A Data Set

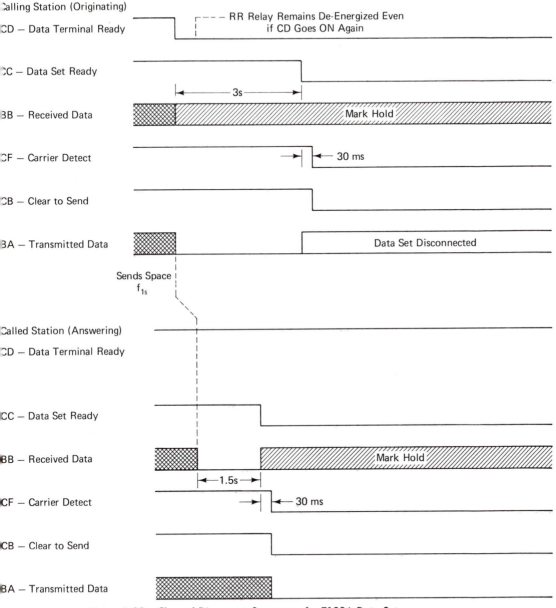

Figure 5-39 Channel Disconnect Sequence of a T103A Data Set

Summary of Channel Establishment (Automatic Answer)

1. Operator presses Talk key and dials the number of the distant modem. Originating modem is on MARK HOLD on Received Data Circuit.
2. The answering modem turns on the Ring Indicator.
3. The answering modem turns the Received Data Circuit to MARK HOLD.
4. The answering modem turns on the Data Set Ready Circuit. The data button lights at the modem.
5. After 1.5 s of delay, the answering modem transmits an f_{2m} tone to disable the echo suppressors.
6. The originating modem operator hears the tone and presses the DATA push button. Its Data Set Ready Circuit turns on and the DATA push button lights.
7. After 150 ms of delay, the Received Data goes to a non-hold condition.
8. One-and-one-half s after Data Set Ready has gone on, the originating modem transmits an f_{1m} tone. Clear to Send and Carrier Detect Circuits turn on after a 250 ms delay. This later delay allows sufficient time for the answering modem to lock on the f_{1m} tone.
9. One-hundred-fifty ms after the answering modem has detected the f_{1m} tone, its MARK HOLD on the Received Data is removed.
10. The Carrier Detect and Clear to Send circuits are turned on after a 250 ms delay. With CTS circuits on at both modems, data can be transmitted and received.

Summary of Channel Disconnect Sequence (Automatic Answer)

The called modem will disconnect when receiving a 1.5 s spacing signal.

1. The DTE at the originating station turns off the Data Terminal Ready Circuit for at least 50 ms. The Transmitted Data goes to a 3 s space.
2. After a 3 s delay, the Data Set Ready circuit is turned off.
3. The answering modem turns off the Data Set Ready circuit and places the Received Data in Mark Hold after receiving a 1.5 s spacing signal. This is called long space disconnect.
4. Thirty ms after the DSR goes off, the Clear to Send and Carrier Detect circuits turn off and the Received Data goes to Mark Hold.
5. Thirty ms after the DSR circuit at the originating modem is turned off, its Carrier Detect and Clear to Send circuits are turned off.

Both modems are now in a non-data mode.

5-6.2   Procedures on a 4 W Private Line FDX/HDX Service

In 4 W operation, the direction of data transmission never changes over each pair of wires and no TELCO line turnaround time is encountered. If the Request to Send circuit is used, the Clear to Send delay need only take into consideration the "carrier detect" time of the distant modem. This CTS delay must be greater than 15 ms, as the carrier detect time is typically 15 ms. This savings in line turnaround time is the main reason for using 4 W private lines in HDX operations.

Although some systems use only the Signal Ground, Transmitted Data and Received Data circuits, it is more common to include the Clear to Send, Data Carrier Detect and optionally the Request to Send interchange circuits to note the status of the DTE/DCE and to assure circuit operation. These configurations are illustrated in Fig. 5-40. The EIA denotes the interface configuration as Type D if the RTS circuit is controlled, and as a Type E if it is not. The SG, RD, TD and DCD circuits are not included in the illustration, but are always present. If synchronous modems are employed, the Transmitter and Receiving Signal Element Timing Circuits must also be included.

In the case of the Type E interface, the CTS delay is encountered only once, when the DTE is turned on, as the carrier remains on from that point. The DTE must be suitably strapped to perform this delay.

5-6.3 Procedures on a 4 W Private Line Multipoint Service

The procedures on a multipoint dedicated network, employing a centralized or primary station are similar to that described for the point-to-point operations. A private 4 W line is used to eliminate the long connect times that would be encountered on the switched telecommunication network. The remote stations can only communicate with the primary station and traffic between remote

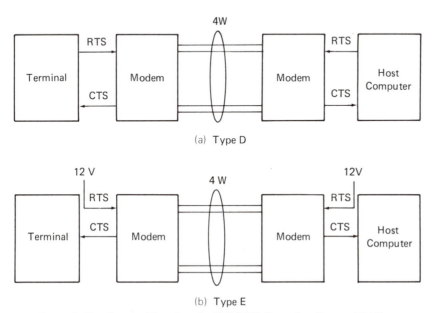

(a) Type D

(b) Type E

Figure 5-40 Standard Interface for FDX/HDX Operation Over a 4 W Line

stations must be relayed via the primary station. The primary station can leave its Request to Send circuit high, as it is the only transmitter on the outbound 2 W channel (type E interface). All the remote stations must control their Request to Send circuits (type D interface) as they must share the inbound 2 W channel back to the primary. The primary station selects a remote station by transmitting the appropriate terminal address. The selected terminal then replies with a response.

The primary only encounters the Clear to Send Delay once, when it is turned on. The secondaries encounter the delay every time they are required to transmit. If there is an exchange of several messages between a secondary and primary station, the secondary station may turn the Request to Send off after each transmission, or it may leave it on until the primary station commands the secondary station to turn the RTS off. The multipoint configuration is shown in Fig. 5-41.

5-6.4 Secondary Channel Turnaround Operation

When employing a half-duplex modem, such as the Bell 202 modem on a 2 W line, the reverse channel can be used to control the direction of data transmission on the main channel. In this operation, the Request to Send signals are controlled by the secondary channel signals. As an example, consider a terminal that is communicating with a host computer, with the host Secondary Request to Send ((S)RTS) circuit controlling the direction of main channel data flow.

As illustrated in Fig. 5-42, when the host is receiving data, its (S)RTS is held high and its Data Carrier Detect is ON as data is being received on the main channel. The RTS and CTS circuits are OFF as the host is in the receiving mode. During this time, the terminal is transmitting data (RTS and CTS

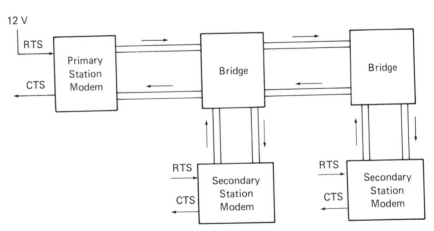

Figure 5-41 Configuration of a 4 W Multipoint Service

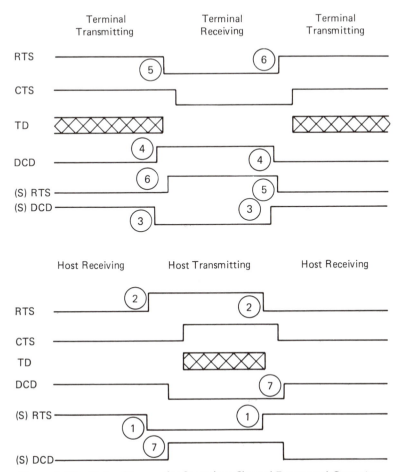

Figure 5-42   Timing Diagram for Secondary Channel Turnaround Operation

are ON) on the main channel, and is detecting a carrier on the Secondary Data Channel Detect circuit (S)DCD. While the terminal is transmitting data, the host may desire to begin data transmission. To initiate the turnaround operation, the host lowers its (S)RTS circuit as shown in the timing diagram. This is detected by the terminal in the removal of the (S)DCD signal. The RTS and the delayed CTS signals at the terminal are then removed. The RTS and CTS circuits are raised at the host, permitting the host to commence sending data.

An alternative method of initiating the turnaround can be by the use of the BREAK key in the terminal. If the BREAK key is pressed while the terminal is transmitting data, SPACES are transmitted on the XMIT DATA line. The host detects the break and lowers its (S)RTS circuit, which in turn is detected at the terminal in the removal of the (S)DCD signal. The same turnaround process then occurs as described earlier.

When the terminal is receiving data, the channel turnaround operation can be commenced by the remote host raising its (S)RTS circuit, thereby raising the terminal's (S)DCD circuit as shown in the timing diagram. At the same time the host's RTS circuit drops and a bit later the DCD signal is lowered in the terminal. The terminal then raises its RTS signal as well as lowering its (S)RTS circuit. After a 250 ms delay the CTS circuit is raised and the terminal can commence transmitting data.

Alternatively, the BREAK key at the terminal can be present while the terminal is receiving data from the host. The (S)RTS circuit at the terminal immediately drops and is detected at the remote host by a drop in its (S)DCD signal. The host may then take action to reverse the channel operation.

5-6.5 Code Turnaround Operation

When operating HDX over a 2 W line, the channel can be turned around by employing either an End of Text (ETX) or an End of Transmission (EOT) command on the main channel. If, for instance, a terminal is receiving data from a remote host computer, the terminal's DCD circuit will be high as a carrier is being detected and its RTS signal will be off. If the host then transmits an ETX or EOT command, this will be decoded by the terminal, indicating that character transmission is complete.

When the remote end drops the line, the DCD at the terminal falls and the RTS is raised. After the CLS is set, the terminal can begin to transmit data.

When the ETX (EOT) code is again detected, this time the result of the command being keyed into the terminal, the RTS (and CTS) signals at the terminal will be dropped. This will be detected at the remote end by a loss of carrier, indicating that the host may now raise its RTS (and CTS) signals and begin transmission. The interface timing for code turnaround is illustrated in Fig. 5-43.

5-7   Echoplex and Local Copy

An echoplex or echo-back system is one in which all the characters received at the receiving end are retransmitted back to the sender. The user can verify that the data is free of errors. If errors are present, the operator can re-enter the erroneous characters. This scheme normally employs full-duplex lines and has the disadvantage of poor line efficiency because of the 100% redundancy. Echoplex is frequently used on time-sharing arrangements.

Local copy is similar to echoplex, but it gives a duplicate copy of the input data to the user *locally* while permitting the data to be transmitted to the remote receiver at the same time. The local "loop back" occurs either in the terminal itself or in its attached data set. It is normally utilized in half-duplex operations. Fig. 5-44 pictorally illustrates the difference between echoplex and local copy.

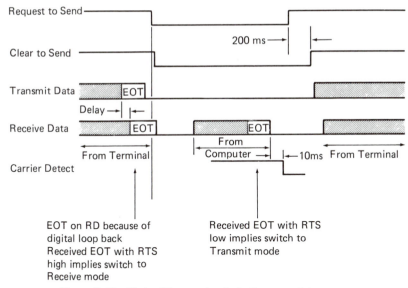

EOT on RD because of
digital loop back
Received EOT with RTS
high implies switch to
Receive mode

Received EOT with RTS
low implies switch to
Transmit mode

Figure 5-43 Timing Diagram for Code Turnaround Operation

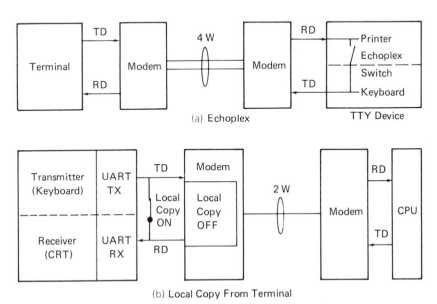

(a) Echoplex

(b) Local Copy From Terminal

Figure 5-44 Echoplex (a) and Local Copy (b) System Arrangement

5-8   Synchronization

5-8.1   Asynchronization Transmission

In low speed data transmission, say 300 b/s or less, asynchronous transmission is usually employed. In this case, each transmitter character is preceded by a start bit, which is a "0" bit or "space" and terminated with one or two stop bits, which is always a "1" bit or a "mark." When the receiving DTE detects the modem switching from a 1 to a 0, it determines that the following bits will make up a character and, knowing how many bits from each character, will be able to decode the bits into a character. Every character will also be followed by one or two stop bits. The actual function of the stop bit(s) is to assure the detection of the next "start" bit.

There can be a variable time interval between the transmitted characters, particularly if an unbuffered TTY device is used. Most devices buffer their messages before transmission and in such cases when a final key depression occurs, the complete message such as shown in Fig. 5-45 is transmitted.

The timing is supplied by the DTE and the receiving and transmitting terminals should be clocking close to the same rate. Typical receiver circuits sample the received data at a rate 16 times the bit arrival rate. Thus a START bit can be detected within $\frac{1}{16}$th of a bit time. After the detection of the start bit, the data stream is sampled approximately midway on each bit; that is, the received signal is sampled at 8, 24, 40, etc. clocking periods. In theory, the sampling of a bit can be off by 50% of a bit time before an error occurs.

As illustrated in Fig. 5-46, after the detection of the START element (a 1 to 0 transition), the center of the STOP element noting the end of a character typically occurs 9½ bits later. This represents 9½ × 16 or 152 clocking periods. The detection of the START bit can be out by $\frac{1}{16}$th of the bit time because it may not be detected until almost a full sample period after the 1 to 0 transition. When sampling the STOP element, we can be off an additional $\frac{7}{16}$th of a bit time or 7 clocking periods before the edge of an adjacent bit is encountered. The signal would then represent the appropriate value for the adjacent bit. The maximum clock deviation may thus by $\frac{7}{152}$ or 4.6%. For a 300 baud

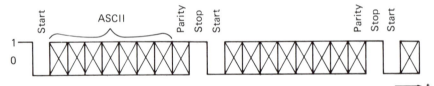

Figure 5-45   Asynchronous Transmission of Buffered Message

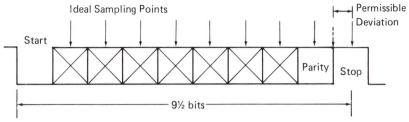

Figure 5-46   Calculation of Permissible Receiving Clock Deviation Using 16x Clock

asynchronous receiver, the ×16 clock could vary from a low of 16 × 300 (1.00 − 0.046) = 4579 Hz to a high of 16 × 300 (1.00 + 0.046) = 5020 Hz.

5-8.2   Synchronous Transmission

In synchronous transmission, synchronization is on a message basis rather than on a character basis. It employs no start and stop bits, but each message is preceded by two or more synchronization or sync characters. These sync characters provide a specialized bit pattern which enables the receiver to lock on to the data characters. Once in synchronization, each set of seven or eight consecutive bits is considered as a character.

Normally the synchronous modem or DCE is responsible for the transmit and receive timing or clock pulses. Both the transmitting and remote receiving modems must clock at exactly the same rate. The receiver modem derives its "receive clock" signal from the analog data signal, a technique referred to as data-derived timing. For this reason, it is imperative that the analog data signal change state frequently enough so that the receive clock can stay in sync with the receive data. As illustrated in Fig. 5-47(a), the modem at each end provides the transmit clock to both the terminal and to itself. Either terminal transmits under the control of its attached modem. As each end has its own master transmit clock, the transmit and receive clocks may be slightly different at either terminal.

Both modems could also be clocked by their associated terminals. This is not frequently used, as modems tend to have more stable clocks. This configuration is shown in Fig. 5-47(b). Again, each modem derives its receive timing from the received data.

If a common clock is required for the entire system, either one terminal or one modem can supply the master clock. Fig. 5-47(c) illustrates one such arrangement, where one modem supplies the clock to its associated terminal, and the data derived clock the transmitted data out of its associated terminal. In this manner the transmit and receive clocks are identical at both terminals.

Since characters must be steadily transmitted in synchronous transmission once message transmission has begun, the terminals must have the ability to buffer the data.

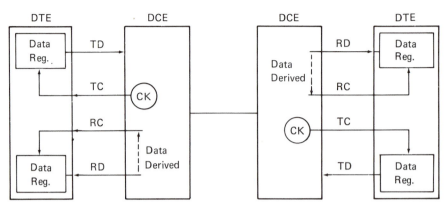

(a) Modem Control at Both Ends

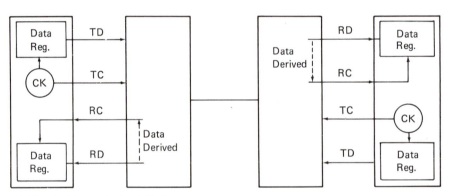

(b) Terminal Control at Both Ends

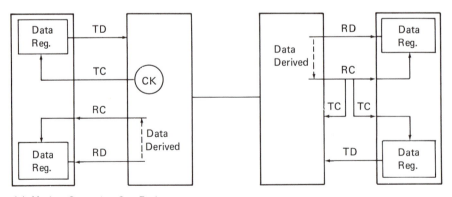

(c) Modem Control at One End

Figure 5-47 Synchronization Configurations: CK = Clock; TC = Transmit Clock; RC = Receive Clock

5-9   Line Circuit and Connection Arrangements

As transmission systems are designed to work with specific signal levels, a reference signal of 1000 Hz is often used to test telephone circuits. All test equipment should have balanced (ungrounded) inputs/outputs or be connected through an isolation transformer. When monitoring a line signal with an oscilloscope it is important not to ground either of the signal lines. All 4 W circuits, such as those used in private lines, have an impedance of 600 ohms, resistive, whereas most of the switched 2 W lines are 900 ohms. Modems usually have selectable 900 ohm or 600 ohm line impedances.

In 4 W circuits the end-to-end loss is held at 16 dB, but in 2 W switched circuits the loss varies from 16 dB to about 30 dB, depending upon the distance the signal travels and the nature of the line characteristics. Fig. 5-48 shows what percentage of calls would obtain a given receive level at the local exchange for various line lengths. The local loop and telephone hybrid loss must be subtracted from this receive level.

Many modems have a Carrier Detect Level Adjustment to control the sensitivity of the carrier detect or circuitry. It determines the minimum signal level at which the modem reliably detects the carrier signal. Its objective is to block the passage of the line "noise" and to allow data to pass through. For private lines, the acquisition level is typically set somewhere in the −25 dBm to −33 dBm range.

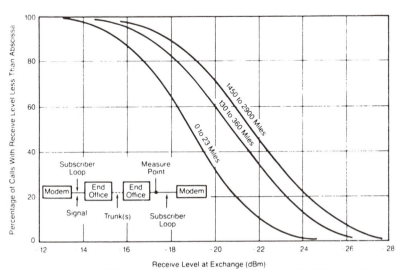

Figure 5-48   End Office Receive Levels on Direct Distance Dial Network. Courtesy of Racal-Vadic.

To illustrate the typical levels on a transmission faculty, consider a simple analog private line channel using loop and carrier facilities such as that shown in Fig. 5-49. Although we are primarily interested in data levels, the levels that would be expected if a 1000 Hz test line was used are also included in the layout. In this example a 0 dBm 0 test tone is assumed; in many facilities −10 dBm 0 is used.

In a carrier or multiplex system, the circuit always has 23 dB of gain and the standard test tone at the Modulation Input is at −16 dBm.

In data circuits all data levels are 13 dB below the Transmission Level Point, i.e., −13 dBm 0. Thus, at the modem, the transmit level is at 0 dBm, at the carrier system Modulation Input the level is −29 dBm, and at the input to the remote modem it is at −16 dBm.

The Data Service Unit (DSU) is used on 4 W dedicated channels to interconnect the modem to the telephone facilities. The tap jack described later is a similar device used on 2 W public switched channels.

Various pads or attenuators are inserted throughout the circuit to ensure proper operating levels. The normally open contacts in the DSU are for analog loop-back testing. Note also the use of unloaded lines in the local loops.

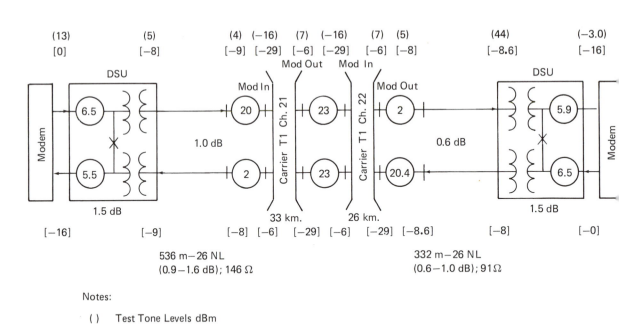

Figure 5-49 An Analog Private Line Channel Using Loops and Carrier Facilities

On dial-up lines, the TELCO usually specifies the signal from the modem to lie in the range of −3 to −9 dBm. Normally, modems have a Transmit Line Signal Level option, which allows for selection of the transmitted signal level, usually from 0 to −15 dBm in 1 dB steps. The particular output level from the modem transmitter depends upon which one of the three industry-approved methods is employed to connect the modem to the telephone Tip to Ring leads. The basic intent of any of these methods is to ensure that the signal arriving at the central office does not exceed −12 dBm. The three arrangement variations are:

1. Permissive Output
2. Programmable Output
3. Fixed Loss Output

5-9.1 Permissive Output

The permissive arrrangement may be used for portable modems where the output signal level need not be optimized. In this configuration, the signal level applied to the Tip and Ring leads is −9 dBm. With a typical loop loss of 3 dB, the −12 dBm at the central office is usually achieved. In some modems the −9 dBm signal level is achieved by the setting of the Transmit Line Signal Level. Some manufacturers prefer to use a special permissible cable containing a built-in resistor to achieve a 9 dB loss, with the modem output set to 0 dBm.

The Universal Service Order Code (USOC) designated jack for permissive applications, the RJ11C voice jack, is identical to the miniature six-pin jack used in recent telephone and extension telephone installations. The equivalent circuit of this jack is shown in Fig. 5-50. The permissive configuration has the advantage of permitting a modem to be carried around and plugged into the standard telephone wall jack; however, it has the disadvantage of achieving less than optimal signal levels at the central office when the local loop loss exceeds 3 dB. A standard telephone hand set may also be bridged across the T-R leads for voice communication.

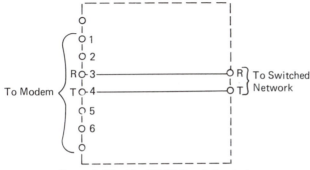

Figure 5-50 The RJ11C Permissive Jack

5-9.2 Programmable Output

The programmable output is an optional arrangement, as the transmit level at the central office is maintained at −12 dBm. In this arrangement, the modem's output level is controlled by a single resistor residing within an eight-pin miniature jack. The value of the external "programming" resistor is determined by the telephone company and is dependent upon the loss of the local loop. The resistor values used for the various line losses have been agreed upon by the industry. With the modem plugged in, it interacts with other resistors within the modem to adjust the transmit output level such that −12 dBm is obtained at the central office.

The programming resistor in the RJ45S programmed jack is attached to the transmitter-attenuator programming terminals, PR and PC, of the associated modem. Fig. 5-51 gives the equivalent circuit of the RJ45S jack, complete with the optional handset. This jack may be used to connect a modem with permissive output level to the telephone line, provided the six-pin modem cable is centered in the eight-pin jack. Pins 3 and 4 of the permissive plug then match up with pins 4 and 5 of the programmable jack.

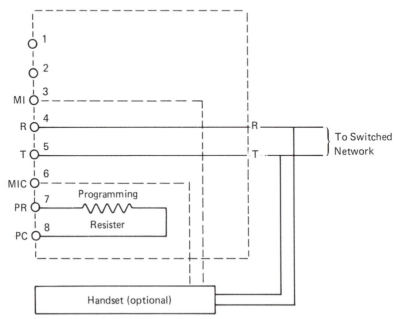

Figure 5-51 The RJ45S Programmable Jack (MI = Mode Indication; MIC r Mode Control)

5-9.3   Fixed Loss Output

Modems registered as fixed-loss loop (FLL) devices have their outputs fixed to −3 dBm. The telephone company provides whatever padding is required to obtain −12 dBm at the central office. These FLL modems are connected to the telephone line with the RJ41S Universal data jack shown in Fig. 5-52. This is an eight-pin miniature keyed jack, usable for permissive, fixed-loss loop or the programmed arrangement. The switch in the jack is closed for the FLL arrangement and open for either the programmed or permissible arrangement. The switch also selects the proper keying to prevent the attachment of the wrong cable. As with the programmable jack, when the permissive plug is connected to the universal jack, it must be centered in the RJ41S jack.

To determine the local loop loss, the central office applies a 1000 Hz tone at 0 dBm on the loop and a transmission test set with a 600 ohm/900 ohm termination monitor at the T to R signal level at the subscriber's station. The numerical reading indicates the loss of the loop in dB. For example, −5 dBm on the meter is equal to a 5 dB loop loss. This loop loss is often written on the data jack label.

A typical modem installation with an auxiliary telephone is shown in Fig. 5-53.

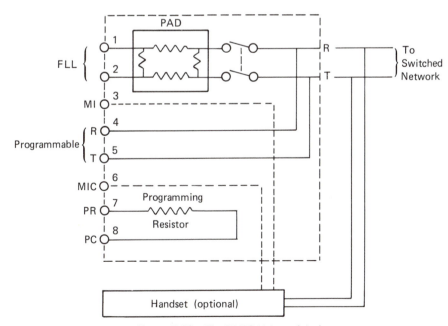

Figure 5-52   The RJ41S Universal Jack

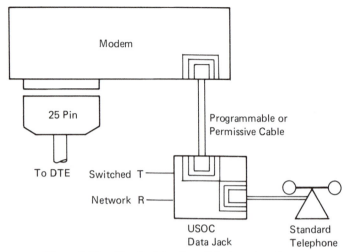

Figure 5-53 Standard Auxiliary Telephone Installation

5-10 Modem and Circuit Diagnostics

When trying to isolate a fault on a communications link, two basic types of tests are performed to determine if the problem exists on the phone line or in one of the modems. Both these are "loopback" tests. The *local* loopback test verifies the send/receive circuitry of the modem whereas the *remote* loopback test verifies the operation of both modems and the telephone line.

The test can loop around the information either in its digital form, called digital loopback, or it can loop around the information on the analog communications lines where it is then called analog loopback. Modem analog self-test should first be made, then the analog local loopback test, next the end-to-end self-test, and finally one of the remote loopback tests. In performing these tests, various push buttons are set and several readout lights are observed from a typical modem to indicate proper operation.

5-10.1 Analog Local Self-test

In the analog self-test, an internally generated test pattern within the modem is transmitted and looped from transmitter back to the receiver. As shown in Fig. 5-54, a mixer is used to convert the transmitted analog frequencies to the proper receive frequency band. The received pattern is compared with what was transmitted, and any errors are indicated by a lamp. The local self-test provides a check of the modem transmitter and receiver circuits.

5-10.2 Analog Local Loopback Test

As illustrated in Fig. 5-55, in the local analog loopback test, the data terminal generates the transmitted data and tests the received data which has been

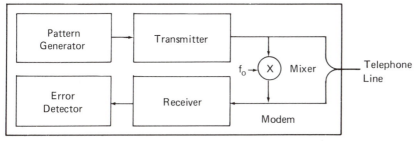

Figure 5-54   Analog Local Self-Test

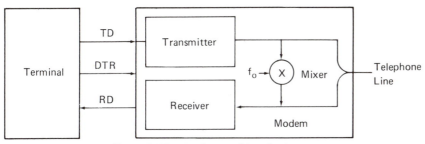

Figure 5-55   Analog Local Loopback Test

looped through the modem transmitter, mixer, and receiver. During this test, the terminal, data set, and the terminal-to-data-set interface are checked out. Both of these tests should be performed on the modems at each end.

5-10.3   End-to-End Self-Test

As shown in Fig. 5-56, this test checks the data sets, the telephone line, and the data-set-to-line interfaces at both ends. Each data set transmits an internally generated test pattern to the other end.

5-10.4   Analog Remote Self-Test

This test is similar to the analog local self-test, but the analog signal is looped around at the far end. After mixing, a 16 dB amplifier must be used at the far end to make up for the additional signal loss as the signal travels over a line twice as long as it was designed to operate over. Because of this additional expense, remote digital loopback is more commonly employed.

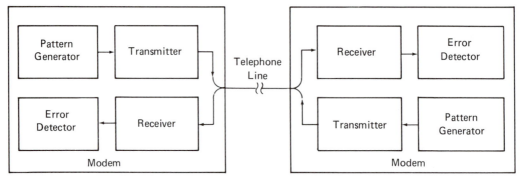

Figure 5-56   End-to-End Self-Test

5-10.5   Remote Digital Loopback Test

As shown in Fig. 5-57, the terminal generates data which is received and re-transmitted back to the terminal. This test checks out the terminal, lines, interface between the terminal and the data set, and both data sets.

An internally generated test pattern from the testing site modem could also be used, but this would not check the interface between the terminal and the modem.

In performing these tests, one should be able to narrow down to which portion of the system is at fault. If the analog local self-test or the analog local loopback test fails, the modem should be suspected. If these tests have passed and the end-to-end self-test or the remote digital loopback test fails, this would indicate a problem with the telephone line and the telephone company or the carrier should be contacted.

Some caution with the latter test should be observed. If an out-of-tolerance timing condition occurs when the two modems are communicating with each other, it could be a modem problem and not a line problem. In addition,

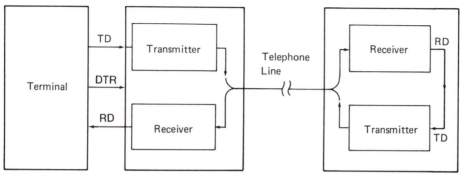

Figure 5-57   Remote Digital Loopback Test

the user must assure himself that the problem lies in the hardware and not in the software. Often the telephone line company is accused of having a faulty line, when in actuality there is a problem with the software. If, however, there is a strong suspicion that there is a line problem, inform the telephone company personnel and describe to them the specifics of the problem.

In all likelihood, the telephone company will initially run a continuity test to see if the connection is broken somewhere. If the line still appears to be faulty, the TELCO personnel will make level checks to ensure that the proper power levels are present at the various points in the circuit.

If faulty communication still persists, they may test the amplitude and envelope delay responses of the network. They may also perform the noise measurement test described in Section 3-2.4 to see if noise may be the problem. For the higher speed data transmissions, tests such as impulse noise, gain hits, drop outs, phase hits, and phase jitter may be performed. These are described later in Chapter 8.

5-11   Multiplexers

In a typical computer system, several remote terminals or work stations communicate with a central computer. In order to avoid the high cost of one telephone line per work station, more than one such device is placed on a telephone line. If the terminals support one of the higher level protocols such as HLDC, SDLC, BSC, etc., they can be multidropped and polled on a shared telephone line. A dumb or nonintelligent terminal, on the other hand, cannot work in such an environment, but can share a single telephone line by multiplexing. The device that takes the output from a number of terminals and combines the various data streams into one composite output signal is called a multiplexer. An identical multiplexer demultiplexes the composite signal at the receiving end. The multiplexer should not in any way interfere with or interrupt the flow of data and the interface signals which are passed across the EIA cable. That is, the multiplexer should be "transparent" to the data stream and, as a result, neither the terminal nor the computer equipment and software need to be modified when installing a multiplexer.

The three major forms of multiplexing are (a) frequency division mutiplexing (FDM); (b) time division multiplexing (TDM); and (c) statistical time division multiplexing (STDM).

FDM divides the telephone line bandwidth into a number of subchannels, with each terminal assigned to a unique subchannel. It is a rather inflexible system in that a change in a channel speed or in the number of channels requires that several of the other channels' center frequencies be reallocated.

TDM interleaves bits or characters from each of the attached channels and transmits them at a higher speed down a wideband transmission facility. Each channel is assigned a unique time slot that contains a predefined number of bits or characters. An elementary picture of TDM transmission is given in Fig. 5-58. For a more detailed description of TDM, please refer to Chapters 6 and 7.

Since the number of bits or characters per time slot is fixed, if a terminal is not transmitting, that time slot remains unused. These time slots are usually filled with *idle* characters, although they are occasionally used for sending remote loopback commands, diagnostic status information, terminal speed indication, etc.

Some time-division multiplexers scan each channel in order, with the result that each channel supplies an identical number of bits upon each sampling and thus occupies the same amount of bandwidth on the high speed line. Thus the highest speed terminal attached to the multiplexer determines the bit rate and thus the bandwidth that must be allocated to every channel on the high speed line. To counteract this inefficiency, many TDMs use variable channel scanning. In this case, one channel may be sampled more frequently than another channel. For instance, if a 300 b/s terminal and a 110 b/s terminal were to be multiplexed, the 300 b/s channel should be sampled three times more often than the 110 b/s channel.

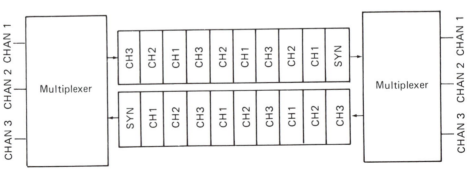

Figure 5-58 TDM Transmission

If a terminal does not transmit continuously, idle time slots appear in the high speed line. If FDM is used, this results in an unused frequency subchannel. Consequently, TDM and FDM are inefficient multiplexers when it comes to the use of line bandwidth.

The statistical time division multiplexer improves the transmission efficiency by using a variable length slot length and by allowing a terminal to vie for any free slot space. It employs a buffer memory which temporarily stores the data during periods of peak traffic. The transmission of variable length data according to the activity of each individual channel allows the STDM to waste no high-speed bandwidth or line time with inactive channels. The modem SMUX is an intelligent device, employing the use of one or more microprocessors. The user, via a control terminal or a keypad on the multiplexer, programs into the multiplexer's configuration memory the network parameters and protocols that are unique to each channel. The microprocessor uses a real-time clock which enables the user to analyze the data that flows into and from each channel and to provide network management statistics. It can also be programmed to run diagnostic tests and measure such parameters as peak loads and degradation of lines.

Many multiplexers can now communicate with each other when on the same network. Each is given an address and generally operates in a master-slave relationship. As each vendor's SMUX has its own communicating protocol, one should not intermix multiplexers from different vendors.

A typical SMUX using a single line is shown in Fig. 5-59. In this scheme, eight 2400 b/s data channels are combined into one 9600 b/s data stream. Note that if TDM was used, an $8 \times 2400 = 19,200$ b/s high speed data bit rate would be required.

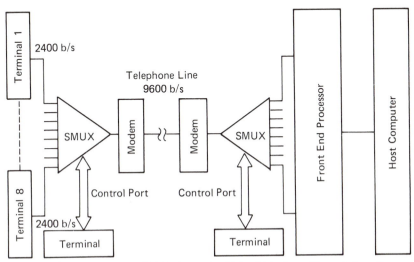

Figure 5-59 A Statistical Multiplexer Combining Eight Data Channels into One Composite Data Signal for Transmission over a Single Line

The front end processor (FEP) or communications controller has the job of removing the data communications control function from the computer, freeing it to devote most of its energy to data processing. The FEP is also well equipped to handle the change and growth that characterize data communications. It readily incorporates different types of lines and devices with different data speeds without modifying the host processor or its software.

A more detailed layout of a statistical multiplexer circuit using analog facilities at 9.6 k b/s is shown in Fig. 5-60. It employs two 209 Data Sets and a Timeplex M 2407 STAT MUX. This MUX is a traffic balance model that incorporates two data links. Each of the dual data links is connected to a similar unit at the remote site; they automatically back each other up if required by heavy traffic or faulty transmission conditions. Balancing traffic between the separate 9600 b/s data channels ensures data integrity against almost all contingencies. The M 2407 incorporates such features as

- any data format 5-8 bits plus parity
- down line programming of remote mux
- real time operating statistics
- automatic hardware and software diagnostics
- traffic flow control, etc.

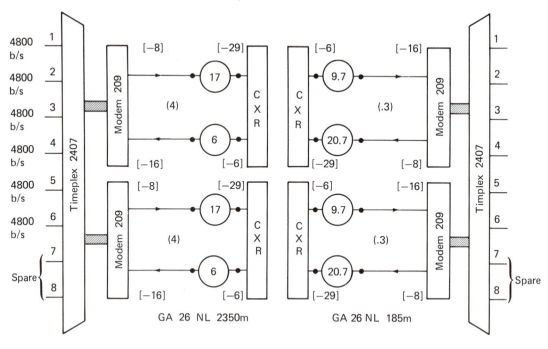

( ) Attenuation at 1kHZ
[ ] Data Signal Level dBm

Figure 5-60 A Statistical MUX Circuit Using Analog Facilities at 9.6 k b/s

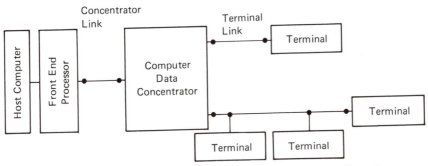

Figure 5-61   Connection of a Computer Data Concentrator to a Computer

A device very similar in nature to the time division multiplexer is the computer data concentrator. It works in concert with the computer and is employed when the number of channels on the terminal end is greater than the number of digital links or circuits to the main computer. The system requires the use of only one computer port (see Fig. 5-61) and employs only one data concentrator, rather than the two when employing SMUX's. The data concentrator is more expensive than the SMUX because of the complexity of controlling the multiplexing and demultiplexing of the various data inputs/output via the software in the main computer.

5-12 Universal Synchronous/Asynchronous Receiver/Transmitter (USART)

Serial data transmission is generally used to transmit data over relatively long distances. The device which converts the parallel bits from a central processing unit (CPU) or some data terminal into a continuous serial data stream, or vice-versa, is known as a Universal Synchronous/Asynchronous Receiver/Transmitter (USART). Either synchronous or asynchronous operation is selected. If the device permits only asynchronous operation, it is known as a UART. This section will briefly consider the COM2502/H UART device.

COM 2502/H UART

This MDS/LSI monolithic circuit is a 40 pin device that transmits and receives asynchronous data in either the half-duplex or full-duplex mode. The transmitter accepts parallel data and converts it into a serial asynchronous output. The receiver and transmitter can operate simultaneously and at different baud rates. A transmit/receive clock must be provided at 16 times the baud rate.

The UART is programmable in the sense that the data word length, parity mode, and number of stop bits can be selected by the control lines. There may be 5, 6, 7, or 8 data bits; even/odd or no parity; and 1, 1.5, or 2 stop bits.

Figure 5-62 illustrates the format of a typical serial character.

Both the receiver and transmitter have double character buffering so that one character time interval is always available to exchange a character with the external device. Figure 5-63 gives the pin configuration of the UART, and Table 5-10 describes each pin function.

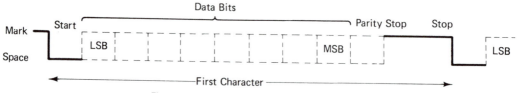

Figure 5-62 Format of Typical Serial Asynchronous Character

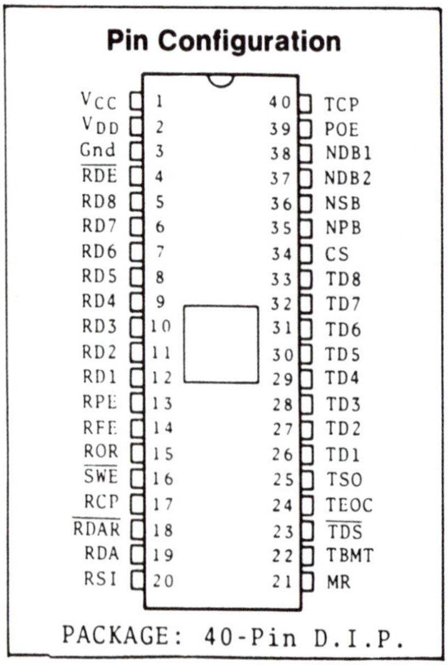

Pin Configuration

V$_{CC}$	1	40	TCP
V$_{DD}$	2	39	POE
Gnd	3	38	NDB1
$\overline{RDE}$	4	37	NDB2
RD8	5	36	NSB
RD7	6	35	NPB
RD6	7	34	CS
RD5	8	33	TD8
RD4	9	32	TD7
RD3	10	31	TD6
RD2	11	30	TD5
RD1	12	29	TD4
RPE	13	28	TD3
RFE	14	27	TD2
ROR	15	26	TD1
$\overline{SWE}$	16	25	TSO
RCP	17	24	TEOC
$\overline{RDAR}$	18	23	$\overline{TDS}$
RDA	19	22	TBMT
RSI	20	21	MR

PACKAGE: 40-Pin D.I.P.

Figure 5-63 Pin Configuration of the COM 2502/H UART. (Courtesy SMC Microsystems Corporation.)

Table 5-10   Description of Pin Functions (Courtesy
SMC Microsystems Corporation)

Pin No.	Symbol	Name	Function
1	V_{CC}	Power Supply	+5 V Supply
2	V_{DD}	Power Supply	−12 V Supply
3	GND	Ground	Ground
4	$\overline{RDE}$	Received Data Enable	A low-level input enables the outputs (RD8-RD1) of the receiver buffer register.
5–12	RD8-RD1	Receiver Data Outputs	These are the 8 tri-state data outputs enabled by $\overline{RDE}$. Unused data output lines, as selected by NDB1 and NDB2, have a low-level output, and received characters are right justified, i.e. the LSB always appears on the RD1 output.
13	RPE	Receiver Parity Error	This tri-state output (enabled by $\overline{SWE}$) is at a high-level if the received character parity bit does not agree with the selected parity.
14	RFE	Receiver Framing Error	This tri-state output (enabled by $\overline{SWE}$) is at a high-level if the received character has no valid stop bit.
15	ROR	Receiver Over Run	This tri-state output (enabled by $\overline{SWE}$) is at a high-level if the previously received character is not read (RDA output not reset) before the present character is transferred into the receiver buffer register.
16	$\overline{SWE}$	Status Word Enable	A low-level input enables the outputs (RPE, RFE, ROR, RDA, and TBMT) of the status word buffer register.
17	RCP	Receiver Clock	This input is a clock whose frequency is 16 times (16×) the desired receiver baud rate.
18	$\overline{RDAR}$	Receiver Data Available Reset	A low-level input resets the RDA output to a low-level.
19	RDA	Receiver Data Available	This tri-state output (enabled by $\overline{SWE}$) is at a high-level when an entire character has been received and transferred into the receiver buffer register.

Table 5-10   (Continued)

Pin No.	Symbol	Name	Function
20	RSI	Receiver Serial Input	This input accepts the serial bit input stream. A high-level (mark) to low-level (space) transition is required to initiate data reception.
21	MR	Master Reset	This input should be pulsed to a high-level after power turn-on. This sets TSO, TEOC, and TBMT to a high-level and resets RDA, RPE, RFE and ROR to a low-level.
22	TBMT	Transmitter Buffer Empty	This tri-state output (enabled by $\overline{SWE}$) is at a high-level when the transmitter buffer register may be loaded with new data.
23	$\overline{TDS}$	Transmitter Data Strobe	A low-level input strobe enters the data bits into the transmitter buffer register.
24	TEOC	Transmitter End of Character	This output appears as a high-level each time a full character is transmitted. It remains at this level until the start of transmission of the next character or for one-half of a TCP period in the case of continuous transmission.
25	TSO	Transmitter Serial Output	This output serially provides the entire transmitted character. TSO remains at a high-level when no data is being transmitted.
26–33	TD1-TD8	Transmitter Data Inputs	There are 8 data input lines (strobed by $\overline{TDS}$) available. Unused data input lines, as selected by NDB1 and NDB2, may be in either logic state. The LSB should always be placed on TD1.
34	CS	Control Strobe	A high-level input enters the control bits (NDB1, NDB2, NSB, POE and NPB) into the control bits holding register. This line may be strobed or hard wired to a high-level.

Table 5-10   (Concluded)

Pin No.	Symbol	Name	Function
35	NPB	No Parity Bit	A high-level input eliminates the parity bit from being transmitted; the stop bit(s) immediately follow the last data bit. In addition, the receiver requires the stop bit(s) to follow immediately after the last data bit. Also, the RPE output is forced to a low-level. See pin 39, POE.
36	NSB	Number of Stop Bits	This input selects the number of stop bits. A low-level input selects 1 stop bit; a high-level input selects 2 stop bits. Selection of two stop bits when programming a 5 data bit word generates 1.5 stop bits from the COM2017/H.
37–38	NDB2, NDB1	Number of Data Bits/Character	These 2 inputs are internally decoded to select either 5, 6, 7, or 8 data bits/character as per the following truth table:
39	POE	Odd/Even Parity Select	The logic level on this input, in conjunction with the NPB input, determines the parity mode for both the receiver and transmitter, as per the following truth table:
40	TCP	Transmitter Clock	This input is a clock whose frequency is 16 times ($16\times$) the desired transmitter baud rate.

NDB table (pins 37–38):

NDB2	NDB1	data bits/character
L	L	5
L	H	6
H	L	7
H	H	8

POE table (pin 39):

NPB	POE	MODE
L	L	odd parity
L	H	even parity
H	X	no parity

X = don't care

Description of Transmitter Operation

A simplified block diagram of the UART Transmitter is shown in Fig. 5-64. When the power is turned ON and the master reset (MR) is pulsed, the Transmitter Buffer Empty (TBMT) line will go high indicating that the transmitter holding register is empty and that the control bit may be set. A high-level signal on the control strobe enters the control bits NDB1, NDB2, NSB, POE and NPB.

The data is loaded into the transmitter holding register by applying a low-level input to the data strobe. If the transmitter shift register is empty, the data is automatically transferred to the shift register. TEOC goes low and TBMT goes high, indicating that the transmitter buffer register has been loaded into the shift register and that the buffer register is available to receive new data. The stop and parity bit are then added to the data, and serial transmission is started. The start bit appears first, followed by the data bits, the parity bit (if selected), and the stop bit(s). TEOC goes high after the last stop bit has been on for one time bit interval. If the transmitter buffer register is loaded (TBMT, low), a new character is transferred to the shift register to be transmitted. If TBMT is high, transmission is stopped until a new character is loaded into the transmitter holding register. New control bits may also be entered prior to the next data transmission.

Description of Receiver Operation

A simplified block diagram of the UART receiver is shown in Fig. 5-65. The receiver is under the same type of control as the transmitter.

A parity check circuit checks for even or odd parity if parity was selected at the transmitter stage. If parity does not check, the parity error signal (RPE) goes high. The logic circuit also checks the first stop bit of each character. If the stop bit is not a logic one as it should be (see Fig. 5-62) the framing error signal (RFE) goes high.

With the power turned ON and the master reset (MR) pulsed, the receiver data available signal (RDA) goes low, indicating that the receiver buffer is empty. The control bits are then entered with the control strobe.

Serial asynchronous data is sent to the serial input line (RSI). The start bit detects circuit searches for a high to low transition (MARK to SPACE) on the input line. If this transition is found, a counter is reset and allowed to count until the center of the start bit is reached. If the input is still low (SPACE), the signal is assumed to be a valid start bit and the sampling continues at the center of all consequent data and stop bits. The character is thus assembled in the receiver shift register. The parity error and framing error signals are put into the status word buffer register. After the stop bit is detected, the received character is transferred to the holding register and the data received line (RDA) is set high, which indicates to the external device that the output data in the buffer register may now be sampled. After the external device samples the output, it should strobe the Receiver Data Available Reset ($\overline{\text{RDAR}}$) with a low, thus resetting the RDA line to a low. If the RDA line is not reset before the next char-

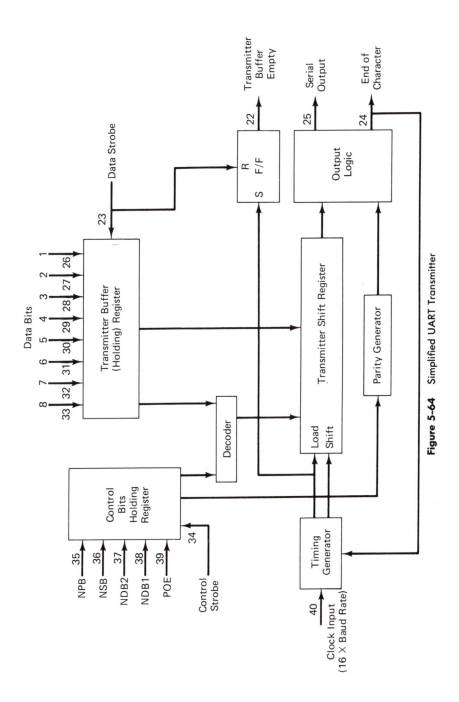

Figure 5-64   Simplified UART Transmitter

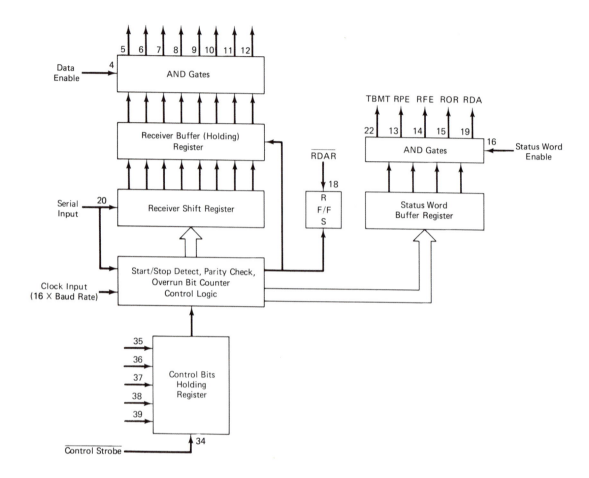

Figure 5-65 Simplified UART Receiver

acter is moved into the receiver holding register, the overrun line will be set high, indicating that a character is lost.

If the UART is to be interfaced with a modem, the transmitter serial output (TSO) and receiver serial input (RSI) must be converted to RS-232 levels.

5-13 Modem Chips

Many of the modem functions are now available in chip form, including the low-speed Motorola MC6860 and MC14412 modems. The MC6860 device shown in Fig. 5-66 includes a modulator, a demodulator, and supervisory con-

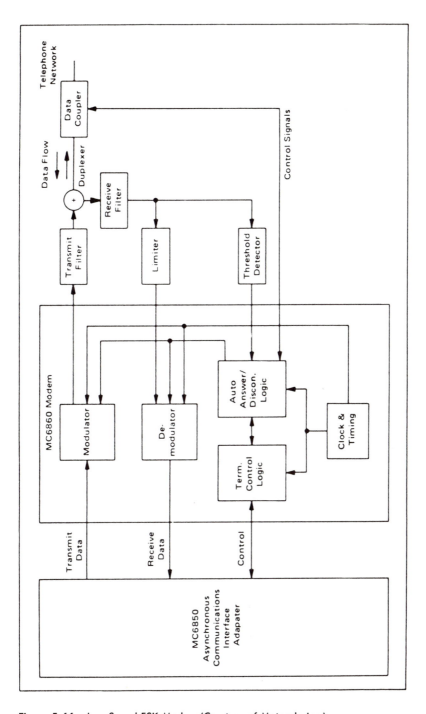

Figure 5-66 Low-Speed FSK Modem (Courtesy of Motorola Inc.)

trol. The modulator produces an eight-step synthesized sinewave which is cleaned up by a bandpass transmit filter. If the modem is in the originate mode, the transmit filter must pass frequencies of 1070–1270 Hz. If it is in the answer mode, it must pass frequencies of 2025–2225 Hz.

The filters should have steep roll-offs when going into the stopband region and should provide linear phase and flat group delay in the passband.

If the attenuation and group delay characteristics are not flat in the passband, the various frequencies which make up the signal experience different attenuations and delays, resulting in distortion. Refer back to Sec. 3-2.6 and 3-2.7 for a more complete explanation of this type of distortion.

The duplexer in the modem schematic is a circuit which causes the transmitted signal to follow a path different from that followed by the received signal, even though both may be simultaneously present on a portion of the path. In the modem under consideration, the signal from the transmit filter reaches the data coupler but not the receive filter. A strong transmit signal is not desirable in the local receiver since the receiver can become overloaded, thus causing large intermodulation products. The received signal from the data coupler can reach both the transmit and receive filters.

Telephone companies have historically required that customer-owned data sets be connected to the telephone network through a data coupler furnished by the telephone company. These data couplers prevent hazardous voltages from occurring on the network by providing data access arrangements (DAA) and sufficient electrical isolation between the customer's device and the telephone network. In addition, the couplers provide suitable signal and control voltage levels which will protect the equipment and will also prevent disturbances from occurring on neighboring telephone circuits. Although requirements for renting a data coupler are changing, the customer must still meet strict coupling regulations. The data coupler in the modem illustrated in Fig. 5-66 meets the requirements of the Bell system CBS and CBT units.

Problems

1. a. State the nature and causes of the skin and proximity effect.
   b. Give the reasons why skin and proximity effect cause pulse distortion.
   c. What is radiation resistance, and under what electrical conditions must it be taken into account?
   d. Define the electronic term dispersion.
2. What is intersymbol interference, and what effect does it have on maximum pulse rates over a transmission medium?
3. What are the functions of the following devices:
   a. Line drivers
   b. Modems
   c. Acoustic couplers
   d. UART

4. Under what conditions can a direct EIA RS-232-C connection be made to interface a DTE to a DCE?

5. What is the purpose of a null modem?

6. Line drivers require
a. FSK modulators/demodulators
b. DC continuity
c. Loaded cable pairs
d. A minimum of three EIA RS-232-C circuits

7. A metallic circuit
a. Has DC continuity
b. Requires FSK modulators/demodulators
c. Is a conducting line that need not carry DC
d. Normally employs carrier equipment

8. A modem eliminator
a. Avoids the use of modems
b. Requires no clocking signal
c. Increases the allowed distance between terminals to several km
d. Reduces back-to-back modems to a single modem

9. Describe with the aid of waveforms the following modulation techniques:
a. AM
b. FSK
c. PSK
d. DPSK

10. a. Describe the difference between full-duplex and half-duplex modem operation. Relate these to the Bell System 103 and 202 modems.
b. What is the function of the reverse channel in the Bell System 202 modem?
c. Describe the difference in operation when a 103 modem is in the originate and the answer mode. What determines the mode a modem undertakes?

11. In "soft carrier turnoff," the carrier
a. Is reduced to some intermediate frequency before being turned OFF
b. Is reduced to some intermediate power level before being turned OFF
c. Turn off is delayed by about 100 ms from the time the RTS circuit is turned OFF
d. Is turned off prior to the removal of the CTS signal

12. On a long distance line, the line turn-around time includes
a. The time to turn around the echo suppressors
b. The time for the modem receiver to turn on the DCD load after the carrier has entered the receiver
c. The transmission delay for the carrier to travel from the modem Tx to the remote modem Rx
d. All of the above

13. Receiver squelch
    a. Enables the receiver while data is received under poor S/N conditions
    b. Is used primarily on 4 W operation
    c. Prevents the detection of the ringing from a pulse signal
    d. Disables the receiver when no carrier is being received

14. The Bell 212A modem
    a. Can operate at 300 b/s or 1200 b/s
    b. Can operate HDX or FDX
    c. Operates in the asynchronous format at low speeds and synchro-nous/asynchronous at high speeds
    d. All of the above

15. What is the chief difference between the Bell 201 and Bell 208 modems?

16. Why is it not feasible to increase DPSK beyond eight phase shifts?

17. Sketch the eye diagram which displays the phase patterns on an oscillo-scope of a 4ϕ DPSK signal (refer to Fig. 5-16) when accompanied by:
    a. Amplitude jitter only
    b. Phase jitter only
    c. Both amplitude and phase jitter

18. a. Describe the purpose of amplitude and delay equilizers within a modem.
    b. What is an "adaptive" equalizer?
    c. Under what operational conditions is it advantageous to employ equal-izers in the transmitting modem?

19. a. Sinusoidal multiplication results in the sum and difference frequencies. Sketch the frequency spectrum of the following kHz pulsed carrier sig-nal.

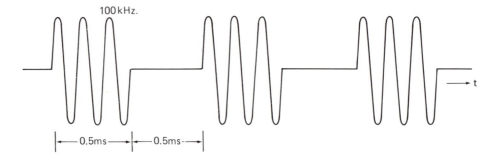

    b. If this waveform is considered an overmodulated AM signal, what effect does overmodulation have on the frequency spectrum of an AM signal?

20. Sketch the frequency spectrum of the following FSK signal.

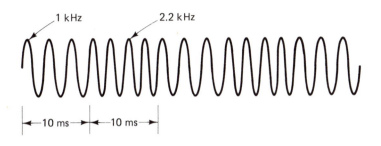

21. a. What is a dibit? Using the waveforms shown in Fig. 5-28, what is the phase of the output signal when the dibit is an 11? A 10?

b. What is QAM?

c. Describe the nature of the Gray code and give its chief advantage.

22. How can the carrier be extracted in a BPSK demodulator?

b. What is the purpose of the LPF in the demodulator of Fig. 5-24?

23. Explain why DPSK is frequently used rather than 4ϕ PSK on digital radio systems.

24. Give the truth table of the exclusive NOR gate.

25. Which of the following limits the baud rate (to 300 baud) in full duplex operation?

a. Noise and crosstalk

b. Bandwidth requirements and constraints

c. Operation on same frequency

d. Transmission in one direction at a time

e. None of the above

26. FSK allows for higher baud rates than PSK because

a. Of better noise immunity

b. It takes less bandwidth

c. On the contrary, PSK is faster, for the above reason

d. PSK operates on two frequencies

e. None of the above

27. a. What is the purpose of the level shift block in Fig. 5-30?

b. If the reference digit is a "0", obtain the encoded message for the same data as given in Table 5-5.

28. A signal element $A \cos \omega_c t$ (representing a 1) succeeded by a signal element $A \cos (\omega_c + \pi)$ (representing a 0) is applied to the DPSK demodulator of Fig. 5-31. Give the analytical expression for the signal during the second signal element at the following locations:

a. Prior to the filter
b. After the LPF
c. At the comparator output
   Assume all gains to be unity.

Note: Questions 29 to 34 refer to the T103A data set block diagram (Fig. 5-34).

29. What is the state of the following contacts?

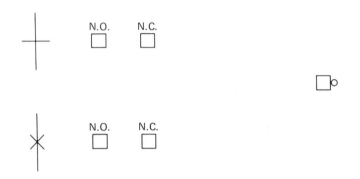

30. What relays must be activated to obtain the following conditions?

Relay(s)

a. DATA LAMP ON   _____
b. DATA SET READY   _____
c. RING INDICATOR   _____
d. TRANSMITTED DATA LEAD CONNECTED   _____

31. a. What is the condition of the locked oscillator when no signal is present?
   b. What type of signal operates the R relay?
   c. If the CR relay is not activated, under what two conditions will the C relay be activated?
   d. What is the purpose of the capacitor across the MD relay?

32. a. Why are there two discriminators circuits, DISC 1 and DISC 2, in the DATA SET?
   b. What effect does the DATA gating circuit have on the multivibrator block?
   c. What activates the AN relay? Is the AN relay activated in both the originating and answering modems?
   d. The AN relay affects the gating circuit to the multivibrator, the choice of discriminator, and free running locked oscillator frequency. Give the center frequency of each block, with AN relay activated and deactivated.

33. a. Under what relay condition(s) is the transmitted Data led (Pin 2) connected to the M.V. gating circuit?
   b. Under what relay condition(s) is the Received Data lead (Pin 3) connected to the slicer output?
   c. What is the function of the slicer block?
34. a. What is the ⬜○ circuit across the T-R input lines? Why must a capacitor be placed in series with this circuit?
   b. What is the purpose of the varistors across the transformer windings?
   c. Under what condition does the DATA lamp turn ON?
35. a. With the use of the phase curve shown, define in words and by an equation the group or envelope delay.

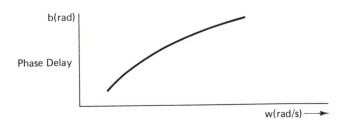

   b. Sketch a desired phase delay characteristic on the figure of (a).
   c. Sketch an ideal attenuation vs. frequency characteristic of a telephone network.
36. State the functions of the following:
   a. duplexer
   b. DAA network
37. In an EIA-D-type interface
   a. The SG circuit is absent
   b. Only the TD, RD, and DCD circuits are present
   c. The RTS circuit is controlled
   d. The RTS circuit is strapped high
38. On a 4 W private line multipoint line
   a. The primary can leave its RTS circuit high
   b. The secondary station must control their RTS circuits
   c. The second stations may transmit only when given the permission to do so by the primary stations
   d. All of the above
39. The difference between asynchronous and synchronous transmission is
   a. The way the synchronization is provided
   b. The way the beginning of a block of data is detected
   c. The way the beginning of a character is determined
   d. All of the above
   e. None of the above

40. Asynchronous data transmission requires a clock
   a. At the transmitter end
   b. At the receiver end
   c. At both ends
   d. At neither end
   e. All of the above

41. The television network is an example of _____ operation.
   a. Simplex
   b. HDX
   c. FDX
   d. TDM

42. Explain with the aid of a diagram how FDX is obtained in the
   a. 4 W circuit
   b. 2 W circuit

43. Multipoint circuits generally
   a. Use leased lines
   b. Use dial-up lines
   c. Give equal status to all sites
   d. Use HDX lines

44. The Bell 103/113 modem is
   a. An HDX 1200 b/s asynchronous modem
   b. An FDX 4800 b/s synchronous modem
   c. An HDX 1200 b/s asynchronous modem
   d. An FDX 300 b/s asynchronous modem

45. Line turnaround procedures apply to
   a. Half duplex lines, where both DTE's wish to transmit at the same time
   b. Full duplex, where the two DTE's must alternate transmission direction
   c. Half duplex lines, where both DTE's cannot transmit at the same time
   d. Full duplex lines, where usage of the primary channel must be shared between the two DTE's

46. The reverse channel on a Bell 202 modem can be used to
   a. Control the direction of data flow on the main channel
   b. Inform the transmitting terminal that data has been correctly or incorrectly received
   c. Maintain the echo suppressors in the active state when the main channel carrier has dropped for a short period of time
   d. All of the above

47. The BREAK key is located in
   a. The transmitting modem
   b. The receiving modem

c. Both terminals
d. The telephone handset

48. Code turnaround emmploys
   a. The secondary channel
   b. The primary channel
   c. The BREAK key
   d. The SOH command

49. Code turnaround is activated by the
   a. SOH command
   b. The ACK or NAK command
   c. The ESC command
   d. the ETX or EOT command

50. In echoplex, the transmitted data is echoed back
   a. From the remote DTE
   b. From the local modem
   c. From the local exchange
   d. Within the local DTE

51. In local copy, the transmitted data is echoed back
   a. From the remote DTE
   b. From the local modem or local DTE
   c. From the local exchange
   d. From the remote modem

52. What is the advantage of buffering the characters transmitted by a TTY?

53. Calculate the minimum and maximum X16 clock rates that are permissible in the detection circuitry of a 300 baud asynchronous system, assuming each character consists of eight bits, plus a single START, STOP and PARITY bit.

54. Sketch the method of obtaining synchronization when clocks are provided within both the receiving and transmitting synchronous modems.

55. 4 W telephone subscriber circuits have a characteristic impedance of
   a. 50 ohms
   b. 75 ohms
   c. 300 ohms
   d. 600 ohms
   e. 900 ohms

56. On a 4 W private line data circuit, the end-to-end loss is
   a. 0 dB
   b. 9 dB
   c. 16 dB
   d. Varies from 16 to 30 dB

57. On a 2 W switched data circuit, the end-to-end loss is
   a. 0 dB
   b. 9 dB

    c. 16 dB

    d. Varies from 16 to 30 dB

58. In 4 w private line data circuits, all received data levels are

    a. At the transmission level point (TLP)

    b. 3 dB below the TLP

    c. 6 dB below the TLP

    d. 16 dB below the TLP

59. In carrier multiplex systems, the standard test tone at the modulation input is at

    a. -16 dBm

    b. 0 dBm

    c. -6 dBm

    d. $+23$ dB

60. The carrier multiplex circuit has a

    a. Loss of 23 dB

    b. Gain of 23 dB

    c. Loss of 16 dB

    d. Gain of 16 dB

61. The basic purpose of specifying the transmit level from a modem is to ensure that the signal level

    a. At the central office is 0 dBm

    b. At the central office does not exceed -12 dBm

    c. At the receiving modem is at least -16 dBm

    d. At the modulation input to the carrier or multiplex system is at -16 dBm

62. The RJ11C permissive jack is employed

    a. To obtain optional signal levels at the central office

    b. In metallic circuit applications

    c. With FDX modems

    d. With portable modems

63. The RJ45S programmable jack

    a. Can be used with the permissive plug

    b. Uses a "programmable" resistor

    c. Maintains a -12 dBm signal at the central office

    d. All of the above

64. In the fixed loss loop arrangement

    a. A "programmable resistor" is employed

    b. FDX modems must be employed

    c. A fixed attenuator is employed

    d. The modem output is adjusted to 0 dBm

65. Describe how the local loop loss is determined.

66. A permissive transmit level sets the transmit level at the T-R point to

    a. -9 dBm

    b. -4 dBm

c. 0 dBm

d. +13 dBm

67. The telephone company selected modem options are/is

a. Transmit level

b. Tip-Ring Make Busy

c. Signal and frame ground

d. All of the above

68. The local self-test checks the

a. Terminal, data set, and terminal-to-data-set interface

b. Data set transmitter and receiver

c. Data set transmitter and receiver and telephone lines

d. Data transmitter and line interface

69. The analog local loop back test checks the

a. Terminal, data set, and terminal-to-data-set interface

b. Data set transmitter and receiver

c. Data set transmitter, receiver, and telephone lines

d. Data transmitter and line interface

70. The end-to-end self-test checks the

a. Terminal, data set, and terminal-to-data-set interface

b. The loop back data set, lines, testing data set, testing terminal and interfaces between each system component

c. Local data set, lines, remote set, local terminal, and interfaces between each system component

d. The local and remote data sets, lines, and data-set-to-line interfaces at both sites

71. Analog loop back refers to data being looped around at the

a. RS-232-C interface

b. Telephone line level

c. Terminal

72. Digital loop back refers to data being looped around at the

a. RS-232-C interface level

b. Telephone line level

c. Terminal

73. a. Describe the advantages of a statistical time division multiplexer (SMUX) when compared to a variable scan time division multiplexer.

b. Describe the application of a data concentrator.

NOTE: Questions 74 to 85 refer to the COM 2502/H UART.

74. The (LSB, MSB) is transmitted first from the UART.

75. The transmitter clock is _____ times the desired transmitted baud rate in the UART.

76. The Transmitter End of Character lead goes (Hi, Lo) when a full character is transmitted.

77. The Transmitter Serial Output remains (Hi, Lo) when no data is being transmitted.

78. A low level input strobe pulse may be applied to the Transmit Data Strobe when the Transmitter Buffer Empty lead is (Hi, Lo).

79. With the NPB lead Lo and the POE lead Hi, (even, odd) parity is obtained.

80. If the parity does not check in the receiver, the parity error signal (RPE) goes (Hi, Lo).

81. After the (START, STOP) bit is detected, the received character is transferred to the Receiver Buffer Register.

82. The Receiver Data Available Reset lead ($\overline{RDAR}$) is controlled by the (UART, external device).

83. The transmit serial output and receive serial input are at (TTL, RS-232-C) levels.

84. A high on the Data Received Line (RDA) indicates that the Receiver Buffer register (has, has not) been loaded with data.

85. Which lead enters the control bits in the Transmitter/ Receiver Holding Register? _____

86. State the functions of the following:
a. Duplexer
b. DAA network

87. If a character is made up of seven data bits and is preceded by one START bit and trailed by a PARITY bit and a STOP bit, a 1200 baud modem would transmit
a. 1200 characters/s
b. 840 characters/s
c. 120 characters/s
d. 84 characters/s

88. If each character from a computer requires 10 bits, a fast typist snapping out text at 100 words per minute produces data at approximately
a. 10 characters/s
b. 100 characters/s
c. 10 bits/s
d. 100 bits/s

89. To fill an 80-character by 25-line screen from a computer at 300 baud takes
a. 7 s
b. 67 s
c. 133 s
d. 960 s
Assume 10 bits per character.

90. To fill an 80-character by 25-line screen from a computer at 1200 baud takes
   a. 2 s
   b. 17 s
   c. 34 s
   d. 375 s
   Assume 10 bits per character.

91. To load a 356 k character floppy disk into a computer at 300 baud takes
   a. 8 minutes
   b. 19 minutes
   c. 30 minutes
   d. 3.3 hours
   Assume 10 bits per character.

92. To load a 356 k character floppy disk into a computer at 1200 baud takes
   a. 3 minutes
   b. 33 minutes
   c. 49 minutes
   d. 3.3 hours
   Assume 10 bits per character.

6

Analog Pulse Modulation

6-1 Introduction

This chapter will consider the transmission of information via analog pulses, whereby each sample of the analog-signal (message) can take on a continuous range of values. In digital transmission, which will be discussed in the next chapter, each sample can take on only a limited number of values.

An analog signal that is sampled often enough can be exactly specified, even if it is continuously varying with time, as long as it is band limited. The analog signal can be fully reconstructed from the sampled pulses if no noise or distortion has been added during transmission.

In analog pulse modulation, the samples directly modulate a periodic pulse train with one pulse for each sample. This discussion will focus on pulse amplitude modulation (PAM), which is probably the most common form of analog pulse modulation. A brief discussion of pulse duration and pulse position modulation schemes will follow. Although all these modulation schemes are widely used, the most favored is pulse code modulation (PCM). This will be discussed in Chapter 7.

6-2 Time Sampling

Analog signals representing voice or temperature are continuous functions with time. Time quantization is used to transmit such signals in a pulsed format, whereby the signal is defined or measured at only discrete time instants. If the sampling rate is sufficiently high, information should not be lost by this technique. The signal can be reproduced from these samples through various means.

The samples ideally consist of pulses of zero width having an amplitude equal to the signal amplitude at the instant of sampling. Since ideal sampling is not a physically realizable process and the resultant pulses do not contain any energy, the following types of sampling will be considered: (a) natural sampling and (b) sample-and-hold.

Natural sampling consists of samples which are of short, but not infinitesimal duration, that have amplitudes which briefly follow the signal. Figure 6-1(b) shows such a waveform. Sample-and-hold is undoubtedly the most common technique since it retains the sampled value for a longer period, as shown in Fig. 6-1(c). When applied to the encoder, sample-and-hold allows time for a sequence of operations rather than a single very rapid operation. The sample-and-hold circuit imposes a non-return-to-zero (NRZ) sampling in the input signal shown.

An example of an electronic switch is the MOS-FET circuit shown in Fig. 6-2. The P-channel acts as a low impedance, or closed, switch when its gate voltage goes negative. Thus, when the input to the driver goes low, G_1 goes

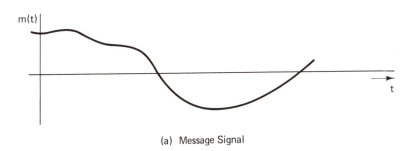

(a) Message Signal

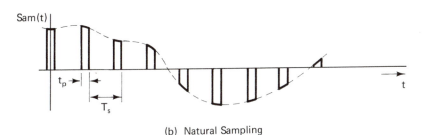

(b) Natural Sampling

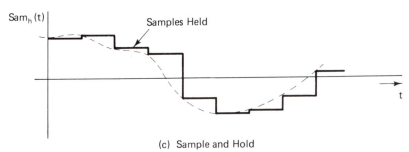

(c) Sample and Hold

Figure 6-1 Sampling Techniques

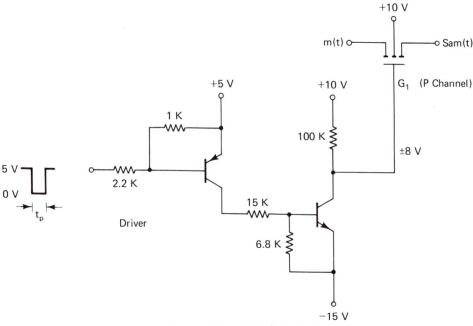

Figure 6-2 MOS-FET Analog Switch

negative, and the switch closes. This allows $m(t)$ to appear at the output. When the driver input goes high, G_1 goes high and the resistance across the drain and source is in the tens of megohms.

Sample-and-hold circuits depend upon the rapid charging of a capacitor to the potential of the input sample amplitude, and then the opening of the switch to retain the charge for the following circuits. The switching in Fig. 6-3 is done by the N-channel FET which presents a fairly low time constant in conjunction with the capacitor when the input sampling pulse goes high. The FET gate opens when the pulse goes to zero and the capacitance charge is maintained except for the leakage of the connecting high impedance circuits.

For high speed sampling, switching diodes in a bridge format can be used. The pulse transformer in Fig. 6-4 isolates the sampling pulse from the signal path, which results in the bridge floating in the signal path.

6-3 The Sampling Theorem: Natural Sampling

Consider a signal $m(t)$ which is band limited and has a highest frequency component of f_{max}. Let this signal be naturally sampled at a pulse rate, f_s, and with

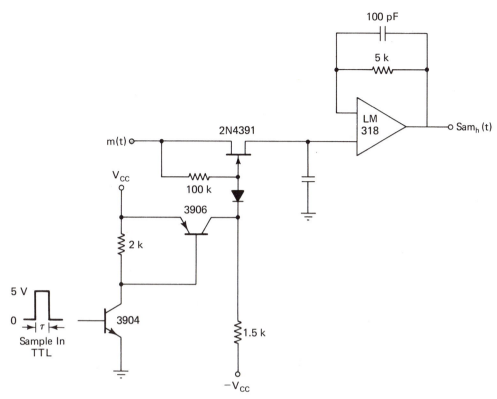

Figure 6-3 Sample-and-Hold Schematic for a Maximum Sampling Rate of About 100 kHz

sampling times, t_p. From Fig. 6-5, observe that the sampled waveform can be expressed as:

$$Sam(t) = m(t) \times S(t) \qquad (6\text{-}1)$$

where $S(t)$ is the sampling function consisting of a rectangular wave train of pulse width, t_p, and period, $T_s = 1/f_s$.

To obtain the frequency spectrum of the sampled waveform, multiply the Fourier series expansion of $S(t)$ by the frequency spectrum of $m(t)$. As derived in Appendix A, the Fourier expansion of a rectangular pulse train of Amplitude 1, period T_s, and pulse duration t_p is given by

$$S(t) = \frac{t_p}{T_s} + \frac{2t_p}{T_s} \sum_{n=1}^{\infty} \operatorname{sinc}\frac{nt_p}{T_s} \cos\frac{2\pi nt}{T_s} \qquad (6\text{-}2)$$

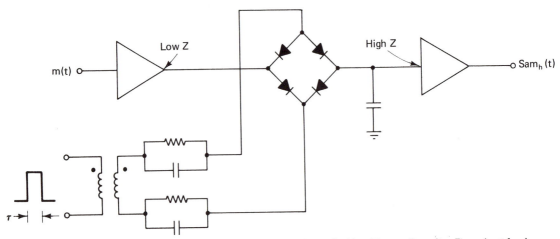

Figure 6-4 Sample and Hold Circuit for Very Narrow Sampling Times ($\tau < 1$ ns)

where
$$\text{sinc } \frac{nt_p}{T_s} = \frac{\sin \pi t_p \, n/T_s}{\pi \, t_p \, n/T_s}$$

and T_s is the sampling time.

This function consists of an infinite number of harmonics of $f_s = 1/T_s$, which have a decaying amplitude given by sinc nt_p/T_s. Since the numerator of the sinc function has a maximum amplitude of 1, the amplitude decays as $1/n$. The sinc function goes to zero whenever its numerator is an integral number of π radians since $\sin n\pi = 0$.

It may appear that the sinc function goes to zero for $n = 0$ since $\sin 0 = 0$; however, the denominator also goes to 0 when $n = 0$, and division by 0 is not allowed. In actual fact, the sinc function has a maximum value of 1 when its argument is zero. This can be checked by substituting an extremely small value for x into $\sin x/x$ and calculating the result.

Thus, $S(t)$ goes through zero whenever

$$\pi t_p \frac{n}{T_s} = m\pi \text{ or when } \frac{n}{T_s} = \frac{m}{t_p}$$

The spectrum of the switching function is shown in Fig. 6-6.

The sampled baseband signal $m(t) \, S(t)$ is thus given by the expression

$$Sam(t) = \frac{t_p}{T_s} m(t) + \frac{2t_p}{T_s} \left[\text{ sinc } \frac{t_p}{T_s} m(t) \cos \frac{2\pi t}{T_s} \right.$$

$$\left. + \text{ sinc } \frac{2t_p}{T_s} m(t) \cos \frac{2\pi 2t}{T_s} + \ldots \right] \qquad (6\text{-}3)$$

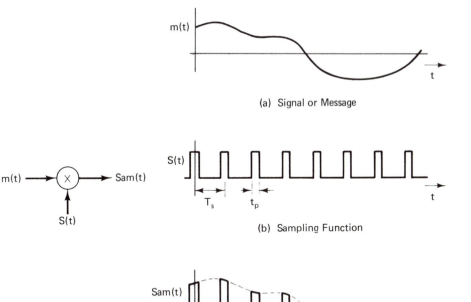

(a) Signal or Message

(b) Sampling Function

(d) Samples of the Message

Figure 6-5   Generation of a Naturally Sampled-Signal

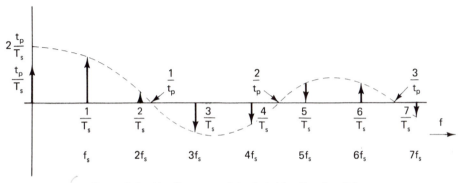

Figure 6-6   Line Spectrum of the Switching Function $S(f)$

The spectrum of the sampled function is now given by the multiplication of the frequency components of the message, $m(t)$, by the spectrum shown in Fig. 6-6. Recall that when two sinusoidal functions are multiplied, such as $\cos \omega_1 t \times \cos \omega_2 t$, *the sum and difference components*, $\frac{1}{2} \cos (\omega_1 + \omega_2)t + \frac{1}{2} \cos (\omega_1 - \omega_2)t$, are obtained. Each message frequency component when multiplied by one of the harmonics of $S(t)$ will result in a lower and upper sideband around each harmonic. Thus, for the message spectrum shown in Fig. 6-7(a), the naturally sampled spectrum appears as shown in Fig. 6-7(b).

It can be observed from Eq. (*6-3*) that the first term in the series is, aside from the constant factor t_p/T_s, the message $m(t)$ itself. This is represented in the frequency domain by the spectrum lying in the 0 to f_{max} range shown in Fig. 6-7(b). Similarly, the second term in (*6-3*) is represented by the first set of sidebands $(f_s - f_{max}$ to $f_s + f_{max})$ shown in Fig. 6-7(b), and so on.

By passing the sampled signal through a low-pass filter, it is possible to extract the signal $m(t)$ if the various spectrum groupings do not overlap. If the sampling frequency is reduced until the spectrum just begins to overlap, as shown in Fig. 6-8, $m(t)$ can still be separated from the rest of the spectrum. As long as f_s is greater than $2f_{max}$, no interference occurs. This is a simple proof of the Sampling Theorem. It can be stated as follows:

(a) Message Spectrum

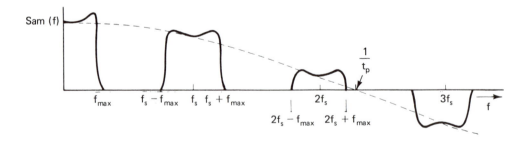

(b) Sampled Spectrum

Figure 6-7 Frequency Spectrum of a Natural Sampled Waveform

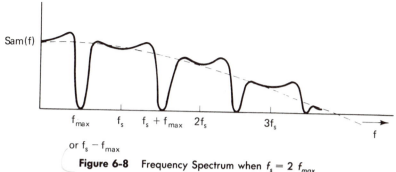

Figure 6-8 Frequency Spectrum when $f_s = 2 f_{max}$

A band limited signal of maximum frequency f_{max}, when sampled at regular intervals at a rate of at least $2 f_{max}$, can be completely reconstructed from these samples with no distortion. That is, a signal waveform can be properly represented if at least two samples are taken for every cycle of the highest significant signal frequency.

CCITT has set a standard of 8000 samples per second for the standard 300 to 3400 Hz voice channel. This complies with the Sampling Theorem, and all information of the original audio signal is recoverable. Thus, a sample is taken every $\frac{1}{8000}$ s or every 125 μs.

Actual analog signals, such as voice, are not so sharply band limited. They roll off gradually, resulting in unnecessarily high sampling frequencies. Filters can be employed to reduce the higher frequencies, but due to their finite roll-off, can never eliminate them. Also, since the spectrum of noise that creeps into any system is broad, noise from the first lower sideband overlaps into the baseband.

If the original signal has frequency components above $f_s/2$, as shown in Fig. 6-9, the lower sideband at f_s will overlap with the baseband and distortion will occur. This effect is termed aliasing, or frequency foldover distortion.

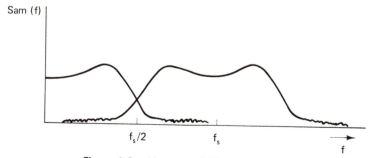

Figure 6-9 Aliasing or Foldover Distortion

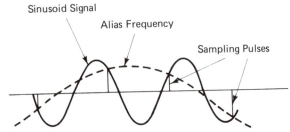

Figure 6-10 Alias Frequency as a Result of Inadequate Sampling Rate

Aliasing can be reduced by improving the filter slope characteristics or by increasing the sampling frequency. Going to a higher pole filter for faster roll-off raises costs. Increasing the sampling frequency increases the bandwidth requirement of the transmission medium.

The effect of an inadequate sampling rate on a sinusoidal signal is illustrated in Fig. 6-10. An alias frequency appears in the recovered signal. This low frequency alias signal, shown by the dotted line, is a result of sampling at less than twice per cycle.

The effect of sampling frequency and rate of filter roll-off on distortion is discussed in the following somewhat simplified case.

For a single pole filter, as shown in Fig. 6-11, the ratio of output to input voltage can be expressed as

$$\frac{V_o}{V_i} = \frac{-j/\omega C}{R - j/\omega C} = \frac{1}{1 + j\omega RC}$$

thus
$$\left|\frac{V_o}{V_i}\right| = \frac{1}{\sqrt{1 + \left(\frac{\omega}{\omega_c}\right)^2}} \tag{6-4}$$

where $\omega_c = \dfrac{1}{RC}$ represents the radial cut-off frequency.

If n such filters are cascaded, each isolated from another, the response becomes

$$\left|\frac{V_o}{V_i}\right| = \left(\frac{1}{\sqrt{1 + (\omega/\omega_c)^2}}\right)^n \tag{6-5}$$

which, for frequencies much above cut-off, approaches the asymptote

$$\left|\frac{V_o}{V_i}\right| = \left(\frac{1}{\omega/\omega_c}\right)^n = \left(\frac{\omega_c}{\omega}\right)^n = \left(\frac{f_c}{f}\right)^n \tag{6-6}$$

In this theoretical development, assume the sharp corner square frequency response holds true, as illustrated by the solid line of Fig. 6-11(b). It is sufficiently accurate for the purpose of this case.

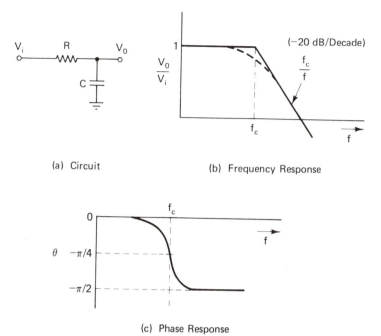

(a) Circuit   (b) Frequency Response

(c) Phase Response

Figure 6-11   Transfer Function Amplitude and Phase Plot of a Single Pole Filter

When considering the ratio of output to input power, the voltage response is squared. Therefore, for $f \gg f_c$

$$\frac{P_o}{P_i} = \left(\frac{f_c}{f}\right)^{2n} \qquad\qquad (6\text{-}7)$$

In logarithmic terms, the roll-off occurs at a rate of

$$10 \log \left(\frac{f_c}{10 f_c}\right)^{2n} = -20 \, n \text{ dB/decade}$$

Consider an audio signal with a flat frequency response of unity amplitude that is applied to an n-pole filter and then sampled at a rate of f_s, as shown in Fig. 6-12(a). The filter will determine the signal baseband bandwidth applied to the sampler. The resulting frequency response after sampling, $Sam\ (t)$, closely resembles the previous sampling spectrum shown in Fig. 6-7. This is illustrated in Fig. 6-12(b).

An examination of Fig. 6-12(b) shows that some of the lower sideband of the first sampled harmonic falls into the band of the baseband signal. This appears as an undesirable signal and distorts the baseband signal. This type of distortion is known as aliasing, or foldover distortion. The aliasing distortion shown by the shaded area X is due to the finite rate of attenuation of the low-pass filter. Since the shaded area Y, which is equivalent to the shaded area X, is

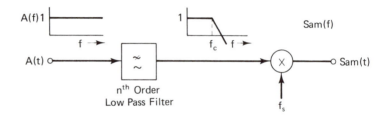

(a) Natural Sampling Scheme Complete with Frequency Plots

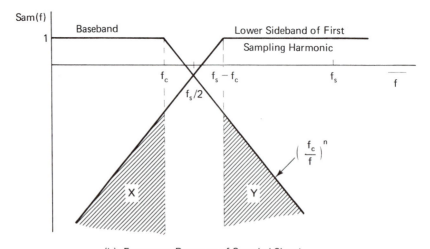

(b) Frequency Response of Sampled Signal

Figure 6-12 Aliasing with n-Pole Filter

the easiest of the two to express mathematically, it will be used to obtain the distorting noise power.

The bandlimited baseband signal amplitude response can be expressed mathematically as:

$$Sam(f) = 1 \text{ for } 0 < f < f_c \qquad (6\text{-}8(a))$$

$$= \left(\frac{f_c}{f}\right)^n \text{ for } f > f_c \qquad (6\text{-}8(b))$$

where n is an integer representing the number of poles of the filter, and f_c represents the filter 3 dB cut-off frequency. For convenience, let the sampling frequency be represented by some constant that is k times the minimum allowable sampling rate of $2f_c$, where the significant baseband signal power is assumed to lie in the frequency range up to f_c.

$$f_s = \text{k } 2f_c \qquad (6\text{-}9)$$

The signal power, S, can thus be obtained by integrating the power spectral response (the square of the $Sam(f)$ response) from 0 to f_c.

$$S = \int_0^{fc} 1^2 \, df = f_c \qquad (6\text{-}10)$$

The corresponding distortion power, represented by the shaded area X or Y, is given by integrating the square of Eq. ($6\text{-}8(b)$) from

$$f_s - f_c \text{ to } f_s$$

$$N = \int_{fs-fc}^{fs} \left(\frac{f_c}{f}\right)^{2n} df$$

$$= \frac{1}{-2n + 1}\left[\frac{f_c}{(2k)^{2n-1}} - \frac{f_c}{(2k-1)^{2n-1}}\right] \qquad (6\text{-}11)$$

Hence the signal power-to-distortion power ratio is given by:

$$\frac{S}{N} = \frac{1}{\dfrac{1}{-2n+1}\left[\dfrac{1}{(2k)^{2n-1}} - \dfrac{1}{(2k-1)^{2n-1}}\right]}$$

which, in decibels, is

$$\left(\frac{S}{N}\right)_{dB} = -10 \log \left\{\frac{1}{-2n+1}\left[\frac{1}{(2k)^{(2n-1)}} - \frac{1}{(2k-1)^{(2n-1)}}\right]\right\} \qquad (6\text{-}12)$$

Table 6-1 and Fig. 6-13 show how the filter roll-off affects the aliasing

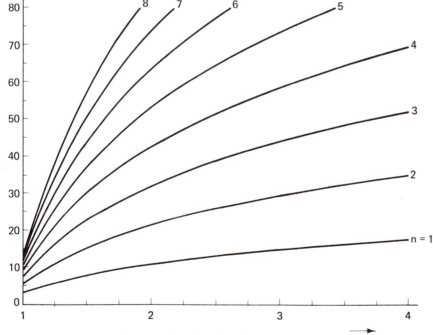

Figure 6-13   S/ N Ratio Due to Aliasing

Table 6-1

N	k	S/N		N	k	S/N
1.00	1.00	3.01		5.00	2.50	64.35
2.00	1.00	5.35		6.00	2.50	77.03
3.00	1.00	7.13		7.00	2.50	89.65
4.00	1.00	8.49		8.00	2.50	102.23
5.00	1.00	9.55		1.00	3.00	14.77
6.00	1.00	10.42		2.00	3.00	29.49
7.00	1.00	11.14		3.00	3.00	44.17
8.00	1.00	11.76		4.00	3.00	58.80
1.00	1.50	7.78		5.00	3.00	73.39
2.00	1.50	15.33		6.00	3.00	87.93
3.00	1.50	22.65		7.00	3.00	102.43
4.00	1.50	29.79		8.00	3.00	116.90
5.00	1.50	36.75		1.00	3.50	16.23
6.00	1.50	43.58		2.00	3.50	32.43
7.00	1.50	50.30		3.00	3.50	48.59
8.00	1.50	56.93		4.00	3.50	64.73
1.00	2.00	10.79		5.00	3.50	80.82
2.00	2.00	21.46		6.00	3.50	96.89
3.00	2.00	32.02		7.00	3.50	112.93
4.00	2.00	42.47		8.00	3.50	128.94
5.00	2.00	52.82		1.00	4.00	17.48
6.00	2.00	63.08		2.00	4.00	34.94
7.00	2.00	73.27		3.00	4.00	52.37
8.00	2.00	83.39		4.00	4.00	69.77
1.00	2.50	13.01		5.00	4.00	87.15
2.00	2.50	25.95		6.00	4.00	104.51
3.00	2.50	38.82		7.00	4.00	121.84
4.00	2.50	51.62		8.00	4.00	139.16

distortion. The curves show that the greater the number of poles in the filter, or the greater the filter roll-off, the less will be the distortion. The sampling frequency also has a significant effect on distortion in that an increased sampling frequency inserts a larger guard band between the baseband and the sideband, thus improving the S/N ratio.

6-4   Sample and Hold: Flat-Top Sampling

More practical sampling systems use the sample-and-hold circuit. This circuit samples the input at precise intervals and holds the input amplitude constant for the duration of the sampling time even though the measured signal is varying. This holding time can extend in time to the next sampling instant, but may be of a shorter duration, as shown in Fig. 6-14(b).

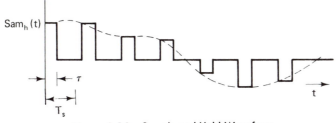

Figure 6-14   Sample-and-Hold Waveform

Convolution must be used to analyze the sampled waveform. Since this technique may be unfamiliar to many readers, this discussion will merely state its results. The frequency spectrum of the sampled signal consists of a baseband around the harmonics of the sampling frequency very much like that shown in Fig. 6-7. However, there is one additional contribution to distortion, called aperture distortion. Aperture distortion depends upon the holding time, and follows the frequency spectrum envelope of a single pulse of time duration equal to the holding time, τ.

The frequency spectrum of a single pulse of duration is given by:

$$\tau \operatorname{sinc} f\tau = \tau \frac{\sin \pi f \tau}{\pi f \tau} \qquad (6\text{-}13)$$

which is graphically illustrated in Fig. 6-15(b).

The spectrum shown in Fig. 6-15 is a continuous one which decays with frequency, as can be seen from Eq. (*6-13*). Furthermore, as τ decreases, so does the energy content of the signal and the various frequency components. Given a flat frequency response for $m(t)$ (shown in Fig. 6-16(a)) which is sampled and held for a duration of τ, results in the frequency spectrum for the sample-and-hold signal shown in Fig. 6-16(b).

The distortion can be more clearly illustrated by expanding the low frequency region that contains the frequencies transmitted after filtering. Thus, from Fig. 6-16(c), the percentage error due to aperture distortion can be calculated by:

$$\% \text{ error} = \left(1 - \frac{\sin \pi f \tau}{\pi f \tau}\right) \times 100 \qquad (6\text{-}14)$$

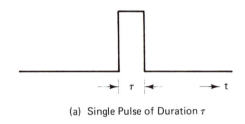

(a) Single Pulse of Duration τ

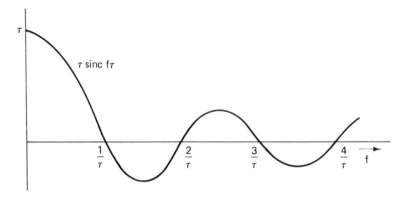

(b) Frequency Spectrum of a Single Pulse

Figure 6-15 Frequency Spectrum of a Single Pulse of Time Duration τ

It can be readily seen from Fig. 6-16(b) that the larger the sampling aperture, τ, the faster the envelope drops off in amplitude and the more attenuated the higher frequency components of the signal become. Obviously, this creates an error in the recovered signal. Figure 6-17 is a useful design chart which shows the percentage error introduced versus the sampling aperture time multiplied by the signal frequency. The rate of sampling does not affect aperture error.

Although it might appear wise, at first glance, to keep the aperture time short in order to minimize error, this results in reduced signal energy content which makes the system more susceptible to noise. As a result, the aperture time is usually maximized, as illustrated earlier in Fig. 6-1(c).

Fortunately, this type of distortion is linear. It can be taken out by inserting an equalizer in cascade with the input or output filter whose transfer function is the inverse of the single pulse frequency response of time duration equal to the sampling time, τ. (See Fig. 6-15(b).) That is, if the equalizer has a response of $1/\text{sinc } f\tau$, the equalizer action, in combination with the aperture effect, will yield a flat overall transfer characteristic between the input baseband signal and the output received signal. Some error always remains because the $u/\sin u$ transfer characteristic cannot be exactly synthesized.

(a) Message Spectrum

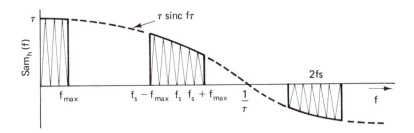

) Frequency Spectrum of Sample-and-Hold
Signal with a Holding Time of τ

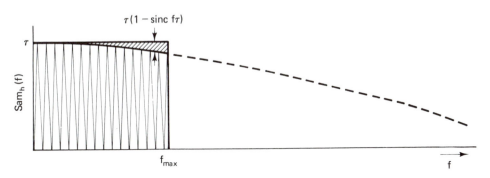

(c) Aperture Distortion Indicated by the Shaded Area

Figure 6-16   Sample-and-Hold Circuit Frequency Response

In PCM telephony, a flat signal is first filtered with a 5-pole filter at -100
dB/decade, and a break frequency of $f_c = 3400$ Hz. The signal is then sampled
at a rate of $f_s = 8$ kHz. Determine the S/N ratio of the received baseband signal.

**Example
6-1**

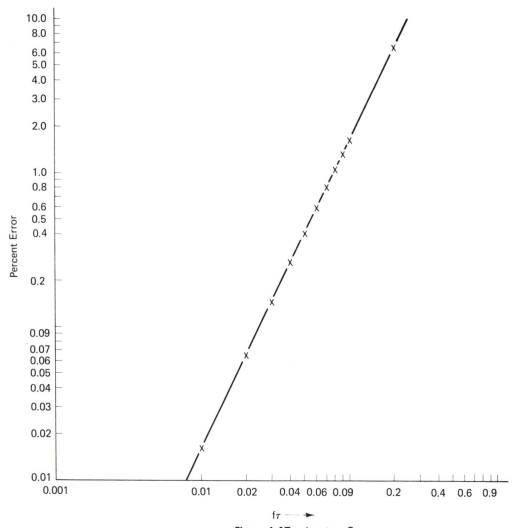

Figure 6-17   Aperture Error

Solution

$$f_s = 2\,k\,f_c$$

$$8000 = 2\,k\,(3400)$$

$$k = \frac{8000}{6800} = 1.176$$

$$S/N = -10\,\log\left\{\frac{1}{-2 \times 5 + 1}\right.$$

$$\left(\frac{1}{(2 \times 1.176)^{(2\times5-1)}} - \frac{1}{(2 \times 1.176 - 1)^{(2\times5-1)}}\right)\right\}$$

$$= 21.4\ \text{dB}$$

The spectrum of the sampled signal is shown in Fig. 6-18. The sidebands around 16 kHz are not shown. Note that the folded data is about 13 dB down at $f_c = 3.4$ kHz.

In the telephone system, the sampling frequency is standardized at 8 kHz. If the error due to sampling aperture cannot be more than ½% at 3 kHz, what aperture time is permitted?

Example 6-2

Using Fig. 6-17, for 0.5% error, $f\tau = 0.055$. Therefore the maximum aperture time is:

Solution

$$\tau = \frac{0.055}{f} = \frac{0.055}{3000} = 18.3 \ \mu s$$

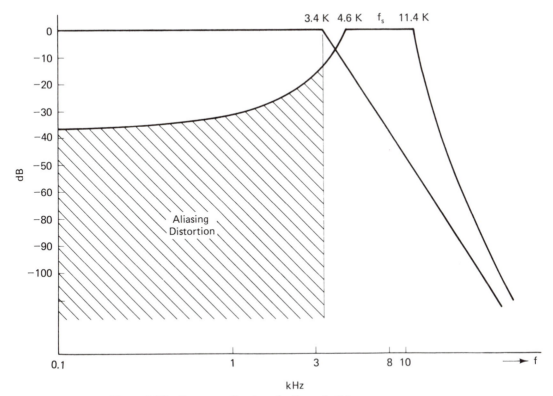

Figure 6-18 Frequency Spectrum for Example 5-1

6-5 Filter Characteristics

Without going into the details of filter design, this section will outline the basic characteristics of the four major classifications of analog filters: namely, the Bessel, Butterworth, Chebyshev, and elliptic filters. The frequency and time delay characteristics of these filters will be sketched in the low-pass mode since the front end pre-sampling, or anti-aliasing, filters have this feature. Bandpass or high-pass filters have similar frequency and phase characteristics.

The passband of the filter is the frequency range in which a signal is transmitted through a filter with little insertion loss. The high attenuation region is called the stopband. The region between the passband and stopband, with somewhat flexible boundaries, is called the transition region, or skirt.

In the ideal filter, the amplitude response and group delay should be constant with frequency in the passband regions. The group, or envelope, delay is related to the phase response by the expression

$$T_g = \frac{db}{d\omega} \quad \text{where } b \text{ is the phase response}$$

Refer back to Sec. 3-2.7 for a more detailed discussion of group delay. Group delay in the stopband is of little interest since this region rejects the undesired signal and noise components.

A typical feedback low-pass Bessel, Butterworth, or Chebyshev filter element is shown in Fig. 6-19. The choice of resistance and capacitance values determines the frequency and phase characteristics of the filter. By cascading several such networks, higher-order filters may be formed.

The Bessel filter has a very linear phase response but a fairly gentle skirt slope, as illustrated in Fig. 6-20. It is well-suited for pulse applications.

The Chebyschev filter has a sharp roll-off but a significant passband ripple. The faster the roll-off, the greater the peak-to-peak ripple in the passband. The phase response is highly non-linear in the skirt region. Such unequal delay of data frequencies in the passband causes severe pulse distortion and thus in-

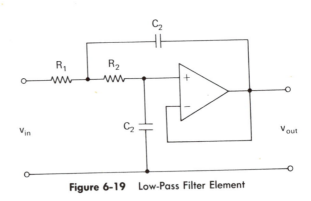

Figure 6-19 Low-Pass Filter Element

creased errors at modem demodulators. This can be somewhat overcome by increasing the bandwidth of the filter so that the more-or-less linear phase region is extended.

The Butterworth filter has characteristics somewhere between those of the former two. It has a moderate roll-off of the skirt and a slightly non-linear phase response.

The elliptic filter has the sharpest roll-off of all the filters in the transition region but has ripple in both the passband and stopband regions. It can be designed to have very high attenuations for certain frequencies in the stopband, which lessens the attenuation for other frequencies in the stopband.

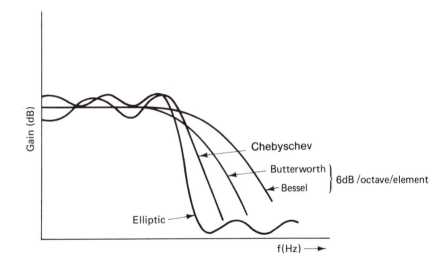

(a) Amplitude Response

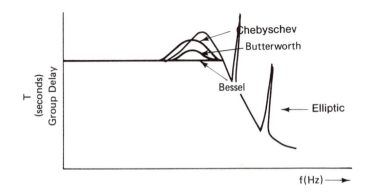

(b) Group Delay Response

Figure 6-20   Filter Characteristics

Table 6-2 presents a summary of the filter's response characteristics discussed.

Table 6-2 Filter Response Summary

Bessel	Butterworth	Chebyschev	Elliptic
Maximally flat time delay in the *dc* vicinity. Monotonically decreasing in amplitude with frequency.	Maximally flat amplitude response in the *dc* vicinity.	Equiripple amplitude in the passband. Fastest roll-off for the same order of filter when compared to Bessel or Butterworth filters.	Equiripple amplitude in both passband and stop band. Sharper transition band and higher attenuation in the lower portion of the stop band.
"Best" phase response; "poorest" amplitude response.	Compromise between amplitude and phase when compared to Bessel and Chebyschev filters.	"Best" amplitude response. (More attenuation in the stop band as ripple is increased.) "Poorest" phase response.	High frequency attenuation lower than that of the other filters. Sensitive to component values.
Attenuation continues to increase as frequency increases			

6-6 Time Division Multiplexing

Any pulse modulation scheme involves translating the audio, or modulating, signal into a series of encoded pulses, sending these pulses over a transmission medium, and reverting the pulses back to an analog signal. Regardless of the encoding used, a PAM signal is always obtained initially.

The audio signal is first sampled at a sufficiently high rate in order to minimize distortion. By keeping the samples short in comparison to the time intervals between them, many different channels can be interleaved in one sampling period. This process, as performed by a commutator, is illustrated in Fig. 6-21. It is commonly called Time Division Multiplexing (TDM). TDM allows each channel the full bandwidth of the transmission medium whenever its signal is transmitted, although each channel is not continuously on the system. This is in contrast to Frequency Division Multiplexing (FDM) in which each channel is continuously on the system but allowed only a limited portion of the system bandwidth. This is clearly illustrated in Fig. 6-22.

After a PAM signal is obtained, it can be transmitted directly or quantized and encoded to provide improved noise immunity. The latter is more commonly done. This will be further discussed in Chapter 7 when considering Delta and Pulse Code Modulations systems.

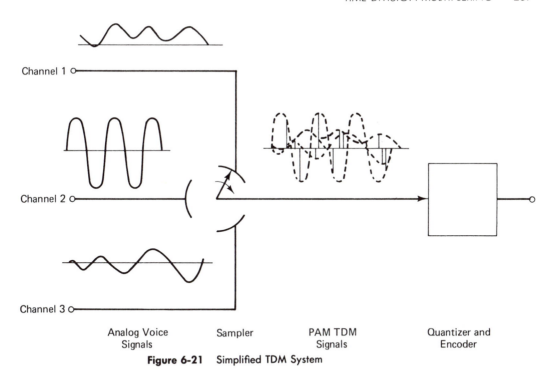

Figure 6-21 Simplified TDM System

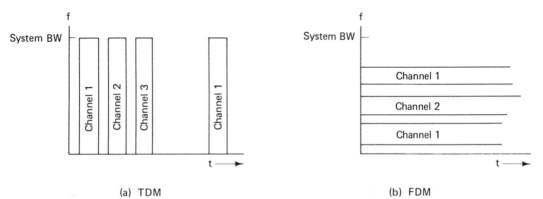

(a) TDM   (b) FDM

Figure 6-22 Bandwidth Utilization of TDM and FDM Systems

An example of a four-channel electronic sampling system is shown in Fig. 6-23. This system employs two flip-flops and MOS P-channel enhancement transistors as switches. When the gate to the MOS P-transistors becomes negative, the P-channels act as a low impedance circuit, with resistance at about 200 Ω. Normally, the resistance across the drain and source is in the tens of megohms.

Flip-flops A and B convert the serial clock input to binary parallel output. The four NAND gates decode this to sequentially drive the switches. When all the inputs to a NAND gate go HI, the output goes LO, thus causing the attached transistor to act as a low resistance path.

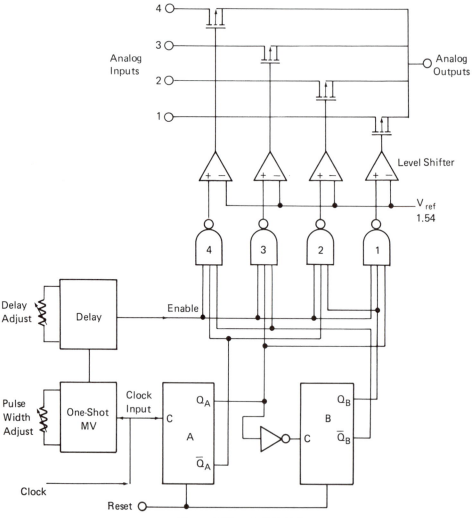

Figure 6-23   PAM-TDM System

Since the two flip-flops provide two digits, the number of commutator switches that can be driven is 2^2 or 4. An eight-channel multiplexer would need three flip-flops; that is, $2^3 = 8$. Figure 6-24 shows the waveforms of the four-channel commutator. Table 6-3 gives the flip-flip conditions for the channel selection.

Table 6-3   Channel Selection Conditions for Fig. 6-23 Circuit (Enable must be HI)

Q_A	Q_B	Channel Selected
LO	LO	4
LO	HI	2
HI	LO	3
HI	HI	1

In order to establish a variable sampling aperture, the clock pulse is fed to a monostable multivibrator with pulse width adjusting control. This will establish the aperture time. A delay circuit then moves the sampling pulses such that the sampling pulse occurs some time away from the gates switching edges.

A typical commercial eight-channel multiplexer is the RCA COS/MOS CD4051A integrated circuit, the functional diagram of which is shown in Fig. 6-25. It has the three binary control inputs of A (LSB), B,C(MSB), and an inhibit input. The three binary inputs select which one of the eight channels is turned ON to connect the input to the output. For instance, with A = 1, B = 1, C = 0, and inhibit = 0, Terminal 3 is connected to output Terminal 12 (Channel 3). Logic $0 = V_{ss}$, and $1 = V_{DD}$. The terminals marked CHANNEL IN/ OUT and COMMON OUT/ IN can be used as input or output terminals.

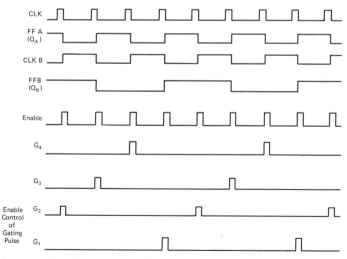

Figure 6-24   Waveforms for the Four-Channel Commutator of Fig. 6-23

When used on a multiplexer, the CHANNEL IN/OUT terminals are the inputs and the COMMON OUT/IN terminal is the output. When used as a demultiplexer, these roles are reversed.

When $V_{DD} - V_{EE} = 15$ V (as when $V_{DD} = 5$ V, $V_{SS} = 0$, V and $V_{EE} = -10$ V), the channel ON resistance is typically 50 Ω. The OFF resistance-input leakage is typically 10 pA for the same supply voltage condition.

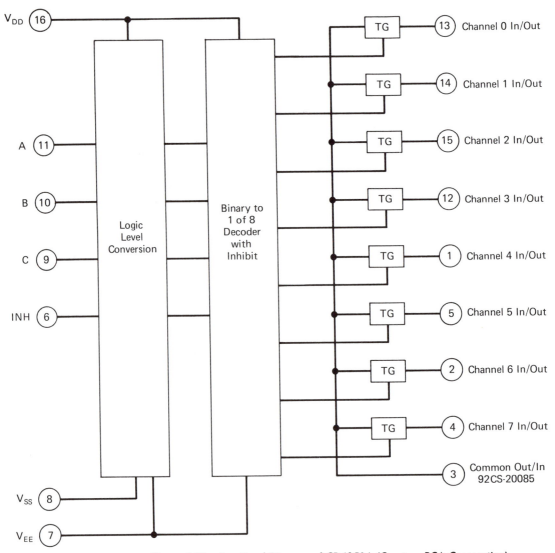

Figure 6-25 Functional Diagram of CD4051A (Courtesy RCA Corporation)

6-7   Pulse Duration and Pulse Position Modulation

In the PAM, the baseband signal modulates the amplitude of a pulse train which is spaced at regular intervals and has fixed time slots. Rather than varying the pulse amplitude, alternative modulation schemes vary either the pulse intervals, called pulse position modulation (PPM), or the duration of the time slots, called pulse duration modulation (PDM) or pulse width modulation (PWM). Figure 6-26 shows the various analog pulse modulations waveforms

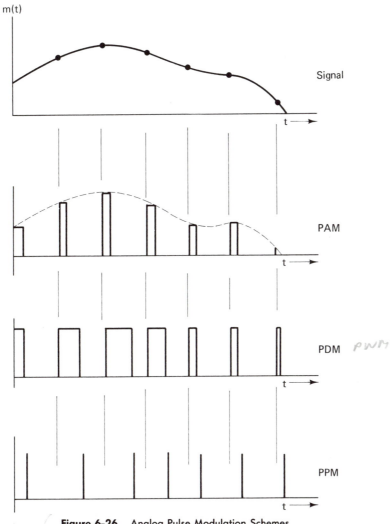

Figure 6-26   Analog Pulse Modulation Schemes

for the same message signal. In PDM, the pulse width is proportional to the amplitude of the modulating signal. In PPM, the pulse delay from some reference point is proportional to the amplitude of the modulating signal.

In PDM and PPM, information is conveyed by a time parameter, or the location of the pulse edges. Thus, these modulation types are referred to as pulse time. In PAM and PDM, the sample values equal to zero are usually represented by nonzero amplitude or duration in order to prevent missing pulses and to preserve a constant pulse rate. This is important for synchronization purposes when time division multiplexing is used.

Pulse position or pulse width measurements are usually based on the leading or trailing edge of the pulse. Usually the uncertainty of the pulse position is stated in terms of the rise and fall time of the pulse, (or time required to go from 10 to 90% of its full amplitude). An analysis of typical low-pass filters shows that the rise time, t_r, is inversely proportional to the bandwidth. As a result, the bandwidths required for PDM and PPM are much greater than for PCM, where only the presence or absence of a pulse is of interest.

For PAM systems, in which the pulse amplitudes are continuously variable, the problems of cumulative noise and distortion in the transmission system can greatly disturb the sampled waveforms. Because of the inherent disadvantages mentioned for PDM, PPM, and PAM, PCM has become the most common digital technique in cable systems. The latter does not require excessive bandwidths, and its amplitude is of little importance except to note whether or not a pulse is present.

To produce PDM, which is often called pulse width modulation (PWM), a circuit employing a 555 timer can be used, as illustrated in Fig. 6-27. The internal block diagram of the 555 chip and the resulting waveforms are shown in Fig. 6-27(b) and (c). When a clock pulse that is going negative is applied to Pin 2, the output (Q_1) of the FF resets to LO. Q_T is cut off and C begins to charge linearly, as illustrated in Fig. 6-27(c). The buffer output remains HI until the capacitor voltage, which appears at Pin 6, the input to Comparator 1, reaches the value of the signal voltage which is applied to the Pin 5 input of Comparator 1. This trips the output of Comparator 1 and sets the FF output HI. Buffer output goes low and the capacitor rapidly discharges through Q_T. This cycle repeats upon the application of the next pulse.

One method of transmitting PDM more efficiently is to transmit only the pulse signs. Figure 6-28 shows the arrangement and the waveforms. At the receiving end, the appropriate pulse can be stretched to recover the original pulse.

The circuit of Fig. 6-29 shows the PDM recovery circuit.

It is assumed that the maximum duration of PDM is less than one-third of the signal period, T_s. The one shot will set the flip-flop to zero only when two successive PDM pulses pass through with separations greater than $T_s/3$. This assumes that Pulse 2 triggers the flip-flop to zero and that the next pulse that triggers the flip-flop corresponds to the leading edge of the transmitted pulse.

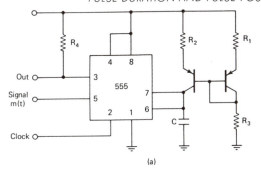

(a)

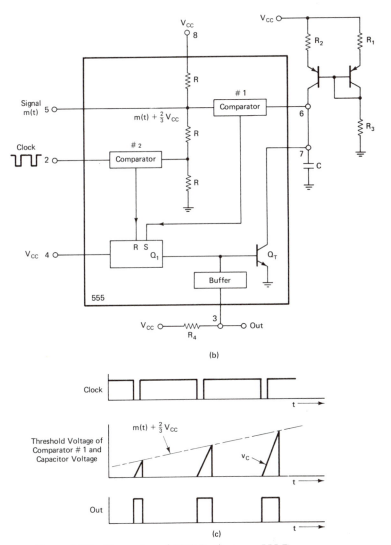

(b)

(c)

Figure 6-27   Generation of PDM Employing a 555 Timer

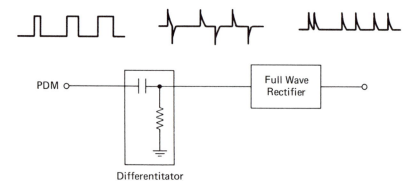

Differentitator

Figure 6-28 Transmission of PDM Using Pulse Edges Only

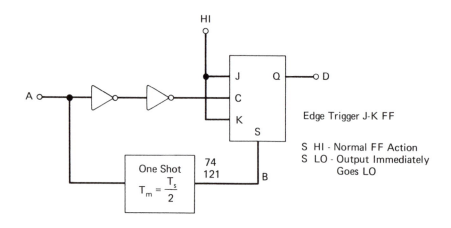

Edge Trigger J-K FF

S HI - Normal FF Action
S LO - Output Immediately
Goes LO

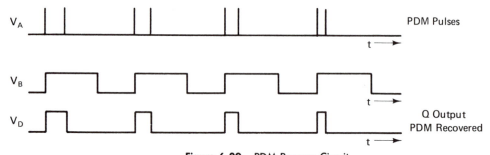

Figure 6-29 PDM Recover Circuit

In demodulating PDM, the best results are obtained when it is converted to PAM and low-pass filtered. A general block diagram of a PDM or PAM is shown in Fig. 6-30. The constant current generator assures a linear voltage rise on the capacitor. Note that the discharging pulses have to be located beyond the widest pulse width. The easiest way to obtain such pulses is to produce narrow pulses by triggering a multivibrator and delaying it with another monostable.

PPM is very closely allied with PDM. In fact, it can be derived directly from PDM, as is shown in Fig. 6-31. The PDM is differentiated, then rectified and shaped. Remember that in PDM it is the position of the pulse edges that carry the information and not the pulse itself. Therefore, PPM carries exactly the same information as long as the position of the clock pulses (leading edge) is well defined in the received signal.

It is easy to see that PPM is superior to PDM for message transmission since the wide pulses of PDM require more energy than PPM when transmitted. Specifically, PPM is suited for communication is the presence of noise.

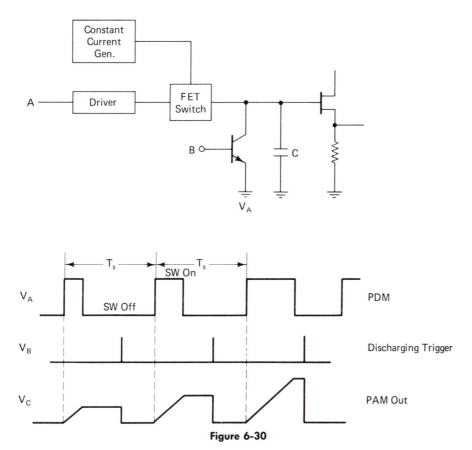

Figure 6-30

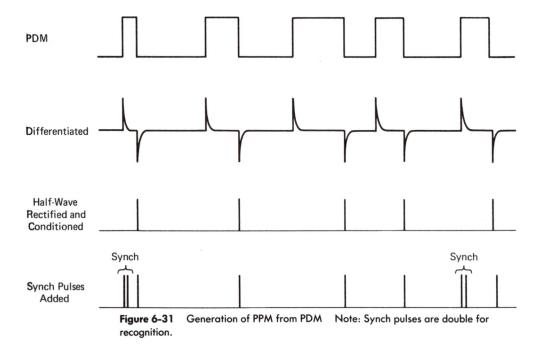

Figure 6-31 Generation of PPM from PDM   Note: Synch pulses are double for recognition.

Very high peak narrow pulses can be transmitted, and the pulse position can be determined even when the noise level is high. Note, however, that transmitting very narrow pulses requires a large bandwidth. When light is used as the media for transmitting analog signals, PPM or PCM are the most suitable types of modulation because the maximum power output in the modulated light source, such as LED or LASER, is achieved when it is pulsed at a very low duty cycle.

Mathematical analysis of PPM is much more cumbersome than PDM. The easiest way to demodulate is to convert PPM to PDM and follow with low-pass filtering. The PDM could also be converted to PAM and followed by low-pass filtering. The latter method is superior. In transmitting PPM, it is necessary to transmit a series of sync pulses at a much lower repetition rate than sampling pulses so that they will not interfere with the original signal and/or minimize the number of pulses transmitted in order to conserve transmission power. Sampling pulses can be recovered from the sync pulses by a phase lock loop frequency synthesizer, or by synchronizing an oscillator. In PPM, a series of double pulses are used for synchronization since the synch pulses must be narrow and at the same time distinguishable from the signal pulses.

Problems

1. With the aid of graphs, describe both the time domain and frequency domain characteristics of:
   a. natural sampling
   b. Flat-topped sampling (also known as the sample-and-hold technique) Assume a flat baseband frequency response.

2. What is the aliasing, or foldover distortion? Give two techniques for reducing aliasing.

3. State the Sampling Theorem.

4. In a PCM telephony circuit, the voice signal is filtered with a 4-pole filter with a break frequency of 3300 Hz. If the signal is sampled at an 8 kHz rate, what is the approximate S/N ratio? Draw the sampled signal spectra and note the signal power and noise power by appropriate shading. Since voice spectra is not flat, would you expect a higher S/N ratio on an actual system? Explain.

5. Derive Eq. (*6-11*) from the integral given.

6. What is aperture time and aperture distortion? Give two techniques for reducing this type of distortion.

7. Why is flat-topped sampling more commonly used than natural sampling?

8. If the error due to aperture is to be less than 1% at 3.3 kHz, what maximum aperture time is permitted when the signal is sampled at an 8 kHz rate?

9. a. With the aid of a typical filter response curve, indicate the passband, stopband, and transition regions.
   b. With the aid of a gain plot, the differences in charcteristics of an elliptic, Chebyshev, and Butterworth filter.
   c. What effect does non-linear envelope delay have on a rectangular pulse train?

10. With the aid of diagrams, define the following:
    a. PAM
    b. PDM
    c. PPM

11. Design a two-channel sampling system which employs N-channel FET's rather than the P-channel FET's shown in Fig. 6-23.

12. List some of the disadvantages of PDM, PPM, and PAM.

13. A natural sampling system with $f_s = 12$ kHz, 4-pole low-pass filter has an input bandlimited at 4 kHz. Its output S/N in dBs is: (use graph in fig. 6-13)
    a. 40
    b. 60
    c. 30
    d. 65

14. Match the filter type to the characteristic.

a. Chebychev   _____ Ripple in stop-band and passband

b. Bi-quad   _____ Slowest roll-off in transition region

c. Butterworth   _____ Ripple in pass-band only

d. Bi-quartic   _____ No pass-band ripple and moderate roll-off

e. Elliptic

f. Circular

g. Bessel

7

Digital Modulation

7-1 Introduction

Up to this point in the text, either the amplitude or the timing of the transmitted pulse modulated waveform was made to vary continuously with the amplitude of the modulating signal. These types of modulation systems are analog in nature. This chapter will consider two of the most common types of systems in which the encoded signal consists of binary digits. The two systems are Delta Modulation (DM) and Pulse Code Modulation (PCM).

7-2 Simple Delta Modulation

The less frequently used of the two systems is Delta Modulation (DM). Currently, it is being applied to satellite communications, television, and subscriber telephone loops. It is less complex and therefore less costly than PCM, more tolerant to transmission errors, and does not need the synchronization requirements of PCM. On the other hand, it is sensitive to slope overload (which will be described in more detail later) and is unsuitable for time sharing as an encoder/decoder among multiple channels.

The delta modulator transmits binary output pulses whose polarity depends upon the difference between the modulating signal and the feedback signal corresponding to the history of the signals previously sent. In the simple Delta Modulator of Fig. 7-1, the integrated feedback $\tilde{m}(t)$ is compared to the input signal $m(t)$.

If $m(t) > \tilde{m}(t)$, a positive pulse forms the output, $\Delta(t)$, and if $m(t) < \tilde{m}(t)$, a negative pulse forms the transmitted signal. This output pulse train also forms the approximation $\tilde{m}(t)$ after being integrated.

The following waveform sketches will clarify the operation of this modulator. As illustrated in Fig. 7-2, a series of trangular waveforms results for $\tilde{m}(t)$ when there is no signal applied to the Delta Modulator. This is because integration of a constant gives a ramp. (See left-hand side of Fig. 7-2.) In this idling condition, the transmitted signal consists of alternate positive and negative pulses. The difference between the original signal, $m(t)$, and the reconstructed signal, $\tilde{m}(t)$, results in an error signal, which is often called granular or quanti-

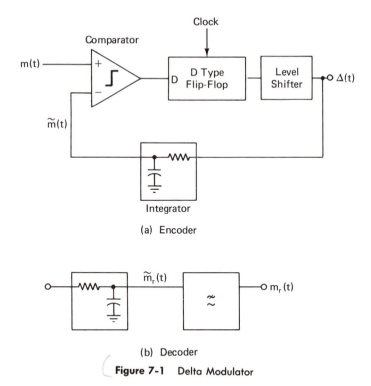

(a) Encoder

(b) Decoder

Figure 7-1   Delta Modulator

zation noise. This noise can be decreased by either decreasing the magnitude of the step size "a" or by increasing the sampling frequency.

If $m(t)$ exceeds the feedback signal, $\tilde{m}(t)$, the transmitted wave train remains positive. When $\tilde{m}(t)$ begins to overshoot $m(t)$, the encoder output begins transmitting a negative pulse. If the modulating signal changes more rapidly than the encoder can follow, slope overload occurs. This slope overload, shown towards the right of Fig. 7-2, causes waveform distortion and is one of the severe limitations of Delta Modulation. The integrator is unable to closely track large amplitude, high frequency signals.

When in slope overload, the difference between $\tilde{m}(t)$ and $m(t)$ is called slope overload noise, a noise which greatly exceeds quantization noise. As a function of input signal power, the S/N ratio initially increases with power until slope overload is reached, as illustrated in Fig. 7-3. This increase is due to the signal power term increasing while the N-term, chiefly quantization noise, remains fixed. Upon a further increase in signal power, slope overload occurs and the S/N ratio suffers.

Given a modulating signal of $m(t) = A \sin 2\pi f_m t$, the maximum rate of rise would be $dm(t)/dt|_{max} = 2\pi f_m A$. For a step size "a," as shown in Fig. 7-2, and for a sampling rate of f_s, the rate of rise of the reconstructed signal $\tilde{m}(t)$ is $\frac{a}{1/f_s} = af_s$. Equating these to determine the maximum amplitude-modulating frequency product allowable produces

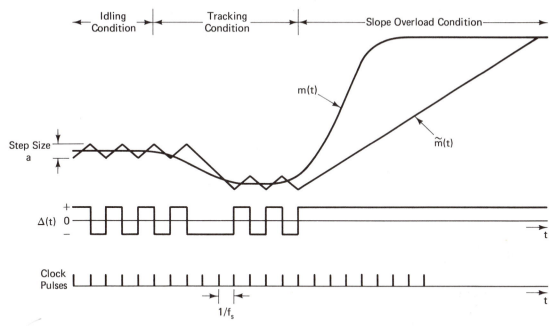

Figure 7-2 Delta Modulation Waveforms Illustrating Idling, Tracking, and Slope Overload

$$2\pi f_m A = af_s$$

or

$$Af_m = af_s/2\pi \qquad (7\text{-}1)$$

Thus, given a maximum of f_m, a maximum amplitude can be determined. As noted in Eq. (*7-1*), slope overload can be delayed by increasing the sampling frequency f_s or increasing the step size "a." This, of course, either forces an increase in the bandwidth or increases the quantization noise.

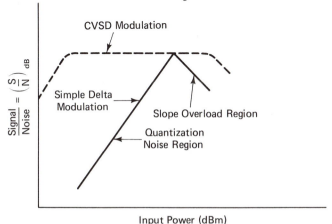

Figure 7-3 Signal-to-Noise for Simple and Adaptive Delta Modulation

7-3 Companded Delta Modulation

Studying the S/N response of the simple Delta Modulation Scheme shown in Fig 7-3 shows that the S/N ratio varies considerably with the signal power. At low signal power, the S/N ratio is much worse than at high signal powers as long as it remains within the quantization noise region.

In order to maintain a more constant S/N ratio, the step size is made adaptable to the input signal amplitude. For small input signals, $m(t)$, the step size is kept small but increased as the input signal is increased. This makes the noise power vary with signal power, maintaining a constant S/N ratio over a fairly large range of input power levels. In addition, slope overload is less likely to occur since the step sizes are the largest when nearing this region.

For telephony, the type of adaptive delta modulation used is Continuous Variable Slope Delta Modulation (CVSD). A block diagram of such a modulator is shown in Fig. 7-4. The squaring circuit causes the gain of the feedback

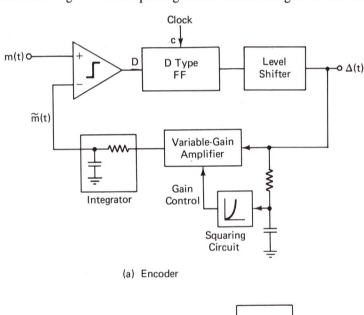

(a) Encoder

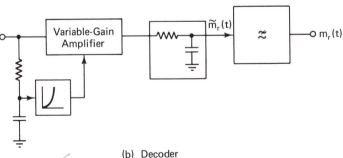

(b) Decoder

Figure 7-4 Variable Slope Delta Modulator

amplifier to increase with increase in signal amplitude, regardless of the polarity of the voltage. In addition, by squaring the quantization noise, power is made to vary linearly with input signal variations.

Single-chip CVSD encoders/decoders are now available. The Harris HC-55516/55532 and the Consumer Microcircuit of America FX209 are typical examples.

One chief advantage CVSD has over companded PCM is that it can handle voice at about half the bit rate. For good voice intelligibility, companded PCM requries about 64k bit/s whereas CVSD requires only 32k b/s for similar quality.

On the negative side, unlike PCM, CVSD is not suitable for time-sharing among multiple channels. Also, at 32k b/s rate, CVSD cannot handle tone and phase encoded modem transmission as well.

7-4 Pulse Code Modulation

The most common form of digital transmission is Pulse Code Modulation (PCM). PCM performs the three functions of

1. Sampling the analog signal,
2. Quantizing the sampled amplitudes, and
3. Encoding the quantized sample into a digital signal.

The principle area of PCM application has been in linking relatively close terminals or nearby local exchanges by wire or cable. In time, PCM will be employed in very large networks spanning great distances.

Consider a simplified version of a 24-channel PCM system that is generally used throughout North America. Since it often displaces the old FDM on VF cable, it is necessary to take certain precautions when the transfer is made. Voice usage on a cable requires consistent characteristics out to about 4 kHz whereas PCM requires consistent performance out to a much higher frequency of 2.5 MHz. As a result, the alterations or precautions must be taken in order to prevent the cable from behaving as a low-pass filter with a cut-off frequency at around 4 kHz. Loading coils, which act as low-pass filters, must be removed. Building-out networks, which are added to make a cable appear electrically longer in order that standard loading plans can be used, should be removed since they severely attenuate the high frequency components in a signal. In addition, taped circuits and cross-split pairs should be removed. The latter is caused by improperly sorting out continuous pairs during splicing. Unused taped circuits are often left in place when a customer's connecting loop is cut to disconnect him from the line. These loose ends can cause excessive phase distortion because of the added capacitance. All are illustrated in Fig. 7-5.

In order to adequately understand the concept of sampling and to obtain the bandwidth required by a PCM signal, review the Sampling Theorem which states: For a signal bandwidth limited to B, all the information in this signal will be retained by the sample if the signal is sampled at regular intervals of

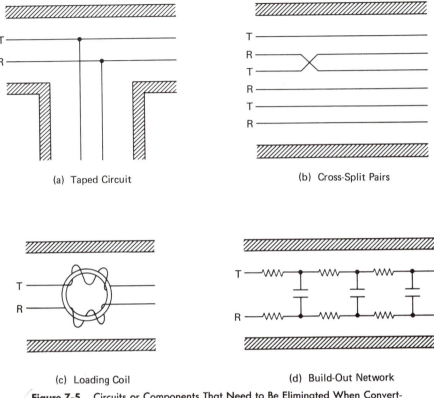

(a) Taped Circuit

(b) Cross-Split Pairs

(c) Loading Coil

(d) Build-Out Network

Figure 7-5 Circuits or Components That Need to Be Eliminated When Converting to Digital Transmission

time and at a rate higher than 2B. For a 4-kHz audio signal, for example, the sampling rate must be at least $2 \times 4000 = 8000$ samples/s, or one sample must be taken every 125 μs (1/8000) for no information to be lost. The 125 μs interval between samples is taken up by 24 channels, signaling pulses, and sync pulses. Because the samples are short compared to the time intervals between them, many different channels can be interleaved and transmitted in one sampling period. This process is known as time division multiplexing (TDM).

Figure 7-6 shows a basic block diagram of a PCM speech terminal. When transmitting, the audio signal is first bandlimited to minimize aliasing or foldover distortion and then sampled to form a PAM signal. At this point, the signals from all channels go to common equipment where the coding and additions of sync are performed. Since each individual channel requires very little equipment and the more exotic switching and coding is done in common circuitry, PCM terminals can be inexpensive for a large number of channels.

With the declining cost of Analog-to-Digital (A/D) converters, the future trend is to have an encoder on each channel and then combine the 24 channels with a digital multiplexer. In this manner, processing such as reducing redundancy before it is multiplexed, can be performed on the sampled signal.

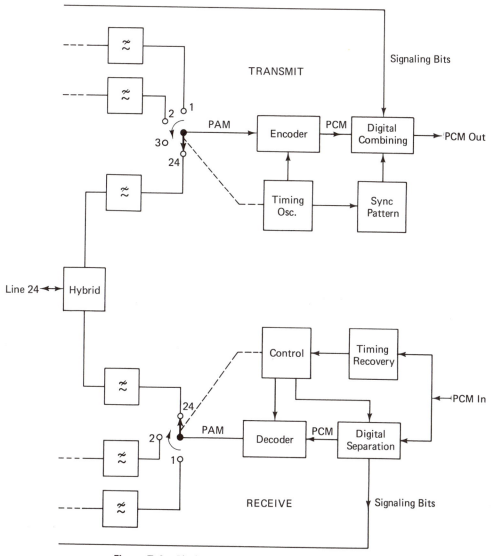

Figure 7-6 Block Diagram of PCM Terminal

After sampling, the voltage amplitude is "quantized" or is forced to take on certain discrete voltages. This new signal forms an approximation to the old one. Because the two forms are not identical, an inherent error occurs. The difference between the old and new signals is called quantizing error, and is referred to as quantizing noise when heard or measured. The error signal has frequency components extending from above the voice band. Because of aliasing, these higher frequency components fold back into the voice band and result in an error signal at the receive-filter output. In PCM transmission, this quantizing noise is one of the dominant sources of impairment, and it remains significant even when other noise contributions are minimized. The transfer function of a linear quantizer is shown in Fig. 7-7 along with the quantizing of a sampled signal.

The quantizing noise is reduced as the spacing between the quantized steps are reduced or, in other words, as the number of levels are increased. For voice communication, 256 levels are commonly used, which can be represented by an eight-digit code. In general, for an n-digit code yielding 2^n equally spaced steps, the rms S/N ratio in decibels is given by:

$$\frac{S}{N} = 1.8 + 6n \ dB \qquad (7\text{-}2)$$

which for 8 digits gives a $\dfrac{S}{N}$ ratio of 49.8 dB

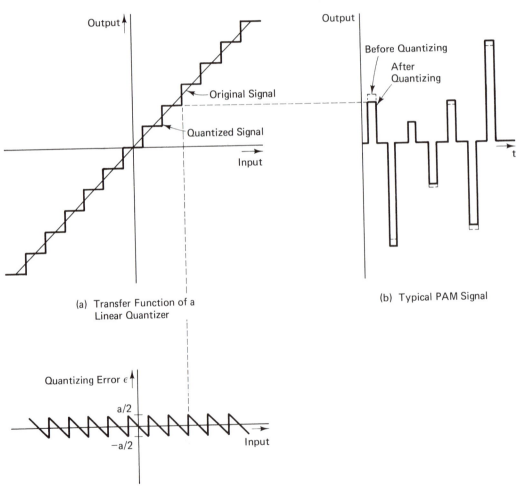

(a) Transfer Function of a
Linear Quantizer

(b) Typical PAM Signal

(c) Quantizing Noise

Figure 7-7 Quantizing Process

Equation (*7-2*) assumes that the maximum amplitude signal is obtained just before clipping occurs. In practice, low level signals are more probable than high level ones. Hence, the real S/N ratio can be much smaller than that shown in Eq. (*7-2*) since the quantizer noise is essentially independent of signal level for uniform step sizes. To obtain a constant S/N ratio for all signal levels, non-linear or piece-wise linear encoders are employed. This will be discussed in Sec. 7-7.

Once the level of quantization distortion is selected, immunity from channel noise and distortion can be obtained. As long as signal distortion is kept within certain bounds, identification of the proper signal level can be maintained. After amplitude quantization, the sample is converted into a code which consists of either seven or eight binary bits for the most common PCM system.

The signal is now discrete in both amplitude and time. If eight bits are transmitted for each sample as shown in Fig. 7-8, a fresh train of elements equivalent to the old ones can be regenerated with very little impairment. Although the signal is subject to noise, distortion, and bandwidth limitation, if these flaws are small enough, the regenerated signal will correspond to the original one. Thus, with suitable spacing, a series of regenerative repeaters will pass along a discrete signal without any accumulation or impairment along the route.

At the sampling instant, the regenerator compares the signal with the threshold decision level, or the level midway between the two expected signal levels. If positive, a positive pulse is generated; if negative, a negative pulse is generated. As can be seen from Fig. 7-8 (c), fairly large impairments can be encountered before the signal is lost. This is one of the chief advantages that digital transmission has over analog transmission.

PCM has great immunity to interference and noise. This noise immunity is not gained without compromise, but at the expense of increased bandwidth. To transmit the pulse stream of 8 bits/sample shown in Fig. 7-8 (e), $8 \times 2 \times 4000 = 64k$ b/s must be transmitted for a 4 kHz bandwidth limited signal which represents a minimum baseband bandwidth of 32 kHz/s. Thus, practical PCM systems require 32k/4k, or 8 times the bandwidth of their analog counterparts.

In addition to speech signal information, signaling information such as ON and OFF hook and dialing must also be transmitted. In addition, synchronizing information (framing code) must be generated so that the receiver can sort out the code words and signaling bits into the proper channels. Alarm code can also be added to inform the user of local trouble.

In addition to PCM's immunity to noise, which gives transmission characteristics almost independent of distance, the following factors also favor this mode of modulation:

a. It can easily carry a mixture of traffic such as telephony, telegraphy, data, and encoded video information, provided the medium has sufficient capacity.

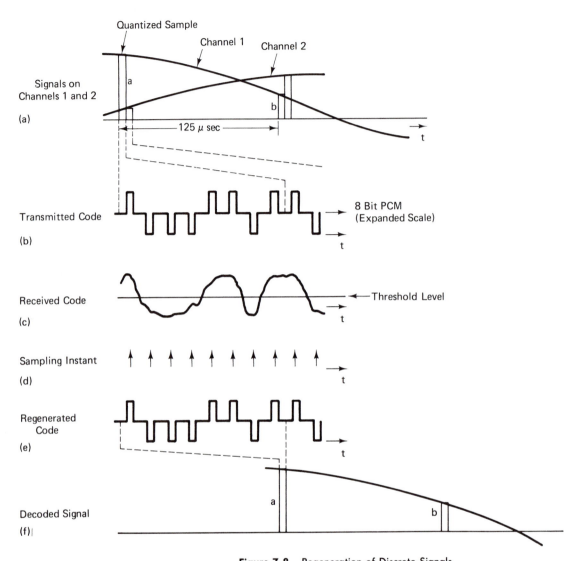

Figure 7-8 Regeneration of Discrete Signals

    b. It can increase the capacity of single telephone channels over cable pairs by multiplexing.

    c. It can lend itself to such novel facilites as crytography, storage, and other forms of digital processing.

    d. It is more suitable for the newer types of transmission media such as light beams in optical fibers and multiple access satellites.

    e. Its signal characteristics allow easy access to electronic switching in which groups of digits are selected to be switched in turn onto various highways.

7-5   Aperture Time

When sampling an analog signal, the signal usually is changing in amplitude during the sampling process. In order to maintain a maximum limit on the amplitude uncertainty, or error, in the measurement of a signal which is varying with time, the aperture time or sampling time window must be of a limited time duration. This maximum aperture time is related to the rate of signal amplitude change.

For an A/ D converter, the aperture time is the conversion time. For a sample-and-hold circuit, it is the signal averaging time during the sample-to-hold transition. A sample-and-hold circuit placed before an A/ D converter effectively reduces the aperture times since the sample-and-hold can take a very fast sample of the analog signal and then hold the value while the A/ D operation is performed. In addition, the position of the sample point to sample point may vary because of slight variations in the sampling times. This manifests itself as "jitter" on the sampling waveform. This aperture uncertainty is normally much less than the aperture time, but can be significant in sample and hold applications.

As shown in Fig. 7-9, the maximum amplitude uncertainty occurs when the slope of the signal is a maximum. If the maximum amplitude error is "a," it will be related to the maximum rate of change with time of the input signal by the relation:

$$a = \tau \left. \frac{dv(t)}{dt} \right|_{max} \qquad (7\text{-}3)$$

Consider the case of a sinusoidal signal, $A \sin \omega t$. The maximum amplitude uncertainty will occur at the zero crossing of the waveform, that is,

$$a = \tau \frac{d}{dt} (A \sin \omega t)|_{t=o} = \tau A\omega \cos \omega t|_{t=o}$$

$$= \tau A\omega \qquad (7\text{-}4)$$

Let the step size of the encoded signal be equal to "a," since this is the size of the worst amplitude uncertainty. If the encoder has n bits of resolution, it

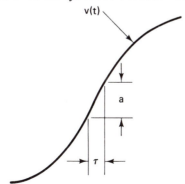

Figure 7-9   Illustration of Aperture Time (τ) and Amplitude Uncertainty (a)

can distinguish between 2^n levels. The maximum peak to peak voltage that can thus be obtained is a $(2^n - 1)$ volts, where the -1 takes care of the beginning level which has no amplitude, therefore for a peak amplitude A,

$$A = \frac{a(2^n - 1)}{2} \qquad (7\text{-}5)$$

Substituting $(7\text{-}5)$ into $(7\text{-}4)$ we obtain

$$\tau = \frac{1}{(2^n - 1)\,\pi f} \qquad (7\text{-}6)$$

Equation $(7\text{-}6)$ is graphically plotted in Fig. 7-10. Note that for a 4 kHz sinusoid with 8 bits of resolution, the maximum aperture time is 0.312 μs. Within this time, the A/D converter must encode the analog signal into 8 binary digits. For several channels sharing the same A/D converter, this time must be even further reduced.

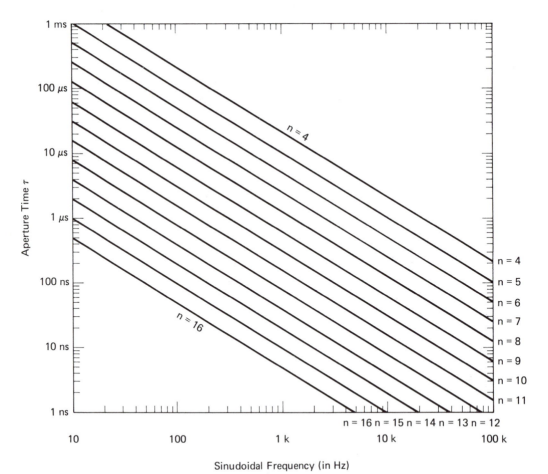

Figure 7-10 Aperture Error as a Function of Sinusoidal Frequency

7-6   Amplitude Sampling

Because the encoder in the transmitter section of the PCM system converts continuous amplitude into a discrete set of amplitude levels, an error waveform known as quantizing noise results. Quantizing noise exists only when a signal is present, and appears much as white noise. Even when no signal is present, some idle channel noise will still exist at the encoder output because of residual noise such as thermal and shot noise at the encoder input. This residual noise will occasionally exceed the thresholds of the first few encoding stages at the low voltage levels, resulting in random output pulses. Excessively large applied signals should also be avoided since they will be clipped when they exceed the maximum threshold level and will introduce large amounts of distortion.

By maintaining uniform encoder step sizes, the quantizing noise would essentially be kept constant for all signal levels. The maximum possible error is the same for all sample amplitudes for equally spaced or linear quanta. This is fine for strong signals, but it results in very poor signal-to-noise ratios for the weaker signals.

In order to improve the S/N ratios for the weak signals, which are much more probable to occur in the case of speech signals, non-uniform quantization can be used. Rather than maintaining the same maximum possible error for all sample magnitudes, the spacing of decision levels, or step sizes, is reduced for the smaller sample magnitudes but increased for the larger sample magnitudes. That is, the step sizes are made approximately proportional to the signal amplitude, resulting in a nearly constant S/N ratio over the speech volume range. Figure 7-11 illustrates a nonuniform encoder transfer characteristic which shows how quantizing noise can be increased with signal size to keep a constant signal-to-mean quantizing noise ratio.

Another technique commonly employed to maintain a more or less constant S/N ratio over the dynamic signal range is companding. The companding process is called compressing at the transmitting end and expanding at the receiving end. In compressing, the higher amplitude signals effectively obtain a lower gain than the weaker signals. In expanding, the reverse takes place.

The compandor devices are called compressors and expandors. Using this method, the signal is instantaneously compressed, and the higher amplitude signals are compressed more than the lower amplitude signals. The companded signal then enters an encoder which performs uniform quantization on the signal prior to coding.

At the receiver, expansion is performed after decoding is completed. Companding can improve the S/N ratio in voice circuit by up to 25 dB. It is of little value in data communications where a signal is fully OFF or ON, at 0, or in saturation. Each condition is equally probable and thus there is no advantage to using companders. The compander, because of its nonlinear characteristics, only adds intermodulation distortion.

The D2 or D3 PCM-channel banks commonly in use throughout North America employ a compressor with a logarithmic gain characteristic. It follows

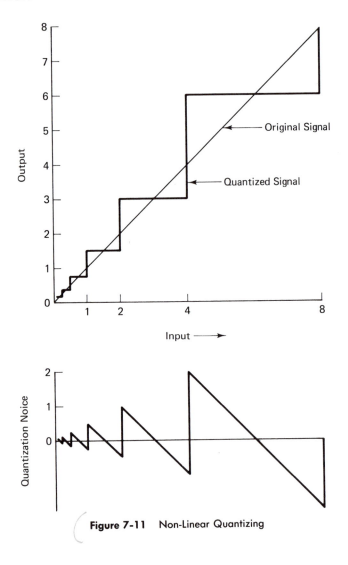

Figure 7-11   Non-Linear Quantizing

an encoding law known as the mu law, Eq. (*7-7*), where the value $\mu = 255$

$$Y = \frac{V \ln(1 + \mu v/V)}{\ln(1 + \mu)}, \quad 0 \le |v| \le V \qquad (7\text{-}7)$$

v = input voltage
V = maximum input voltage = maximum output compressed voltage

The dotted curve in Fig. 7-12 illustrates the companding effect of the mu law.

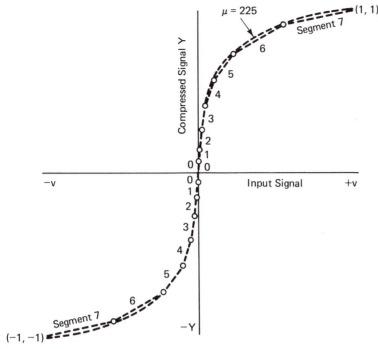

Figure 7-12   Fifteen-Segment CompressionCharacteristic for the Mu Law ($\mu = 255$) Where the Maximum Voltage is Equal to 1

$$Y = \frac{\ln (1 + \mu v)}{\ln (1 + \mu)}$$

In Europe, the CCITT's "A law" on the companding curve is followed. It is given by the expression

$$Y = \begin{cases} \dfrac{Av/V}{1 + \ln A}, & |v| \le \dfrac{V}{A} \\[2ex] \dfrac{1 + \ln Av/V}{1 + \ln A}, & \dfrac{V}{A} \le |v| \le V \end{cases} \tag{7-8}$$

where v = input voltage

      V = maximum input voltage = maximum output compressed voltage

where A = 87.6 for a 13 segment approximation

Rather than follow a smooth logarithmic curve, the D2 or D3 channel banks assume a segmented approximation to the continuous function. A total of 15 linear segments, as illustrated in Fig. 7-12, make up the approximation: seven positive segments, seven negative segments, and one segment near zero, which consists of two co-linear segments on either side.

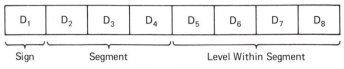

Figure 7-13 Digit Pattern

Within each segment there are 16 steps of equal size. The step sizes increase by a factor of 2 when moving from one segment to the higher adjacent one. In total, there are 255 discrete voltage levels for the D_2/D_3 type system, which require eight bits for coding.

The 8-bit code representing the signal is broken up into three groupings. As illustrated in Fig. 7-13, the first bit represents the sign (1 is positive; 0 is negative); the next three bits indicate which segment contains the sample; and the last four bits specify the level, up to 16 possible, within the segment. The various segments are counted sequentially from the lower levels towards the higher ones.

As an example, consider the coding scheme used by the encoder found in the Lenkurt 9002B system. The code word is split up as shown in Fig. 7-13 where:

a. The sign bit is 1 when the signal is positive, and 0 when the signal is negative.

b. The segment code is represented by a sequential binary code in reversed form; that is

$$\text{segment } 0 = 111$$
$$1 = 110$$
$$\cdot$$
$$\cdot$$
$$7 = 000.$$

c. The level within the segment is represented by a sequented binary code in reversed form; that is,

$$\text{level } 0 = 1111$$
$$1 = 1110$$
$$\cdot \qquad \cdot$$
$$15 = 0000$$

A pattern given by 0 111 1100 represents a negative voltage on Segment 0 at Step 3. Note that the segment, or step number, can be obtained by subtracting the actual code digits from 7(111) or 15(1111). Thus, for a segment binary representation of 4(100), the actual segment is $7 - 4 = 3(011)$. Although this type of coding, frequently referred to as folded binary coding, may appear strange since the maximum number of 1's appear near the no-signal level, it does give a higher pulse pattern for the more probable lower speech levels where there is a danger of losing timing recovery in the receiver. If few pulses are transmitted, synchronization could well be lost.

Number the levels from 0 to +127 for the positive signals and from 0 to −127 for the negative levels, with level 0 representing the zero voltage level. An expansion of an encoded signal at low voltage levels is illustrated in Fig. 7-14. Note that 0 volts is represented by the 0 levels of both the + and −0 segment; that is, by the codes 11111111 and 01111111. This results in a total number of 255 permissible levels for the 8-bit code. Also note that Segment 1 has twice the step size as Segment 0.

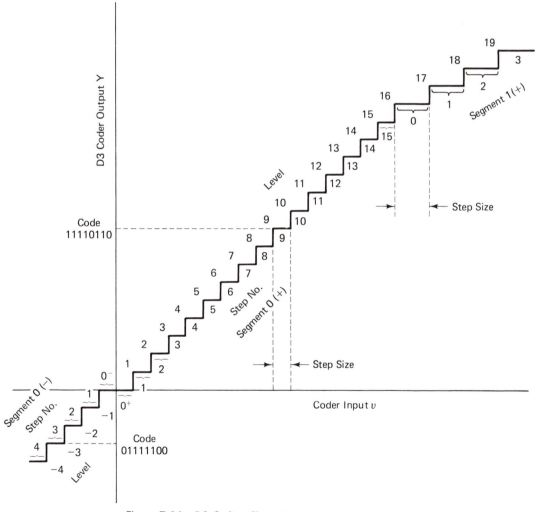

Figure 7-14   D3 Coding Characteristic

For the positive samples, when the input signal resides between level n and $n + 1$, the code word output can be obtained by expressing $255 - n$ in binary form. For Level 9, for instance, the code number is $255 - 9 = 246$. This is represented by the code 1 111 0110.

This can also be interpreted as a positive signal (first bit is a 1) on Segment 0 $(7 - 7)$ at Step 9($15 - 6$). Similarly, for negative examples the output code word is given by $127 - n$, expressed in binary form. Thus, Step 3 on the 0 negative segment is represented by $127 - 3 = 124(01111100)$.

If all the coding steps were to be of the same height as the smallest step for all the segments, a linear code requiring 13 bits would result. Codes in which step sizes increase by multiples of the smallest step are readily convertible from the compressed form to the linear form. The linear form is necessary if processing is to be done on the digital signal. For example, if two digital speech signals are to be added together, as in a conference call, each digital signal must first be converted to the linear counterpart, added, and then converted back to the compressed non-uniform code.

Table 7-1 gives the increment step sizes and the break points for the various segments. Table 7-2 tabulates some typical 8-bit PCM words for the Lenkurt 9002B channel bank. Figure 7-15 illustrates the corresponding positive section of the fifteen segment compression characteristic.

As an example, find the voltage, sign, segment, step, and level for the following code word

1 101 1101

Step = 15 (1111) − 13 (1101) = 2 (0010)

Segment = 7 (111) − 5 (101) = 2 (010)

Sign = 1 (Positive)

The voltage at Step 2 of Segment 2 from Table 7-1 is $(23 + 2 \times 1.93)\text{mV} = 28.86$ mV. The voltage is on level $255 - 221(11011101) = 34$.

Table 7-1   Increment Step Size and Breakpoints for the Lenkurt 9002B Channel Bank

Segment	Step Increment (mV)	Breakpoint (mV)	
0	0.475	16 × 0.475 =	7.6
1	0.963	7.6 + 16 × 0.963 =	23.0
2	1.93	23 + 16 × 1.93 =	53.9
3	3.88	53.9 + 16 × 3.88 =	116
4	7.69	116 + 16 × 7.69 =	239
5	15.4	239 + 16 × 15.4 =	486
6	30.9	486 + 16 × 30.9 =	981.0
7	63.7	981 + 16 × 63.7 =	2000

Table 7-2   Folded Binary Code
8 Bit PCM Word

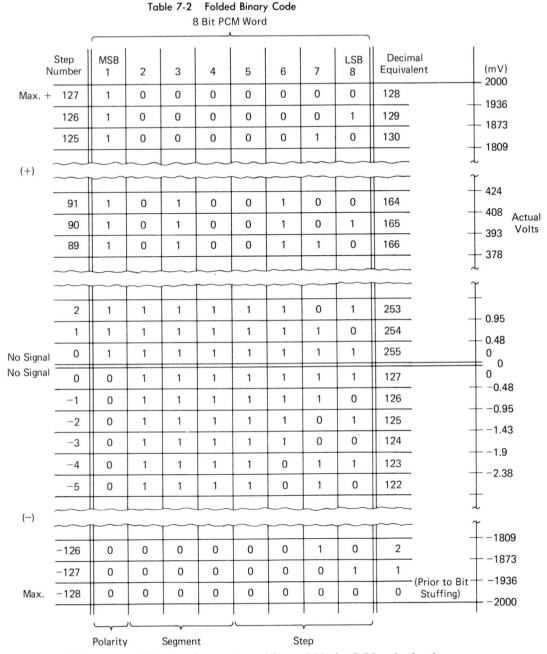

	Step Number	MSB 1	2	3	4	5	6	7	LSB 8	Decimal Equivalent	(mV)
											2000
Max. +	127	1	0	0	0	0	0	0	0	128	1936
	126	1	0	0	0	0	0	0	1	129	1873
	125	1	0	0	0	0	0	1	0	130	1809
(+)											424
	91	1	0	1	0	0	1	0	0	164	408
	90	1	0	1	0	0	1	0	1	165	393
	89	1	0	1	0	0	1	1	0	166	378
	2	1	1	1	1	1	1	0	1	253	0.95
	1	1	1	1	1	1	1	1	0	254	0.48
No Signal	0	1	1	1	1	1	1	1	1	255	0
No Signal	0	0	1	1	1	1	1	1	1	127	0
	−1	0	1	1	1	1	1	1	0	126	−0.48
	−2	0	1	1	1	1	1	0	1	125	−0.95
	−3	0	1	1	1	1	1	0	0	124	−1.43
	−4	0	1	1	1	1	0	1	1	123	−1.9
	−5	0	1	1	1	1	0	1	0	122	−2.38
(−)											−1809
	−126	0	0	0	0	0	0	1	0	2	−1873
	−127	0	0	0	0	0	0	0	1	1	−1936
Max.	−128	0	0	0	0	0	0	0	0	0 (Prior to Bit Stuffing)	−2000

Actual Volts

Polarity   Segment   Step

For 8-bit systems following the mu law with $\mu = 255$, the S/N ratio that is C-message weighted never exceeds 38 dB and remains at this level for about a 40 dB variation in signal level. Figure 7-16 shows the theoretical performance of such a terminal.

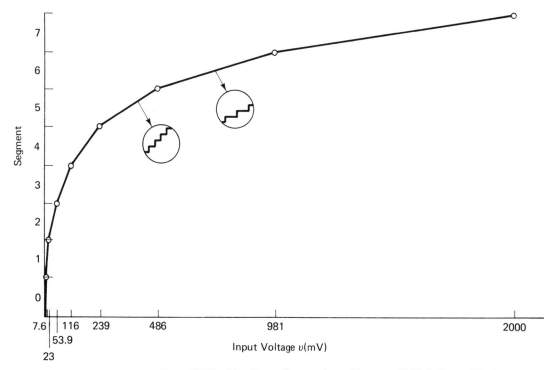

Figure 7-15 Non-Linear Companding of Lentcurt 9002 B-Channel Bank

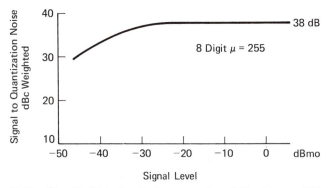

Figure 7-16 Signal-to-Distortion Performance of an 8-Bit with $\mu = 255$ Compression Characteristic

7-7   Encoders and Decoders

After filtering, time quantizing (sampling), and amplitude quantizing the analog signal, the result is encoded into a binary signal. (See the block diagram of a PCM terminal in Fig. 7-6.) Over the years, a large array of encoders have been invented, which for the purposes of this text, can be broken down into three main categories:

1. Parallel encoders, which generate a whole character at a time
2. Serial encoders, which generate a digit at a time
3. Hybrid encoders, which generate several digits at a time but not a whole character

Parallel Encoder

The parallel encoder has the advantage of being very fast and able to readily utilize any code. Its chief drawback is that for a large number of amplitude levels, it requires a large number of comparators and crosspoints. For n-bits resolution, it requires $2^n - 1$ comparators. A 4-bit converter, for instance, requires only 15 comparators, but an 8-bit converter requires 255.

A basic format of a parallel quantizer is shown in Fig. 7-17. Latch, D-type flip-flops store the binary outputs from the comparator. The comparator threshold points are spaced 1 LSB apart for the series resistance chain and voltage references. For an applied analog voltage, all comparators biased below the threshold turn ON and all those above it remain OFF. In addition, a serial register is added which is clocked at $n f_s$ to transfer the encoded signal to a serial train of two-level pulses.

Serial Encoder

One type of serial, or sequential encoder, uses a successive approximation technique which generates one bit at a time, with the most significant bit (MSB) determined first.

For example, take a 4-bit encoder with a range of 16. The MSB would represent one-half of the full range scale, or 8. The next MSB would represent one-fourth of the full scale range, or 4, and so on.

The successive approximation process is controlled by a programmed register as shown in Fig. 7-18.

The sample-and-hold circuit is inserted to hold the input signal constant while the encoding is in process. The analog input is initially compared with one-half of the full range scale by setting the MSB in the register to 1. If the analog input, X, is greater than this approximation, the first bit of the transmitted character is 1. This bit is also held in the programmed register and this weighting of 8 is held on the summing network.

At the beginning of the next digit period, the next weight of 4 is applied to the summing network. This causes the input to be compared to 8 + 4 or 12.

On the other hand, if the input is less than 8, the first character bit is 0. During the second digit period, the input would be compared to 4. This process

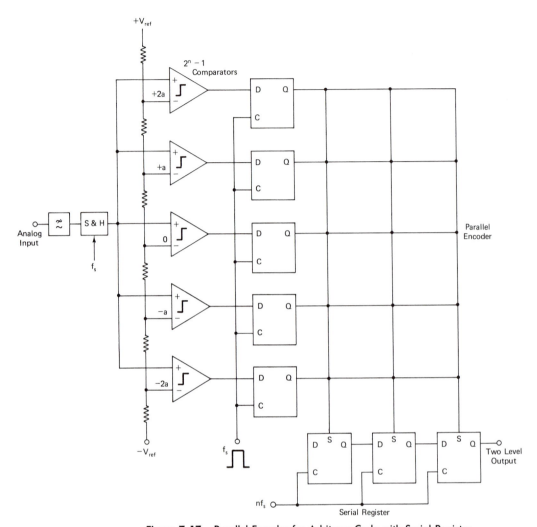

Figure 7-17 Parallel Encoder for Arbitrary Code with Serial Register

is continued until the least significant bit is determined. A typical cycle, with the analog voltage setting at 10½, is illustrated in Fig. 7-19.

Another type of serial encoder is one that scans the signal range one level at a time. The encoder shown in Fig. 7-20 employs an up/down counter which is clocked at a rate of $(2^n - 1)f_s$, where 2^n represents the number of levels for n bits. This assumes the counter resets at an extreme positive or negative level. The clock rate could be halved if the counter was reset at the midway level at the beginning of a cycle. A 1 must be subtracted from the total number of levels, 2^n, because the extreme positive (or negative) level is one of the 2^n levels. Therefore, there are only $2^n - 1$ steps above (or below) the extreme for the A/D converter. If, for instance, an 8-bit code represents an analog signal sampled at

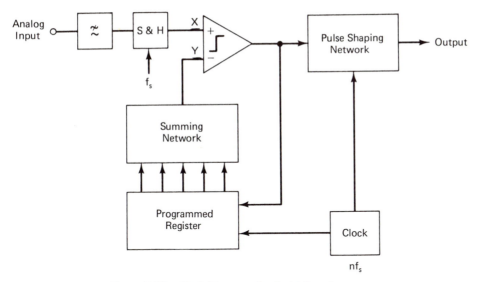

Figure 7-18   Block Diagram of a Serial Encoder

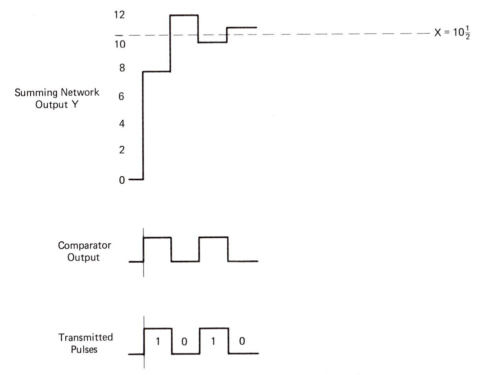

Figure 7-19   One Operational Cycle of a Serial Encoder

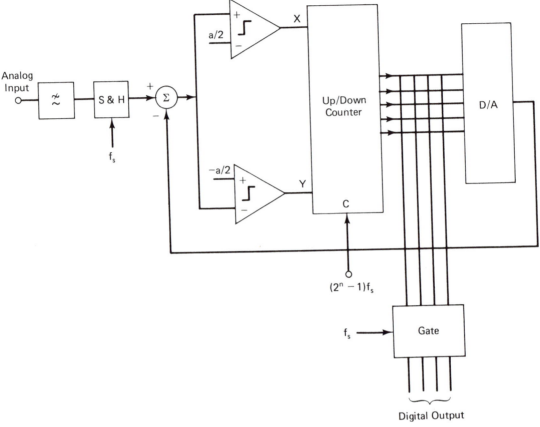

Figure 7-20 Serial Encoder Employing a Counting Technique

a 8 kHz rate, the counter must operate at a $(2^8 - 1)8k$ or 2.04 MHz rate. The output of the comparators, X and Y, controls the direction in which the counter must run or whether the counter output should remain stationary. If X = 1, the counter output should increase; if Y = 1, the counter output should decrease; and if both X and Y are 0, the counter output should remain fixed.

The digital output from the counter is applied to a D/A converter which in turn is applied to a difference amplifier at the encoder input. This negative feedback causes the output of the D/A converter to track the input signal and thus the counter output represents the encoded signal when a steady-state condition is achieved.

As the number of distinguishable levels increases, the clock rate of the up/down counter rapidly rises. This is one of the chief disadvantages of a serial encoder. On the other hand, very few digital chips are needed, which results in a very reasonably priced encoder-decoder. Non-uniform quantizing, or companded PCM, can be generated by changing the weight for each digit of the circuit shown in Fig. 7-18. The weight of each digit is a function of the current

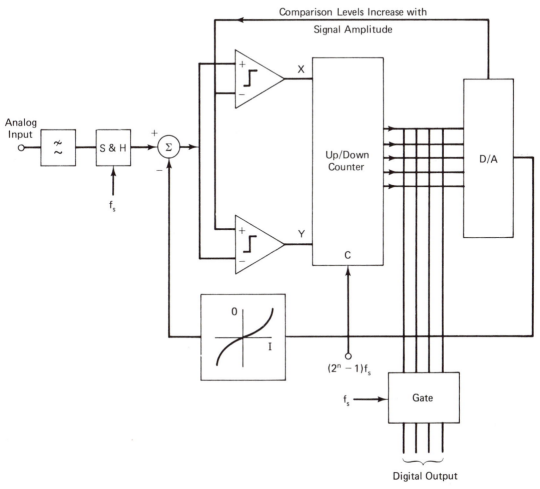

(a) Non-Uniform Serial Encoder

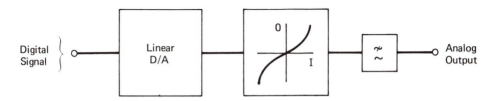

(b) Decoder for Non-Uniform Quantization

Figure 7-21 Block Diagram for the Encoding and Decoding of Companded PCM

binary decision together with the information stored about the previously determined digits. Figure 7-20 can be transformed into a non-uniform encoder by
inserting a non-linear transfer device after the D/A converter and forcing the
distance between adjacent levels to increase as the signal level increases. This is
illustrated in Fig. 7-21, along with a corresponding decoder circuit.

Hybrid Encoder

A hybrid encoder is a combination of the parallel and serial encoders
with expected accompanying compromises. It requires less equipment than a
parallel encoder and can operate at lower clock frequencies than the serial encoder. One technique it uses is the subranging converter shown in Fig. 7-22
whereby k bits are determined simultaneously and the entire n bits are obtained in n/k steps.

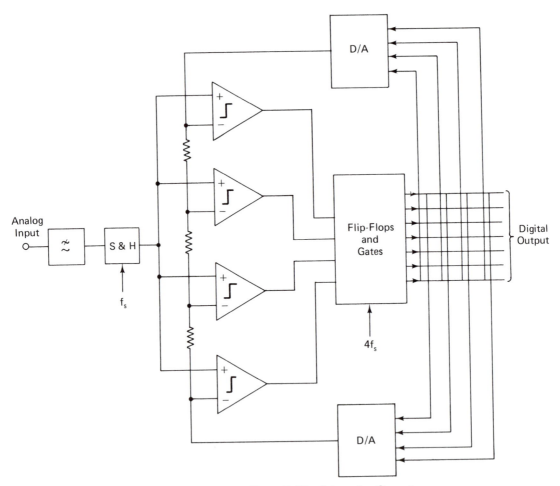

Figure 7-22 Subranging Converter

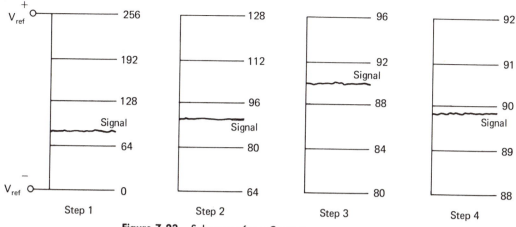

Figure 7-23   Subranges for a Converter

To illustrate this technique, consider a system which has four subranges per step as shown in Fig. 7-23. During the first step, the input voltage, which here is represented as a number equivalent to a level number, is compared with the various boundaries to determine between which boundaries it lies. The selected subrange is then divided into four more subranges and the process repeated. It takes 2 bits to determine one of four levels. After 4 steps, the 8 bits are determined in the case of a 256-level system.

The comparators are referenced at equally spaced intervals in the range between the two D/A converters. Initially, the lower D/A is set at zero and the upper D/A at the maximum of 256 V. The output of the comparators indicates in which subrange the signal lies; in this example, between 64 and 128 V. The reference voltage of 64 V is then applied to the lower D/A and 128 V are applied to the upper D/A. This process is repeated until the smallest range, in this case a range of 4 V, is attained.

Decoders

The device which converts a digital word into its analog counterpart is known as a digital-to-analog converter. D/A converters are also referred to as DAC's and decoders. For each input code word applied to a decoder, a discrete analog voltage is produced at the output.

The most popular decoder in use is the weighted current source D/A illustrated in Fig. 7-24. Each transistor current source has a different emitter resistor with binary related values of $R, 2R, 4R, \ldots 2^n R$, where n refers to the length of the code word. These current sources are switched ON or OFF by the binary signals applied to the control diodes connected to the emitters. If the input signal is low, no current flows. If the input is high, there is a current contribution towards the output summing line.

Assuming a 0.6 volt-drop from emitter to base, the total current contribution from all the transistor collectors is:

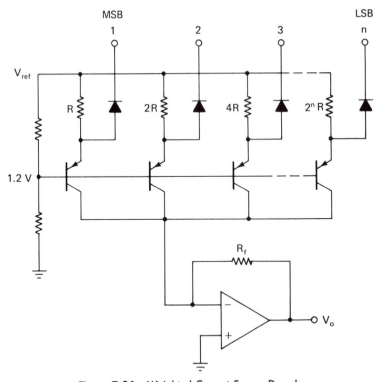

Figure 7-24 Weighted Current Source Decoder

$$I_o = \frac{V_{ref} - 1.8}{R} + \frac{V_{ref} - 1.8}{2R} + \frac{V_{ref} - 1.8}{4R} + \ldots \frac{V_{ref} - 1.8}{2^n R}$$

Since the output voltage from the ideal current to voltage amplifier is $-I_o R_f$, the resultant output voltage will be

$$V_{out} = \frac{-(V_{ref} - 1.8)}{R} R_f \left[1 + \frac{1}{2} + \frac{1}{4} + \ldots \frac{1}{2^n} \right] \qquad (7\text{-}9)$$

This can be expressed in the form of a digital word as

$$V_{out} = A \left[\frac{b_0}{2^0} + \frac{b_1}{2^1} + \frac{b_2}{2^2} + \frac{b_n}{2^n} \right] \qquad (7\text{-}10)$$

where   A is the constant $\dfrac{-(V_{ref} - 1.8)}{R} R_f$ and $b_1, b_2, \ldots b_n$

are the bit coefficients which are quantized to be either 1 or 0. From Eq. (7-9) we can see that the most significant bit (MSB) should be applied to Terminal 1, and the LSB applied to Terminal n.

One concern of this network is the accuracy required of the resistors, particularly for the higher resolutions. If, for example, a typical R value of 10k is used in an 8-bit converter, and R is out by 1% (9.9 kΩ), the MSB error contributions would be more than that contributed by the LSB. To overcome this difficulty, a R-2R ladder network such as that shown in Fig. 7-25 is employed. The inputs operate single-pole-single throw switches which tie the shunt resistors either to ground or to the current summing line. Such a network employs only a narrow range of resistances which need to be matched and have identical thermal tracking properties. The absolute values of the ladder resistances are not critical.

As can be observed, the input resistance at every junction is the same and equal to R. Also, the current splits into equal parts at every junction, giving the desired binary division of currents, or binary-weighted currents, to either the summing line or ground. The net output voltage is thus:

$$V_o = -I_oR_f$$
$$= -\left(\frac{V_{ref}}{2R} + \frac{V_{ref}}{4R} + \frac{V_{ref}}{8R} + \frac{V_{ref}}{2^nR}\right)R_f$$
$$= \frac{-V_{ref}R_f}{R}\left(\frac{1}{2} + \frac{1}{4} + \frac{1}{8} + \frac{1}{2^n}\right) \qquad (7\text{-}11)$$

As before, this circuit gives the proper weighting to each binary digit of the input code word

7-7.1   Codecs

Monolithic integrated circuit chips that convert voice are now being produced on a massive scale. These chips, which are called CODECS (Coder-Decoders), are becoming so inexpensive that they are being installed in the individual channel units rather than as a single encoder/decoder unit that is time shared

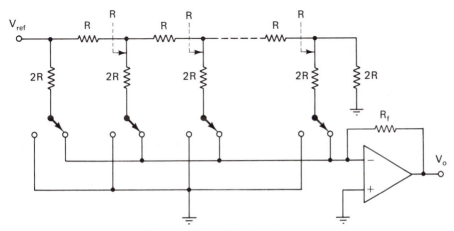

Figure 7-25   R-2R Ladder Decoder

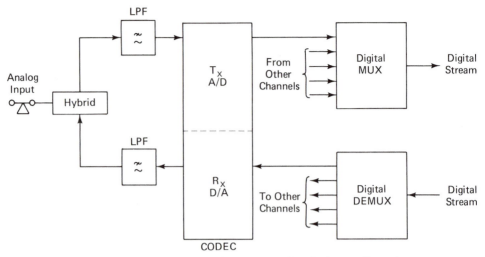

Figure 7-26 PCM System Using One Codec per Channel

by all the channels. Thus, as illustrated in Fig. 7-26, each channel contains a coder for sampling, quantizing, and encoding. The digital outputs are then passed to a digital multiplexer where they are combined to form a single serial bit stream for transmission. The inverse process occurs in the receiving section of the codec.

Single-channel codecs have the advantage of

1. Being less catastrophic than a time-shared system because the loss of a codec results in the loss of a single channel rather than all the channels in an entire channel bank, and
2. Employing digital multiplexing which tends to be simpler and easier to implement than analog multiplexing. By the proper combining of digital pulse streams, analog baseband signals of various bandwidths can be digitized and carried over the digital system.

An example of a codec is the INTEL2910A PCM CODEC. This chip contains sample-and-hold circuits, an A/D converter (successive approximation type), a D/A converter, as well as some other logic and interface circuitry which make it suitable for full-duplex PCM operation. The block diagram of this Codec is shown in Fig. 7-27.

The codec provides two major functions:

1. Encoding and decoding of analog signal (VOICE, dial tone, busy tone, ring-back tone)
2. Encoding and decoding of the signaling and supervision information (OFF or ON HOOK, rotary dial pulses, etc.)

On a non-signaling frame, the Codec encodes the incoming analog signal at the frame rate, FS_x, into an 8-bit PCM word which is sent out on the D_x lead

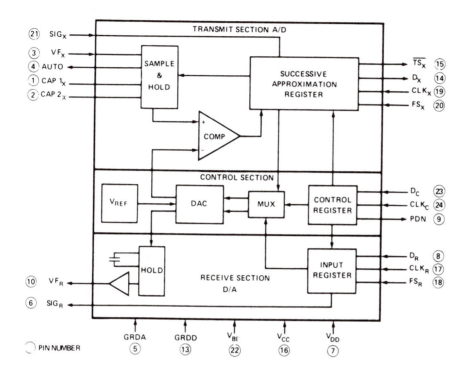

Figure 7-27 Block Diagram and Pin Names for 2910A Codec (Reprinted by permission INTEL Corporation, Copyright 1979. All mnemonics, Intel Corporation, copyright 1979).

at the proper time. A frame consists of a single sample of all channels, which will be further discussed in the next section. Similarly, the Codec fetches an 8-bit PCM word from the receive highway, D_R lead, and decodes an analog value which will remain constant on lead VF_R until the next receive frame. Transmit and receive frames are independent. They can be asynchronous (transmission) or synchronous (switching) with each other.

On a signaling frame, which occurs every sixth frame, the Codec transmit side will encode the incoming analog signal as previously described and substitute the signal present on lead SIG$_x$ for the least significant bit of the encoded PCM word. Similarly, on a receive signaling frame, the Codec will decode the seven most significant bits and will output the least significant bit value on the SIG$_R$ lead until the next signaling frame. Signaling frames on the send and receive sides are independent of each other, and are selected by a double-width frame sync pulse on the appropriate channel.

The 2910A Codec is intended to be used on line and trunk terminations. The call progress tones, which include the dial tone, busy tone, ring-back tone, and re-order tone, and the pre-recorded announcements, can be sent through the voice-path. Digital signaling, which includes off-hook and disconnect supervision, rotary dial pulses, and ring control, is sent through the signaling path.

Circuitry is provided within the Codec to internally define the transmit and receive timeslots. In small systems, this may eliminate the need for any external timeslot exchange. This feature can be bypassed and discrete timeslots sent to each Codec within a system.

In the power-down mode, most functions of the Codec are directly disabled to reduce power dissipation to a minimum. The Codec can be portioned into three major sections, as noted in Fig. 7-26: the middle or control section, the top transmit or A/D section, and the bottom receive or D/A section.

7-8 24-Channel System Frame Alignment

In a practical multi-channel PCM system, it is important for the receiving terminal to properly interpret which bits in the incoming bit stream belong to which channel, which bits are signaling bits, and which bits are synchronizing bits. The signaling bits relay signaling information such as ON HOOK or OFF HOOK condition, dialing, and BELL ringing. Since signaling is rather slow, it needs to be sampled less often than the speech information.

In the Lenkurt 9002B system, each of the voice frequency (VF) channels is sampled at 8000 times per second (8 kHz rate), and every sample is encoded into an 8-bit digital word. A frame consists of a single sample of all 24 channels, hence 24 digital words or 192 bits. Because frame synchronization is required to determine which bits make up Channel 1, 2, and so forth, a single synchronization bit or frame bit (s-bit) is added to give a total frame of 193 bits. This yields a total bit rate of 1.544b M/s (total bits/frame × sampling rate). In terms of timing, each channel is sampled at an 8 kHz rate, yielding 125 μs ($\frac{1}{8000}$) frame time or time needed to sample all 24 channels. Each frame consists of 193 bits (192 channel bits + 1 framing bit), which gives a time slot of 648 ns for each bit or 5.18 μs for each 8-bit channel word. Figure 7-28 indicates the relationship of the multiframe, the frame, channel words, and bit sequences along with the synchronized information bits.

Two signaling channels are made available by stealing the eighth bit from Frames 6 and 12. This causes about a 2 dB reduction in the theoretical

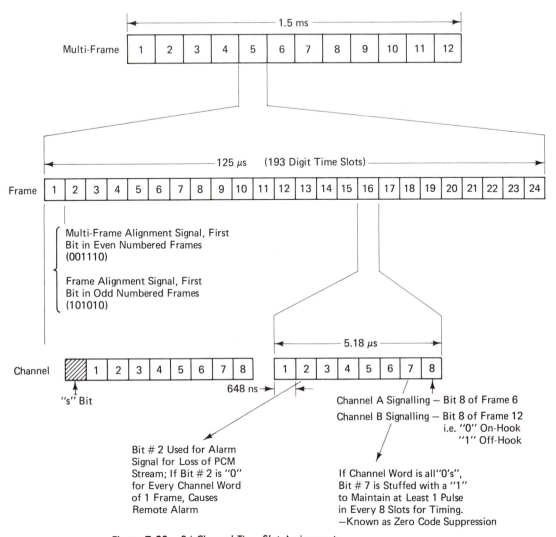

Figure 7-28 24-Channel Time Slot Assignment

S/N ratio. Each signaling channel thus transmits at a rate of 666 b/s (¹⁄₁.₅ ms), or one twelfth of the channel signal rate. This is quite adequate for monitoring the slower signaling information.

The following list gives the various time allotments in abbreviated form:

1 channel time slot = 8 digit timeslots

sampling rate of each channel timeslot = 8 kHz

$$1 \text{ frame time} = \frac{1}{8000} = 125 \ \mu s$$

1 frame = 24 × 8 × 1 = 193 timeslots

1 multiframe = 12 × 125 μs = 1.5 ms

$$1 \text{ digit timeslot} = \frac{125 \ \mu s}{193} = 648 \text{ ns}$$

channel A and channel B signaling rate
= 1 bit/multi-frame/channel
= 666 b/s/channel

(combined signaling rate = 2 × 666 = 1333 b/s/channel)

A study of Fig. 7-28 shows that the s-bit must provide a recognizable frame synchronizing pattern to indicate which is Channel 1 in the sequence of channels. Its second function is to indicate when a signaling frame (A signaling channel or B signaling channel) occurs and which one is occurring. This is done by the interleaving of two separate logic patterns to produce the unique framing sequence of s-bits; that is,

```
1   0   0   0   1  [1]  0   1   1   1   0  [0]  1   0   0   0   1  [1]
              Frame # 6              Frame # 12             Frame # 6
                s-Bit                  s-Bit                  s-Bit
```

In this sequence, every sixth frame bit is, alternately, 1 or 0. The A signal channel is enabled during the time slot of Bit 8 for every channel word occurring during a signaling frame (every sixth frame) which follows a frame s-bit of 1. Similarly, the B signal channel is enabled during the Bit 8 slot of every channel word during a signaling frame following a frame s-bit of 0. Hence, signaling information occurs only during the channel words of every sixth frame, or Frames 6, 12, 6, 12, and so forth. From the pattern of s-bits in a 12-frame sequence (called a superframe), a terminal framing pattern identifies the locations of the time slot of Bit 1 of Channel 1 in each frame, and a signaling frame pattern identifies Frames 6 and 12 in which signaling information occurs.

Terminal framing has an alternating 1 and 0 s-bit pattern (101010) in odd-numbered frames between Frame 1 and Frame 11. Signal framing has an s-bit pattern of 001110 in even-numbered frames between Frame 2 and Frame 12. The signaling s-bit changes 0 to 1 at Frame 6 and from 1 to 0 at Frame 12. Since one signaling bit occurs every sixth frame or 2 bits occur in 12 frames (1.5 ms), the total signaling bit rate is 2 b/1.5 ms or 1333 b/s. Thus, each of the two signaling channels, A and B, has a bit rate of 666 b/s.

System supervisory information regarding operational status is also determined by examining Bit 2 of every channel word in every frame. If this bit is 0 for every word of one frame, an alarm is given (*REMOTE* alarm) to indicate that, most likely, no PCM bit stream is present. Bit 2 is used because, under normal conditions, this bit will contain a 1 less than 10% of the time. Its absence becomes a good indicator of the absence of PCM bits.

To ensure that timing information necessary for the receiver is not lost by a long string of 0's representing a particular value, Bit 7 in every word can be used as a timing bit. The receiver timing depends upon detecting a 1 on the average in every 8-bit word, thus allowing a maximum of 15 consecutive 0's in any particular event. All bits in each word are monitored at the transmit side. If all the bits in a PCM word are 0, Bit 7 is forced to 1, which sidesteps the previously stated condition. Changing Bit 7 causes the least error in code value.

One of the practical problems associated with frame synchronization is how to regain framing synchronization within the system once it has been lost; that is after the system is turned on or if there is a break in the transmission path. Using the frame scheme shown in Fig. 7-28, once the receiver loses knowledge of the location of Bit 1 of Channel 1, it will remain lost because there is no inherent method to induce the bit stream of the frame to resynchronize. To solve this, an out-of-frame detector is used. This detector is made up of circuitry which examines every second frame (Frames 1, 3, 5, 7, and so on), hence every 386 bits (2 × 193 bits) are examined for the 010101 framing pattern. If a problem has occurred, either two 1's or two 0's will occur in the pattern: (0101001010) or (0101110101). The existence of either of these two situations causes an extra bit to be inserted in the frame after the location of the incorrect pattern. The detector examines the pattern until this condition is found again, and then adds another extra bit. This process continues until the correct pattern is regained. The receive pattern is internally generated in the receiver and is adjusted by the above process until it matches the incoming frame pattern. Thus, the system becomes synchronized. Synchronization status is indicated by the LOCAL alarm on the system.

7-9 PCM Terminal Equipment

The D3 PCM-channel bank can be applied to intertoll service as well as to direct and to toll connecting service, as illustrated in Fig. 7-29.

The Lenkurt 9002B PCM-channel bank is made up of 24 plug-in channel units, a common transmit unit, a common receive unit, and a common alarm unit. The basic block diagram outlining the relationship of these parts as shown in Fig. 7-30. Note that 24 analog signals are time division multiplexed, although only three channels are shown in the diagram.

Recall that any PCM system essentially must have the functions shown in Fig. 7-31.

The Lenkurt channel bank has all these blocks or functions buried in it. Referring again to Fig. 7-30, it can be seen that the handsets and signal source become the analog input signals into the system, usually with voice frequencies of 200 to 3400 Hz. Examination of the channel unit (CU) shows that it performs the basic function of band limiting the input signal through use of bandpass filter, and that it also performs sampling at a rate of 8 kHz, hence multiplexing.

The CU block is followed by the Common Transmit Unit (CTU), which performs quantizing and encoding functions. The resulting pulse stream is fed to the Common Alarm Unit (CAU) and then to the optional Office Repeater, if

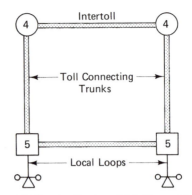

Figure 7-29   Application of D2 Channel Bank

it is present. Basically, the CAU performs the task of setting up standard pulse levels and of detecting basic system failures, which will be seen later. Signals are then passed to the transmission, or span line.

On the PCM pulse stream going into the channel bank, the Office Repeater, if used, regenerates the supposedly low amplitude pulses to make them acceptable to the decoder. Again, the alarm unit basically functions as a signal monitor for obvious system failures, such as loss of pulses. The PCM pulse stream is fed to the Common Receive Unit. At this point, the pulse stream is

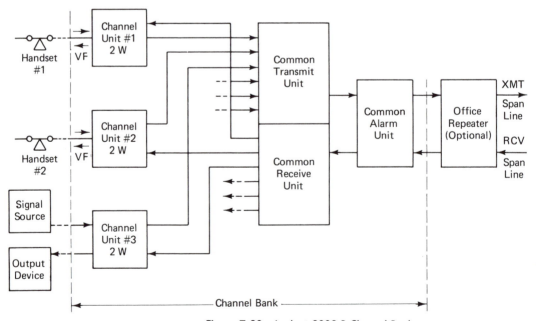

Figure 7-30   Lenkurt 9002 B-Channel Bank

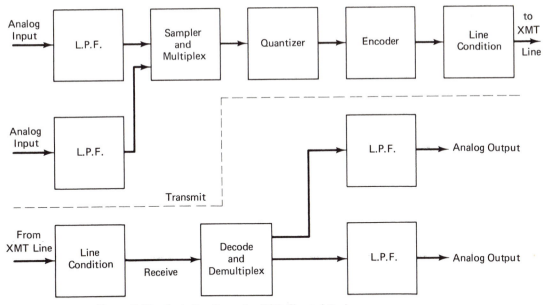

Figure 7-31 Basic Functions of a PCM-Channel Bank

examined to determine timing information regarding the location of each bit (bit timing) and the start of the frame (first channel in the possible set of 24). This information is then applied to the decoder with the PCM pulse stream to produce a quantized analog pulse stream. It should be a replica of that originally produced by the Common Transmit Unit. Hence, the Common Receive Unit functions as a decoder while also generating the appropriate timing information.

The quantized analog pulse stream is passed on the receive-side of the channel unit which demultiplexes, or separates out, the pulses needed to reconstruct its signal. Reconstruction is accomplished by low-pass filtering of the pulses to form a facsimile of the original analog signal applied to that channel. Although the system has been illustrated as a basic PCM system, many additional refinements, techniques, and methods are used to accomplish the needed functions of timing, synchronization of frames, transmission of additional information (such as signaling and supervisory), and detection of system faults.

7-10   Line Pulse Format

Many different types of line codes are used to represent the 1's and 0's of a digital signal on a transmission link. Four of the more common formats are illustrated in Fig. 7-32. Fig. 7-32(a) shows the simplest of the code types which uses a "unipolar" waveform, a term used because it is not symmetrical about 0 V. If the waveform is symmetrical about 0 V, the "polar" code of Fig. 7-32(b) results. Both of these codes are also called Non-Return-to-Zero (NRZ) codes.

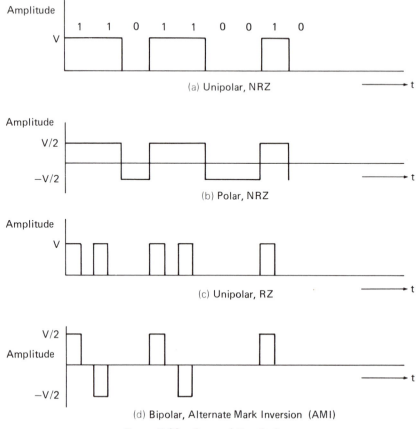

Figure 7-32 Types of Line Coding

The 9002 B system uses the Return-to-Zero (RZ) format of Fig. 7-32(c) with 50% duty cycle since it requires timing information on the receive end to ensure that each information bit is examined in its correct time interval. This allows the receiver to track any small changes in transmitter bit rate without causing errors. In the NRZ format, a long string of 1's would result in virtually no signal transition; hence, no timing information would be obtained. The RZ format ensures that a signal transition will occur regardless of the content of the information bits. This timing information is also necessary for operation of the regenerators, or repeaters, in the transmission line.

Because most carrier transmission systems do not pass dc, a line code should contain no dc. If a large dc component is present, as in the case of transmitting long sequences of 1's or 0's in polar or unipolar format, the waveform amplitude shifts with respect to zero volts, as shown in Fig. 7-33. When this happens, the normally fixed threshold used for deciding whether a "1" or a "0" is received is no longer optimal and a small amount of noise can cause an

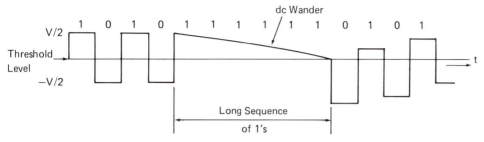

Figure 7-33   DC Wander

error. Special dc restoration circuits can be designed to eliminate the dc wander and restore proper reference levels, but in the 9002 B, the bipolar Alternate Mark Inversion (AMI) of Fig. 7-32(d) is used as the transmission format. The net dc content is zero as successive 1's produce alternate positive and negative level pulses. The removal of the dc component from the unipolar pulses used in the terminal logic circuits is accomplished in the 9002 B system by a bipolar converter in the common transmit unit. The conversion back to the unipolar pulse format from the bipolar line transmission format is performed by a converter in the common receive unit.

In addition to timing information, the bandwidth or frequency content of the pulse stream is important in transmitting pulses. Since the only requirement for PCM signal information recovery is detection of either a 1 or 0, the minimum required bandwidth of a PCM signal is the fundamental frequency component of the pulse stream. This is illustrated in Fig. 7-34(a). Usually the worst, or largest bandwidth, occurs when the string alternates between 1's and 0's, i.e. 101010. This will produce the highest fundamental frequency. For example, a pulse width, t_p, of 324 ns in a RZ-unipolar format has a fundamental frequency (bandwidth) of:

$$BW = \frac{1}{2t_p} = \frac{1}{2(324 \text{ ns})} = 1.544 \text{ MHz}$$

The AMI coding scheme reduces the bandwidth to one-half that of the unipolar situation, as can be observed from Fig. 7-34(b). For the same bit information rate, the fundamental frequency or required bandwidth is:

$$BW_{AMI} = \frac{1.544 \text{ MHz}}{2} = 772 \text{ kHz}$$

The penalty for AMI signal being less sensitive to noise and transparent to transformers and capacitors, due to the removal of any dc content, is increased system complexity to process this format. One additional advantage, however, is that detection of errors due to noise becomes possible. This is done by detecting bipolar violations, or the occurrence of two or more marks, 1's of the same polarity in a sequence.

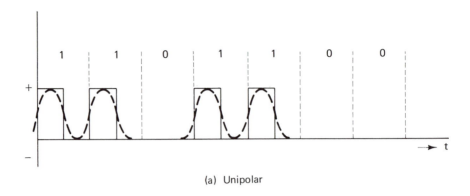

(a) Unipolar

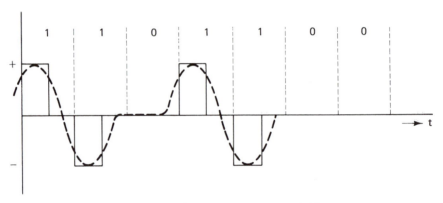

(b) (Bipolar Alternate Mark Inversion; AMI)

Figure 7-34 Line Code Formats

If a large string of zero's is transmitted with the AMI code, no pulses are transmitted and the receiver can lose its timing. To avoid this difficulty, the higher data rate systems substitute a special code pattern that contains mostly "1"s for a long string of "0"s. At the receiver this code pattern is recognized by detecting a bipolar variation and the inserted pattern is replaced with "0"s. These schemes are called Binary N-Zero Substitution Schemes or BNZS.

On the Bell System's T2 transmission line, the Binary 6-Zero Substitution (B6ZS) code is used. Whenever a string of six consecutive zeros is transmitted, a special code given in Table 7-3 is substituted. The code pattern depends upon the polarity of the "1" pulse immediately preceding the string of six zeros. In the substituted code, both the second and fifth substituted pulse cause a bipolar variation.

Table 7-3   B6ZS Substitution

Preceding Pulse Polarity	Substituted Code
$+$	**0 + $-$ 0 $-$ +**
$-$	**0 $-$ + 0 + $-$**

Here are some specific examples:

Binary Code   0 0 1 1 0 0 0 0 0 1 0 0 0 0 0 0 1 0

Substitution   0 0 $-$ + 0 + $-$ 0 $-$ + $-$ 0 $-$ + 0 + $-$ + 0

Binary Code   1 0 1 1 0 0 0 0 0 0 0 0 0 0 0 0 1 1

Substitution $-$ 0 + $-$ 0 $-$ + 0 + $-$ 0 $-$ + 0 + $-$ + $-$

In North America, the original Western Electric T1 carrier system contained the D1-PCM-channel bank. It was suitable for transmission on direct trunks between Class 5 end offices, and between Class 5 end offices and Class 4 offices (toll-connecting trunks). It was, however, of unsuitable transmission quality for intertoll networks. The D2-channel bank was designed for this, and it has recently been upgraded to form the D3-channel bank. The major characteristics of the PCM-channel banks are shown in Table 7-4.

7-11   North American Multiplex Hierarchy

Each type of signal source tends to have its own characteristic digit rate. In telephony, 64k b/s are required whereas in broadcast television about 60M b/s are required. The transmission media over which these signals must be transported also has its own characteristic digit rate. For a balanced two-wire line, 2M b/s is appropriate; a waveguide system may have a capacity in the region of 500M b/s.

Multiplexing is used to match such signals to a transmission path. A number of primary signal sources are combined to form a composite signal suitable for efficient use of the transmission media. By planning a suitable hierarchical multiplex structure, it is possible to combine various primary signals into identical formats that can be carried by the same transmission equipment. Also, signals carried by low capacity transmission systems can be combined to form the basic unit of a higher capacity system.

PCM Multiplex Hierarchies

The Bell System, generally used as a standard in North America, established a multiplexing structure for PCM carrier systems and designated it a T-carrier system. The framework of this carrier system is a multiplexing hierarchy consisting of several levels based on the bit rate of the multiplex line signals. The first level, referred to as T1, is the basic building block of this hierarchy.

Figure 7-35 illustrates the North American multiplex hierarchy. Multiplexer units with lower bit rates and fewer numbers of channels couple to

Table 7-4   Present PCM-Channel Bank Characteristics

Channel Bank Generation	Bell System	Lenkurt System	Characteristics
1st	D1	9001	• 24 channels VF • companding encoder, $\mu = 100$ • 7-bit voice endode (127 levels) • 1-bit signaling encode • pseudo random sampling sequence Time slot: 1 2 3 4 5 6 7 8 Channel unit: 13 1 17 5 21 4 16 7 • used with T1 repeatered line (1.544 Mb/s, RTZ, bipolar)
2nd	D2	9002A	• 24 channels VF • companding encoder, $\mu = 255$ • 8-bit voice encode, 83% of time • 1-bit signal encode, 17% of time • pseudo random sample sequence • uses T1 repeatered line
3rd	D3	9002B	• same as 2nd generation, except for straight sample sequence; e.g., Time Slot 1 is Channel 1.
4th	D4		• 48 channels VF • alternate bit interleaving (3.088 Mb/s plus control bits to give 3.152 Mb/s by multiplexing) • uses T1C repeatered line

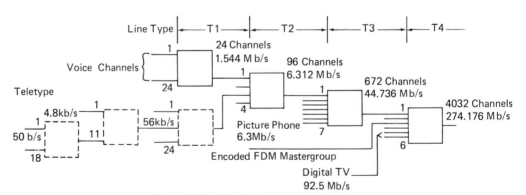

Figure 7-35   North American Digital Multiplex Hierarchy

Table 7-5 Types of Digital Carrier Lines

Line type	Line Signal Standard	Characteristics
T1	DS-1	1.544 Mb/s 24 TDM channels voice Bipolar format
T1C	DS-1C	3.152 Mb/s 2 time division multiplexed T1 lines 48 TDM channels voice Bipolar format
T2	DS-2	6.312 Mb/s 4 time division multiplexed T1 lines 96 TDM channels voice B6ZS format
T3	DS-3	44.736 Mb/s 28 time division multiplexed T1 lines 672 TDM channels voice B3ZS format
T4	DS-4	274.176 Mb/s 168 time division multiplexed T1 lines 4032 TDM channels voice Polar format
T5	DS-5	
T6	DS-6	

MUX's that have higher bit rates and greater numbers of channels. Included in the diagram are a few typical signals that are applied to the network. The analog signals must be digitized before entering the digital network.

Table 7-5 summarizes some of the major characteristics of the North American multiplex systems. It lists their bit rates, the type of line code and the number of multiplexed voice channels that can be carried. The DS-1 signal standard is implemented to handle the newer duobinary PCM system. The T6 level is designed particularly for fiber optic applications.

7-12   Measurement of Quantization Noise

To measure the quantization $(S + N)/N$ ratio, a white noise signal is passed through a bandpass filter that acts as the baseband signal to the PCM channel. The white noise generator and filter are preferred over a sine wave generator since all quantum levels are produced in a random fashion, which closely simulates the normal human voice. After passing through the PCM channel, the re-

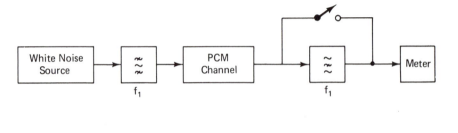

Figure 7-36 Quantization Noise Measurement System

ceived signal and any associated noise are detected with a meter. A filter is then inserted (see Fig. 7-36) to receive the original baseband signal, which leaves only the noise. The ratio between the two measurements, usually expressed in decibels, is the signal-to-noise ratio (actually (S + N)/N ratio).

7-13   Introduction to Information Theory

The basic unit used when discussing information is the bit, an abbreviation of binary digit. It represents the speculation of one of two equally likely alternatives, much like the two sides of a coin in a coin-flip experiment. If one has four equally likely alternatives, as for example four voltage levels, two bits would be required to determine the unique level. The four levels can be represented by the unique pairs 00, 01, 10, and 11.

If the number of levels are doubled to eight, three bits of information would be required. Similarly, sixteen levels would require four bits to determine the unique level. Extrapolating, the information content, H, is found to be:

$$H = \log_2 (\text{number of alternatives}) \qquad (7\text{-}12)$$

Applying Eq. ($7\text{-}12$) to a signal which can assume one of m equally likely voltage levels would result in $\log_2 m$ bits of information. The opposite is also true. If there are n bits of information 2^n different levels can be distinguished. Thus, with eight bits of information, $2^8 = 256$ different voltage levels can be distinguished.

Consider a message with a duration of T seconds over which there are n samples, each sample of which can assume one of m levels. This is illustrated in Fig. 7-37.

In obtaining the total number of possible signal combinations, it can be observed that for the first sample, there are m possible signal combinations; for

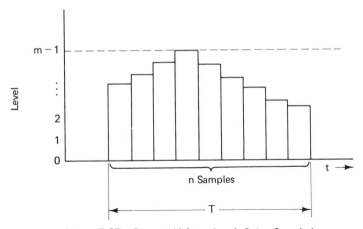

Figure 7-37 Discrete Voltage Levels Being Sampled

the second sample, there are $m(\cdot m) = m^2$ possible signal combinations; and for the n sample, there are m^n possible combinations. The information content of the message over a time T is

$$H = \log_2 m^n = n \log_2 m \qquad (7\text{-}13)$$

This equation indicates that the information content varies directly with the message length, n, as would be expected.

As an exercise, determine the number of bits required to represent the English alphabet, assuming an equal probability occurrence of each letter. If $n = 1$, as shown in Fig. 7-38, the bits needed would be $H = \log_2 26 = 4.7$ bits. Since fractional bits cannot be used, go to the next highest integer, which is 5. This would leave $32 - 25 = 7$ levels for other characters.

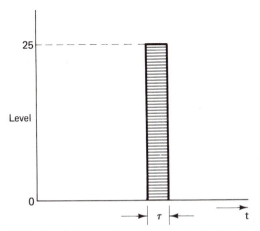

Figure 7-38 Levels Reprinted to Represent the English Alphabet

Increasing the number of levels will increase the information transmitted. However, increasing the number of levels will also cause a decrease in the spacing between them which makes the signal more susceptible to noise and therefore to errors. For this reason, a two-level system is most commonly used. Chapter 9 will consider how the information capacity, or information rate, is related to the bandwidth of the transmission medium.

Coding

A single, multilevel pulse can be broken up into several pulses while keeping the same number of information combinations. Consider a single pulse, 32-level system, which represents an information content of

$$H = \log_2 m^n = \log_2 32^1 = 6 \text{ bits}$$

If this is to be represented by binary pulses, it will require:

$$6 = \log_2 2^n = n \log_2 2 = n$$

$$n = 6 \text{ pulses}$$

If it is to be represented by a ternary system (3 levels) it will require:

$$6 = \log_2 3^n = n \log_2 3 = 1.583 \, n$$

$$n = \frac{6}{1.583} = 3.79 \text{ or 4 pulses}$$

The latter would actually permit 3^4, or 81 combinations.

Refer back to Eq. *(7-13)* to derive an expression for the information rate, or information capacity, of the signal. Since H represents the information content over a time T, the information capacity will be

$$C = \frac{n}{T} \log_2 m \text{ bits} \qquad (7\text{-}14)$$

From this result, it can be seen that the information capacity will be doubled if the number of samples per second are doubled. Doubling the number of distinguishable levels only causes a logarithmic increase in signal capacity. The number of distinguishable levels depends upon the noise power (N) and signal power (S). The greater the S/N ratio, the larger the number of levels that can be resolved.

The number of pulses that can be transmitted per second can be expected to relate linearly to the bandwidth. Chapter 9 will show that for an ideal low-pass transmission path of *bandwidth B, a maximum of 2B b/s can be transmitted* in a two-level system. If Eq. *(7-14)* is simplified to a two-level system where $m = 2$; $C = 2B = n/T \log_2 2 = n/T$, or $n/T = 2B$, then

$$C = 2B \log_2(m) \qquad (7\text{-}15)$$

From this expression we can see that for a given information capacity, number of signal levels (m) can be exchanged for bandwidth.

Example 7-1

In a channel bank the maximum signal-to-noise power ratio is 38 dB. Assuming a fixed noise level and a nominal 4 kHz voice channel, what is the information capacity of one voice channel?

Solution

$$dB = 10 \log \frac{S}{N}$$

where   S and N are the signal and noise powers respectively.

$$dB = 20 \log \frac{V_S}{V_N}$$

and V_S and V_N are the signal and noise voltages respectively.

$$\text{For} \quad 38 \text{ dB} \quad \frac{V_S}{V_M} = \text{antilog} \frac{38}{20} = 79.4$$

Since the noise is approximately ⅟₈₀th of the signal amplitude, 80 levels are distinguishable. The information capacity will be: $C = 2B \log_2 m = 2 \times 4k \log_2 80$ = 50k b/s.

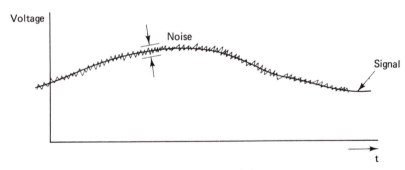

Figure 7-39   Signal with Noise

Example 7-2

A color display contains a grid of 216 × 560 pixels. Each pixel is capable of displaying red, green or blue with 16 levels of intensity for each color. What is the information capacity of the system if the display is refreshed at a rate of 60 Hz?

Solution

Number of alternatives per pixel = 3 × 16 = 48. This represents $\log_2 48$ or 6 bits. Each screen therefore requires

$$215 \times 560 \times 6 = 725{,}760 \text{ bits of information}$$

The information capacity of the system thus is:

$$C = 60 \times 725{,}760 = 43.5456 \text{ M b/s.}$$

Example	In a full color graphics system, each pixel requires 16 bits. If a screen requires

Example 7-3

In a full color graphics system, each pixel requires 16 bits. If a screen requires 1280 × 1024 pixels, how long will it take to paint a screen if a 1200 b/s modem is used to send the character files?

Solution

As each pixel requires 16 bits, a screen of 1280 × 1024 = 1,310,720 pixels requires 20.97152 M bits. Under this circumstance, it would take:

$$20.97152 \times 10^6/(1200 \times 60) = 291 \text{ min or } 4.85 \text{ hours}$$

to paint a full screen of characters.

Problems

1. a. Describe the operation of a simple delta modulator.
   b. Describe slope overload.
   c. Derive Eq. (*7-1*).

2. a. A 1 volt peak audio signal with a maximum frequency component of 3.3 kHz is to be transmitted by a simple delta modulator with a step size "a" of 1 mV. What is the minimum sampling frequency required to assure that slope overload does not occur?
   b. What can be done to decrease the sampling frequency of (a)? What effects would this have on the signal?
   c. What is the chief disadvantage of DM?

3. a. What is companded delta modulation, and why is it employed?
   b. Describe the cause of quantization noise.
   c. What precautions must be followed to prevent a telephone cable from acting as a 4 kHz low-pass filter?

4. Sony manufactures an audio digitizer which samples a signal at a rate of 44 kHz. On a video cassette recorder, it stores 32 bits for each sample taken.
   a. What is the storage rate required of this system
   b. How many bits must be stored for a half-hour program?
   c. What is the advantage of recording digitally?
   d. Why is a *video* cassette required?

5. a. What is a commutator? Illustrate how it can be used to sample a signal.
   b. Sketch a channel-interleaving scheme for Time Division Multiplexing the following channels: Five 4 kHz telephone channels, and one 20 kHz music channel. Note the minimum times required between the samples for the two types of channels.
   c. If 128 distinguishable levels are to be obtained for each signal, what channel capacity is required?
   d. Estimate the minimum system bandwidth required if the minimum bandwidth is half the channel capacity.

6. A telephone T1 carrier consists of 24 voice channels multiplex and PCM modulated. Each signal has a bandwidth of 300 Hz to 3.2 kHz. The encoder digitally encodes to 8 bits, and the rate of sampling meets the CCITT standard.
   a. Determine the bit rate of the T1 carrier.
   b. What baseband bandwidth is required to transmit this signal if the minimum bandwidth is half the channel capacity.
   c. Why does each voice channel require bandwidth limiting?
   d. Give three advantages of PCM over the present analog carrier system.
   e. If high fidelity music is to be encoded over each of the 24 channels, what changes would need to be made in the system? Draw a block diagram of a suitable transmitting section, clearly marking the signal frequencies and/or bit rate at each input/output.

7. a. Show that the rms quantization noise of the quantizing error voltage shown in Fig. 7-7(c) is $a/2\sqrt{3}$. For ease of calculation, assume some time periodicity T for the triangular waveform.
   b. For an n digit code, the number of available quantized signal levels will be 2^n. What is the rms signal amplitude available if the step size is "a"?
   c. Obtain the expression for the rms signal/rms, or quantization to noise ratio. Derive (Eq. 7-2) by converting the expression obtained to decibels.

8. What is the rms S/N or signal to quantization noise ratio for a
   a. 7-bit binary code
   b. 8-bit binary code
   c. 16-bit binary code

9. a. Define aperture time. What effect does a sample hold circuit have on the aperture time when placed ahead of an A/D converter?
   b. A 15 kHz sinusoidal signal is to be digitized to a resolution of 12 bits. What aperture time must be used to give less than 1 bit of error?
   c. A 10 V peak-to-peak signal that changes at a maximum rate of 0.1 V/ms is to be digitized to a resolution of 8 bits. What aperture time must be used to give less than 1 bit of error?

10. a. What is companding, and why is it employed in voice communications?
    b. Why is companding not used when transmitting data?
    c. What type of mathematical characteristics does a compander generally have?

11. a. Why is a code that has a maximum number of 1's near the zero voltage level used in D_2/D_3 channel banks?
    b. Illustrate why 13 bits are required to represent a linear digital version of the 15 segment mu-law curve that is illustrated in Figs. 7-12 and 7-13.

12. Find the absolute voltage, signs, segment, step, and level for the code word 10100101 on the 9002B Lenkurt channel bank.

13. Plot the mu law for $\mu = 255$, and the A law for $A = 87.6$. Use linear graph paper and assume maximum input and output signal voltages of 1.

14. State the advantages and disadvantages of the (*i*) parallel, (*ii*) serial, and (*iii*) hybrid encoders.

15. In order to distinguish 256 levels in a voice channel, determine the minimal frequencies of the various circuit components of the encoded listed in the following table.

	Parallel Fig. 7-17	Serial Fig. 7-18	Non-Uniform Serial Fig. 7-21(a)	Hybird Fig. 7-22
f_s				
Serial of Programmed Register Clock Frequency				
Up/Down Counter				
Flip-Flops and Gates				

16. a. Describe the basic theory behind D/A converters.
    b. What is the advantage of the R-2R ladder decoder over the weighted current source decoder of Fig. 7-24?

17. a. What is a codec?
    b. State two advantages of a single channel codec versus one codec shared by all channels.

18. Describe what is meant by the following terms:
    a. frame
    b. multiframe
    c. signaling
    d. channel

19. a. Obtain the framing sequence (*s*) bits from Fig. 7-28.
    b. Dialing pulses from a telephone handset occur at a nominal rate of 10 pulses/s. If the bandwidth of these square wave pulses is 40 Hz, what bit rate do these pulses represent? Does this put a severe information load on the signaling channel?
    c. Describe how framing synchronization is obtained on the Lenkurt 9002B system.

20. a. Describe the formats of the NRTZ and RTZ signals. What is the advantage of the latter format?

   b. What is the AMI code? Why is it employed on a line rather than the RTZ operation?

   c. Describe the North American digital multiplex hierarchy.

21. A message is sent using five pulses. The first pulse is restricted to two amplitude levels while the remaining pulses can assume any integer voltage level between −4 and +4 (inclusive).

   a. How many different messages can be sent using a single group of five pulses?

   b. If the message is transmitted as a binary sequence, what is the minimum information rate and transmission bandwidth (assuming no redundancy is added to the message) if the maximum transmission time is T seconds?

22. a. Determine the maximum information capacity of a 2-level, 3 kHz bandwidth system.

   b. If the number of levels is increased to three, what bandwidth would be required for the same information capacity?

   c. Which of the two systems, (a) or (b), is most immune to noise?

   d. How long would it take either of these systems to transmit a 10,000 bit message?

23. What is the information capacity of a 3 kHz bandwidth channel which has a 34 dB signal-to-noise ratio?

24. a. How many bits are needed to represent a single letter in the English language which consists of a total of 26 letters. Assume equal likelihood of occurrence for each letter.

   b. How many levels must each of two pulses have in order to obtain the minimal number of combinations that are suggested in (a).

   c. How many levels must each of three pulses have in order to obtain the minimal number of combinations that are suggested in (b).

25. A shaft position encoder is to have a resolution of at least 1°. If it can rotate at a rate of 1200 rpm, what bit rate must be transmitted if it is digitally encoded? Why is the gray code the most suitable for such an application?

26. A noise signal ranging up to 2 V peak-to-peak in amplitude and containing significant components up to a frequency of 15 kHz, is to be transmitted using PCM. If changes in amplitude of 1/64 V are to be detected, what must the information capacity of the channel be?

27. A black and white display is made up of 80 columns and 24 rows of characters. Each character takes up a 5 × 7 dot matrix and is surrounded with a frame of 1 dot on each side to separate the individual characters (see Fig. P7-27). Thus each character effectively takes up a 7 × 9 matrix. What is the information capacity of the system if the display is refreshed at a rate of 60 Hz?

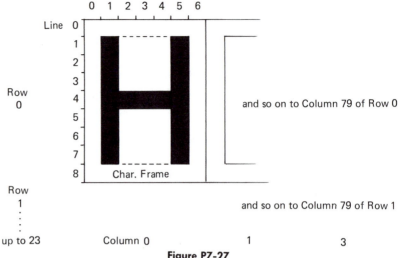

Figure P7-27

28. Complete the B6ZS code pattern for the following binary code word.

0 1 0 0 1 0 0 0 0 0 0 0 0 0 0 0 0 1 1

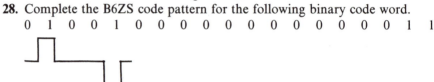

29. In a PCM system the quantizing noise can be reduced by
    a. Decreasing the number of levels
    b. Reducing sampling frequency
    c. Increasing the number of levels
    d. Increasing clock frequency

30. The primary function of a PCM compandor is to
    a. Increase bit rate
    b. Improve S/N ratio
    c. Decrease system bandwidth
    d. Decrease equipment cost

31. The primary function of a PCM CODEC is
    a. Amplitude sampling
    b. S/N enhancement
    c. Clock regeneration
    d. A/D, D/A conversion

32. To increase channel capacity of a PCM system one of the following techniques can be applied:
    a. FDM
    b. SSB
    c. TDM
    d. PPM

33. Natural sampling techniques are not well suited for PCM systems because of
    a. Poor noise characteristics
    b. A/D converters require that input be stable during conversion time
    c. High bandwidth required
    d. Expensive, high quality components are required

34. A PCM system operating at 8 bits/sample (4 kHz baseband) requires a transmission bandwidth of
    a. 16 kHz
    b. 8 kHz
    c. 32 kHz
    d. 64 kHz

35. How many bits of memory are required to store 30 minutes of an LP record? HiFi baseband = 20 kHz, sampling frequency of 40 kHz, each sample = 16 bits.
    a. 10512000 bits
    b. 20152000 bits
    c. 1002000000 bits
    d. 1152000000 bits

36. Slope overload is caused by
    a. Excessive input amplitude
    b. Slow clock rate
    c. Poor comparator performance
    d. The modulating signal changing more rapidly than the encoder can follow

37 Modem A/D converters range up to about 16 bits at conversion rates up to several hundred kHz. What dynamic range in dB does this represent?

8

Effects of Noise and Other Distortions on Digital Transmission

Noise and distortion are always present in any communication system. All extraneous signals appearing at the channel output that are not due to the input signal are considered noise. This noise is due to such factors as thermal agitation, radio frequency interference, crosstalk, and erratic switch contacts. Noise can be broken down into the following categories:

- Quantization
- Thermal
- Transient
- Phase jitter

Keep in mind that the redundancy of human speech and the adaptability of the human ear make noise and distortion less of a problem in voice communication than in data transmission. If a noise burst makes a word indistinguishable, the listener can guess what the word should have been based on the context of the conversation. Machines are unable to do this; therefore, tight controls must be kept on noise levels to minimize errors. When transmitting data, few errors are generally permitted.

8-2   Quantization Noise

Quantization noise is the result of the difference between the signal presented to a Codec and its equivalent quantized value. As suggested by Fig. 7-7, the root-mean-square (rms) quantization noise for uniform quantizing can be obtained by finding the rms voltage of a triangular waveform with a peak-to-peak value of step size "a." Figure 8-1 illustrates the triangular waveform of the quantizing error, assuming a period T.

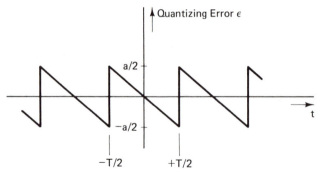

Figure 8-1   Quantization Noise Waveform

The rms quantization noise is thus equal to:

$$\epsilon_{rms} = \sqrt{\frac{1}{T}\int_{-T/2}^{T/2}\left(-\frac{at}{T}\right)^2 dt}$$

$$= \frac{a}{2\sqrt{3}} \qquad (8\text{-}1)$$

Assume that there are M available quantized levels for a single pulse. Thus, the peak signal-to-rms quantization noise will be

$$\frac{V_S}{V_N} = \frac{Ma}{a/2\sqrt{3}} = 2\sqrt{3}\,M \qquad (8\text{-}2)$$

Expressed as a power ratio, this is:

$$\frac{S}{N} = (2\sqrt{3}\,M)^2 = 12\,M^2 \qquad (8\text{-}3)$$

or, in dB

$$\left(\frac{S}{N}\right)_{dB} = 10\log 12\,M^2$$

$$= 10\log 12 + 20\log M$$

$$= 10.8 + 20\log M$$

If the single pulse is represented by a group of n pulses, each having m levels, and if M is replaced by m^n, the following is obtained:

$$\left(\frac{S}{N}\right)_{dB} = 10.8 + 20\log m^n$$

$$= 10.8 + 20\,n\log m \qquad (8\text{-}4)$$

If a sinusoidal signal is assumed, its rms equivalent would be $Ma/2\sqrt{2}$ as Ma represents the peak-to-peak value. Inserting this expression for V_s in equation

(*8-2*) and following the same procedure, we obtain for the rms signal-to-rms quantization noise:

$$\left(\frac{S}{N}\right)_{dB} = 1.8 + 20\,n\,\log m$$

For the binary code $m = 2$, Eq. 7-2 is obtained.

For practical situations, the actual average power is always less than the peak power assumed in Eqs. (*8-3*) and (*8-4*). Thus, an expression developed for the average power would be helpful. As an example, consider a simple case in which each level is equiprobable. For an m-level system, the probability that any particular level is being considered is $1/m$.

For illustration purposes, picture a cubic die which has 6 faces. Since each face has an equal probability of being thrown, each face has a probability of 1 out of 6 or ⅙ of being thrown. If, for instance, a voltage is always at a level a, the average power would be a^2. If, however, the voltage was at that level only ¼ of the time on the average, the average power would be $a^2/4$.

Consider the average power in a single-polarity pulse stream, positive only or negative only, as shown in Fig. 8-2(a). Consider also the average power in a bipolar pulse stream as shown in Fig. 8-2(b), each having four equally likely levels, and level spacings of "a." For the case of the single, four-level pulse stream, the average power is given by

$$S = \frac{1}{4}\,0^2 + \frac{1}{4}\,a^2 + \frac{1}{4}\,(2a)^2\ \frac{1}{4}\,(3a)^2$$
$$= 3.5a^2 \text{ watts} \tag{8-5}$$

For the four-level bipolar pulse stream, the average power is given by

$$S = \frac{1}{4}\left(\frac{-3a}{2}\right)^2 + \frac{1}{4}\left(\frac{-a}{2}\right)^2 + \frac{1}{4}\left(\frac{a}{2}\right)^2 + \frac{1}{4}\left(\frac{3a}{2}\right)^2$$
$$= 1.25\,a^2 \text{ watts} \tag{8-6}$$

As can be seen, the bipolar pulse stream transmits a message with substantially less power than the single polarity pulse stream. Because of this advantage and

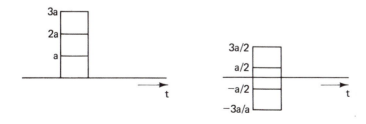

(a) Single Polarity Pulse   (b) Bipolar Pulse

Figure 8-2   Four-Level Pulse Streams

the fact that bipolar pulse streams can be designed without dc components, this examination will concentrate on the bipolar case only.

In general, for the bipolar pulse stream of M levels shown in Fig. 8-3, the average power will be

$$S = \frac{2}{M}\left[\left(\frac{a}{2}\right)^2 + \left(\frac{3}{2}a\right)^2 + \dots\left(\frac{M-1}{2}a\right)^2\right]$$

$$= \frac{a^2}{2M}[1^2 + 3^2 + \dots (M-1)^2]^2$$

The series in the brackets can be expressed by

$$\frac{M/2\ (M-1)\ (M+1)}{3}$$

Thus
$$S = \frac{a^2}{12}\ (M^2 - 1) \tag{8-7}$$

The average signal-to-quantization noise power is thus reduced to

$$\frac{S}{N} = \frac{\dfrac{a^2}{12}\ (M^2 - 1)}{a^2/12} = M^2 - 1$$

or in dB

$$\left(\frac{S}{N}\right)_{\mathbf{dB}} = 10 \log\ (M^2 - 1) \tag{8-8}$$

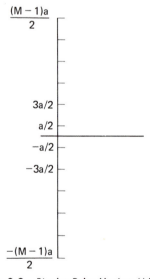

Figure 8-3 Bipolar Pulse Having M Levels

If the single pulse is represented by a group of n pulses, each having m levels, and M is replaced by m^n, the following is obtained:

$$\left(\frac{S}{N}\right)_{dB} = 10 \log (m^{2n} - 1) \qquad (8\text{-}9)$$

For a binary code where $m = 2$, the following is obtained:

$$\left(\frac{S}{N}\right)_{dB} = 10 \log (2^{2n} - 1) \qquad (8\text{-}10)$$

which for 8 bits reduces to 48.2 dB. This represents about a 10 dB reduction in S/N ratio when compared to the peak signal-to-quantization noise power under the same conditions.

8-3   Effects of Gaussian-Type Noise on Digital Transmission

Noise entering or contributed by a system results in errors in the detected signal. The error rate is dependent upon both the signal and noise amplitudes, as well as upon the characteristics of the noise. Noise containing large amplitude spikes causes more errors in PAM systems than, for instance, thermal noise containing the same average power.

For many applications, noise is considered Gaussian, even though in practice this is only an approximate assumption. Noise is Gaussian if it can be considered random and the result of many independent overlapping current or voltage pulses, such as the random motion of free electrons making up thermal noise in a resistor.

Gaussian noise has a probability density function which follows the familiar bell curve illustrated in Fig. 8-4. The larger amplitudes have less likelihood of occurrence than the lower ones. For example, voltage near zero volts has a greater probability of occurrence than the larger positive or negative voltages. When analyzing probable signals, the terms *variance* and *standard deviation* are frequently employed. The standard deviation, σ, is a measure of the spread of observed values. In electronics, it is called the rms voltage of the ac

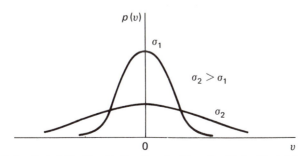

Figure 8-4   Probability Density Function for Gaussian Noise Containing No dc

portion of the signal, or the rms signal voltage with the dc removed. The square of the deviation, that is, σ^2, is called the variance. The larger the variance, the larger the spread, and the more squat the density function, as shown in Fig. 8-4. These curves can be expressed analytically as

$$p(v) = \frac{1}{\sqrt{2\pi}\,\sigma}\,e^{-v^2/2\sigma^2} \qquad (8\text{-}11)$$

where $p(v)$ is the probability density of the voltage v which, when integrated over an interval, gives the probability or the percentage chance that the voltage exists in the interval.

$$\sigma \text{ is the standard deviation}$$
$$\sigma^2 \text{ is the variance}$$

To find the probability, for instance, that the voltage is greater than A volts, the shaded area of Fig. 8-5 must be obtained. Obviously, the area under the entire curve must be unity (1) since the probability that the voltage lies somewhere between $-\infty$ and $+\infty$ is 1, or 100%.

Since it is difficult to integrate Eq. (8-11), tables and graphs have been developed to aid in determining the areas under the Gaussian density function. To analytically determine the probability that the voltage lies between 0 and $k\sigma$ volts, where k is some constant, the following integral must be solved.

$$prob\,(0 < v \le k\sigma) = \frac{1}{\sqrt{2\pi}\,\sigma}\int_0^{k\sigma} e^{-v^2/2\sigma^2}\,dv \qquad (8\text{-}12)$$

This equation integrates Eq. (8-11) over the interval of interest.

When using a graphical or tabular technique, the function $\Phi(k)$ is generally used which represents the white area shown in Fig. 8-6 and is defined as:

$$\Phi(k) = \frac{1}{\sqrt{2\pi}}\int_0^k e^{-u^2/2}\,du \qquad (8\text{-}13)$$

Of special interest in this case is the tail probability, which represents the probability that the noise voltage exceeds $k\sigma$ volts. Since the total area under

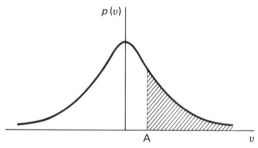

Figure 8-5   Shaded Area Indicating the Probability that $v > A$.

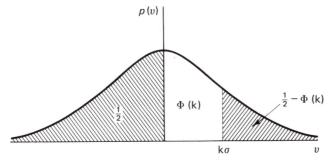

Figure 8-6 Gaussian Density Function

the density curve is unity, and symmetry indicates that the area under either side is one-half, the tail probability must be given by:

$$prob\ (v > k\sigma) = 1 - \frac{1}{2} - \Phi(k)$$

$$= \frac{1}{2} - \Phi(k) \qquad (8\text{-}14)$$

Figure 8-7 gives a plot of $\frac{1}{2} - \Phi(k)$. For $k \gg 1$, this can be approximated by

$$\frac{1}{2} - \Phi(k) \simeq \frac{1}{\sqrt{2\pi}\,k}\,e^{-k^2/2} \qquad (8\text{-}15)$$

What is the probability that a 0.5 volt rms Gaussian noise signal exceeds a magnitude of 1 volt?

The standard deviation is given as 0.5 volts. In this case, the area in the two tails shown in Fig. 8-8 is to be found.

$$\text{Since} \quad k\sigma = 1$$

$$k = \frac{1}{\sigma} = \frac{1}{0.5} = 2$$

Example 8-1

Solution

For $k = 2$, the area under each tail from Fig. 8-7 is $2.3 \times 10^{-2} = 0.023$. Therefore, the probability that the voltage exceeds 1 volt is $2 \times 0.023 = 0.046$ or 4.6%.

Consider the effect of Gaussian noise on a bipolar, or two-level, signal with permissible levels of $A/2$ or $-A/2$ as shown in Fig. 8-9. The decoder assumes that a positive pulse is present if the instantaneous signal exceeds 0 volts and that a negative pulse is present if the signal drops below 0 volts. Assume that either level is equiprobable; that is, each level has a 0.5 probability.

The decoder in Fig. 8-9 responds to the amplitude of the signal-plus-noise at the mid-pulse periods. For the pulse train shown, two errors occur.

To determine the error rate of such a pulse system, assume that Gaussian noise will ride on the two levels, $-A/2$ and $A/2$, as illustrated in Fig. 8-10.

What is the probability that the decision will be in error? When the transmitted voltage is $-A/2$ volts, an error will occur whenever the voltage *exceeds* the decision threshold level of 0 volts, represented by the shaded area of

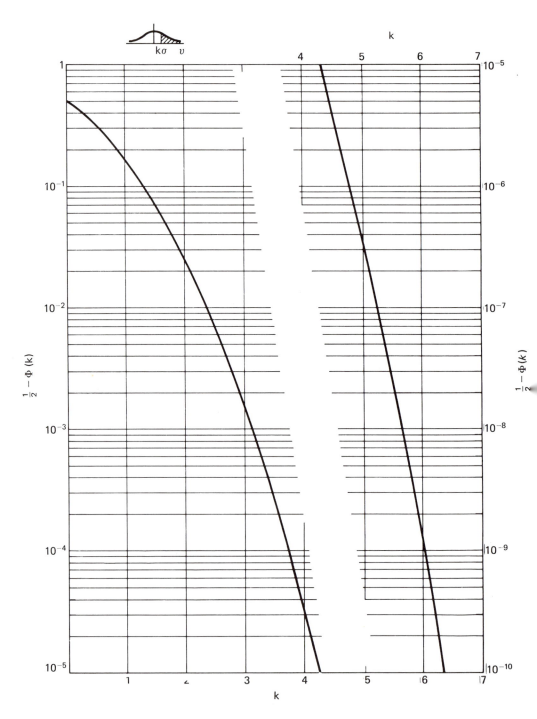

Figure 8-7 Tail Area from $k\sigma$ to ∞ of the Gaussian Probability Density Function

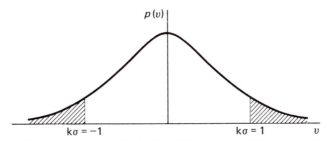

Figure 8-8   Shaded Area Indicate Probability that the Volage Exceed 1 Volt

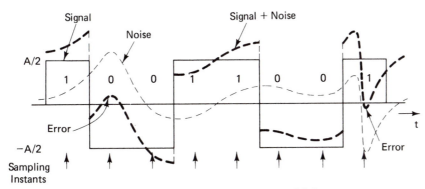

Figure 8-9   Effect of Noise on Two-Level Pulses

Fig. 8-10(b). When the transmitted voltage is $A/2$ volts, an error will occur whenever the voltage drops *below* the decision threshold of 0 volts. This is represented by the shaded area of Fig. 8-10(c). Since either signal is equally probable, and the Gaussian distribution is symmetrical, the probability of error is given by:

> probability of signal at $-A/2 \times$ tail probability
> \+ probability of signal at $A/2 \times$ tail probability
> $= \frac{1}{2} \times$ tail probability $+ \frac{1}{2} \times$ tail probability
> $= 2 \times \frac{1}{2} \times$ tail probability
> $=$ tail probability

The equivalent tail probability has been redrawn in Fig. 8-11.

Since $k\sigma = A/2$, or $k = A/2\sigma$, the error probability can be determined as a function of peak signal ($A/2$) to rms noise voltage (σ). Thus, Table 8-1 can be obtained with the aid of Fig. 8-7.

Figure 8-12 is a graphical representation of the error rate as a function of the peak signal-to-rms noise ratio. As can be seen from the graph, the error rate is very sensitive to the S/N ratio. For a 1 decibel change in S/N in the 16 dB range, the error rate changes by a factor of about 100.

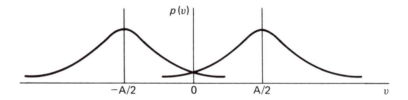

(a) Gaussian Noise on a Two-Level System

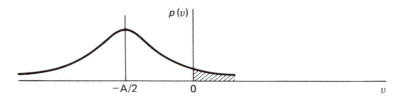

(b) Probability of Error when Signal is at $-A/2$ Volts

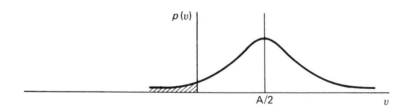

(c) Probability of Error when Signal is at $A/2$ Volts

Figure 8-10   Determination of Error Probability on a Bipolar Pulse Stream

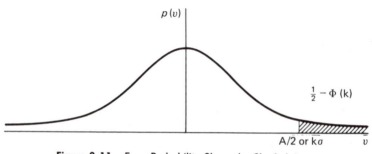

Figure 8-11   Error Probability Shown by Shaded Area

Table 8-1   Error Rate Due to Gaussian Noise

k or $\dfrac{A}{2\sigma}$	$20 \log \dfrac{A}{2\sigma}$	Error rate $(^1\!/_2 - \Phi(k))$
0	$-\infty$	0.5000
0.5	-6.02	0.3085
1	0	0.1587
1.5	3.52	0.0668
2	6.02	0.0227
2.5	7.96	0.0062
3	9.54	0.00135
3.5	10.88	0.00023
4	12.04	0.00003
4.5	13.06	0.00000355
5	13.98	2.97×10^{-7}
5.5	14.81	1.96×10^{-8}
6	15.56	1×10^{-9}
6.5	16.26	4.107×10^{-11}
7	16.90	1.305×10^{-12}
7.5	17.50	3.246×10^{-17}
8	18.06	6.315×10^{-16}
8.5	18.59	9.607×10^{-18}
9	19.08	1.142×10^{-19}

8-4   Transient Noise

The four major types of sporadic, or transient, disturbance in a communication system are impulse noise, gain hits, drop-outs, and phase lists. These noise types account for a substantial portion of errors which occur when digital signals pass through a switched network.

Impulse Noise

Impulse noise consists of irregular noise spikes of relatively high amplitude that tends to occur in sporadic groupings, or bursts. This type of noise is not overly obnoxious to listeners on voice lines, but it can seriously increase error rates on digital systems since it can cause loss of synchronization.

Impulse noise comes from natural causes such as lightning, and from man-made causes such as ignition systems and power lines. It can also come from telephone switching networks as a result of dialing, and of the action of electromechanical switches and relays. Impulse noise is generally measured in terms of the number of times a noise spike exceeds a reference level over a 15 min. time period. This reference level is usually set to 8 dB below the signal level, as shown in Fig. 8-13. Impulse noise pulses that occur within 125 μs of a detected noise pulse are ignored.

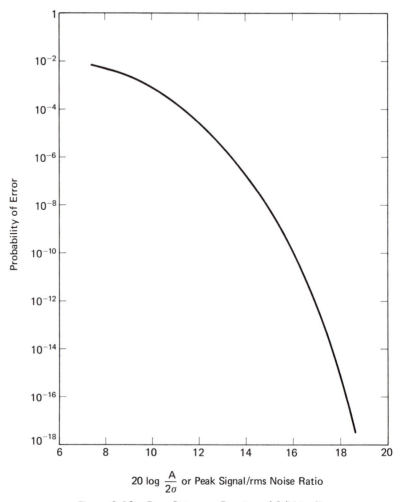

Figure 8-12   Error Rate as a Function of S/N in dB

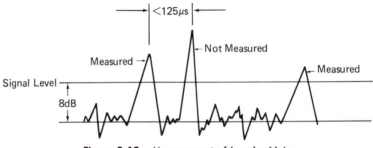

Figure 8-13   Measurement of Impulse Noise

The measurement tests are made at the receiving data locations, and should be averaged out over several days and at different times of the day.

Gain Hits

Gain hits are sudden decreases or increases in recent signal level that cause changes of less than 12 dB in the nominal receive signal level. Because such changes in signal level may appear as data, especially in systems that use amplitude modulation, the error rate of the systems can be significantly increased. A gain hit is registered when the change lasts for more than 4 ms. In the telephone industry, no more than 8 gain hits exceeding 3 dB from the nominal receive level should be detected over a 15 min interval.

Dropouts

Dropouts can be compared to severe gain hits. They are defined as signal decreases that are greater than 12 dB and last longer than 4 ms. Because synchronization can be lost under such circumstances, more data can be lost while trying to resynchronize and automatically equalize. Over a period of 30 min, no more than 1 dropout should occur over the telephone system.

Phase Hits

A phase hit is a sudden change in the received signal phase, and thus frequency, which lasts for durations exceeding 4 ms. Phase hits can appear as false data in PSK and FSK systems. Phase or gain hits occur: (1) when power supplies are changed; (2) when alternate transmission facilities with different propagation times are used; and (3) when the optimum receiver in a space diversity microwave system is used.

Phase Jitter

Variations in the phase of the carrier signal as it passes through a communication network causes a smearing in the zero-crossing of the received data pulses as shown in Fig. 8-16(b). This phase jitter is the traditional undesired angle modulation. For severe jitter, this may result in pulses moving into time slots allocated for neighboring data pulses.

Using the symbols as indicated in Fig. 8-15, the percent of phase jitter is given by $\Delta t/T$ (100%).

Phase jitter is caused by 20 Hz ringing currents in adjacent channels, by the 60 Hz ripple and associated harmonics in the power supplies, and by insufficient filtering of image sidebands. Over the telephone system, phase jitter should not exceed 10° between 20 Hz and 300 Hz and 15° between 4 Hz and 20 Hz.

To obtain a quantitative feeling for phase jitter, consider a single interfacing tone on a carrier signal, as shown in Fig. 8-14. In this illustration the interfacing tone vector A_i rotates around the carrier vector A_c at a rate equal to the difference in their frequencies.

The resultant vector lies somewhere on the dashed circle, indicating that both the amplitude and the phase of the resultant varies. If a clipping circuit is used to remove the amplitude variations, only phase modulation remains. The

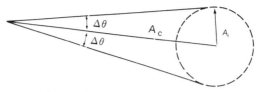

Figure 8-14 Phase Jitter Due to an Interfering Tone

peak phase jitter $\Delta\theta$ is independent of either the carrier or interfering tone frequencies.

The peak phase jitter from Fig. 8-14 is given by:

$$\Delta\theta = \sin^{-1} \frac{A_i}{A_c}$$

Given an interfering tone level, the phase jitter can be calculated. For example, take a -20 dB interference to carrier tone level.

$$-20 = 20 \log \frac{A_i}{A_c}$$

$$\frac{A_i}{A_c} = 0.1 = \sin \Delta\theta$$

$$\Delta\theta = 5.7°$$

$$\text{or} \quad 2\Delta\theta = 11.5° \text{ (peak-to-peak value)}$$

Table 8-2 gives jitter value figures for various interference signal levels.
The tone interference procedure can be used as a calibration check for phase-jitter meters.

Special equipment, such as the Hewlett Packard 4943A Transmission Impairment Measuring Set, is available to measure transient noise and phase jitter. A block diagram of the transient noise measurement is illustrated in Fig. 8-15. A block diagram of the phase jitter measurement is shown in Fig. 8-16(a).

Because phase jitter is a low frequency phenomenon, the front-end, low-pass filter will pass the jitter and, at the same time, reduce the effects of noise

Table 8-2   Jitter Values

Interfering Tone/Carrier	Jitter Values	
	Peak-to-Peak	Percent
−10	36.9	10.2
−15	20.5	5.7
−20	11.5	3.2
−25	6.4	1.8
−30	3.6	1.0

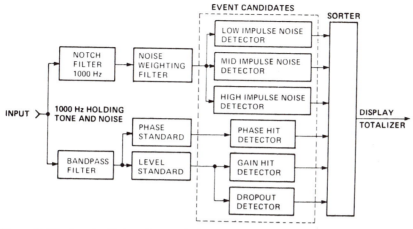

Figure 8-15 Impulse Noise, Gain and Phase Hits, and Dropout Measurement (Courtesy of Hewlett Packard)

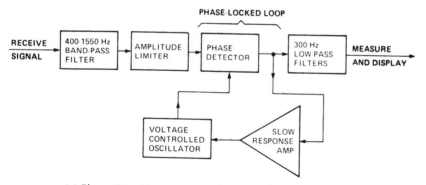

(a) Phase Jitter Measurement (Courtesy of Hewlett Packard)

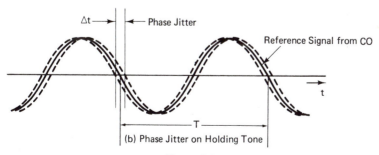

(b) Phase Jitter on Holding Tone

Figure 8-16

and other interference. The voltage controlled oscillator (VCO) is phase-locked to the long term average of the detected signal. It provides the reference signal phase of Fig. 8-16(b). Any phase deviation from this average will be detected by the phase detector and displayed. The phase signal at the 300 Hz low-pass filter output is frequency limited to a band between 20 Hz and 300 Hz. Below 20 Hz, the phase-locked loop is able to track the phase deviation.

In addition to the types of noise just mentioned, many other impairments may also be present in the communication system. In brief, these are:

a. envelope and delay distortion,

b. intermodulation distortion,

c. crosstalk,

d. echo, and

e. frequency shift.

8-5 Peak-to-Average Ratio Test

One measurement which indicates the fidelity of a channel is peak-to-average or (P/AR) measurement. It does not pinpoint the specific impairment which is causing problems, but it gives a good indication of the "health" of the system. This test is sensitive to attenuation distortion, envelope delay, harmonic distortion, and background noise. If there is a decrease in the P/AR from one reading to another, there is a degradation in the system.

The P/AR is determined by applying a complex signal to the channel and then measuring the peak-to-average ratio at the receiver. This measurement is then compared to the P/AR value of 100 which indicates no signal degradation. Figure 8-17(a) shows a complex signal used to determine the P/AR. Figure 8-17(b) shows the frequency spectrum of the signal.

The P/AR values run in the range of 45 for a basic channel, 48 for a C_1-conditioned channel and 78 and 87 for a C_2- and C_4-conditioned channel.

Problems

1. a. Is quantization noise, in actual practice, a periodic triangular wave as illustrated in Fig. 8-1? Sketch the quantization noise for a sinusoidal signal.

   b. What is the peak signal-to-rms quantization ratio (in dB) for a 256-level system?

2. a. Find the average normalized power of a bipolar pulse train having five equally likely levels with spacings of "a."

   b. Find the average normalized power of a single polarity pulse train having five equally likely levels with spacings of "a."

   c. State the advantages of the bipolar pulse train over the single polarity pulse train.

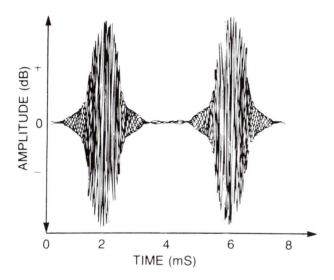

(a) P/AR Transmit Signal Envelope

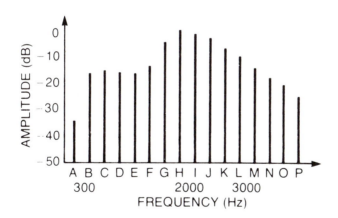

(b) P/AR Transmit Signal Frequency Spectrum

Figure 8-17 P/AR Measurement Test Signal (Courtesy of Hewlett Packard)

3. If the peak voltage of a two-level and a multilevel system is fixed, what approximate noise penalty occurs in an m-level system as compared to a binary system? Figure P8-3 illustrates the levels permitted by each system, both having equal peak amplitudes of A volts. The penalty is determined by the ratio d_b/d_m.

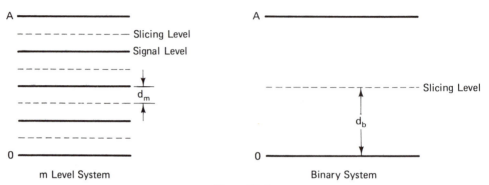

Figure P8-3

4. a. What characteristics does Gaussian noise have, and under what conditions is it produced?
   b. Give an electronic concept of standard deviation and relate it to the Gaussian distribution characteristic.
   c. Why is Gaussian noise so often assumed when evaluating system noise characteristics? Would you consider the noise from a car distributor to be Gaussian?
   d. List some other significant types of noise in cable systems.

5. A two-level system (± 5 volts) experiences a Gaussian noise of 1 volt rms. For a 1-G-bit pulse train, how many bits on the average would be in error? When calculating the tail probability, use both Eqs. (*8-15*) and Fig. 8-7.

6. In order to pass signals in the same frequency band in both directions at once, hybrid transformers, as illustrated above, are used. B_S, which is known as the balance return loss, is the loss experienced due to mismatching. If the loss between the two wire points, A and B, is assumed to be T, the total loss around the loop is

$$\text{Loss} = 2 \ (T + B_s) \ \text{dB}$$

In order to prevent the circuit from oscillating or singing, the loop loss should be greater than 6 dB.

Thus
$$2 \ (T + B_s) \geq 6 \ \text{dB}$$

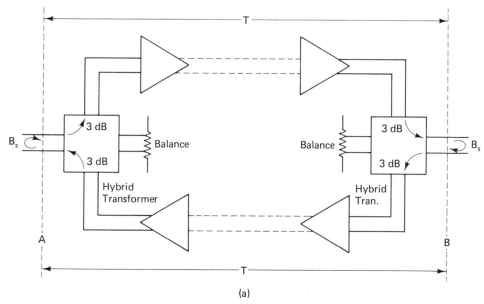

(a)

Figure P8-6 (a)

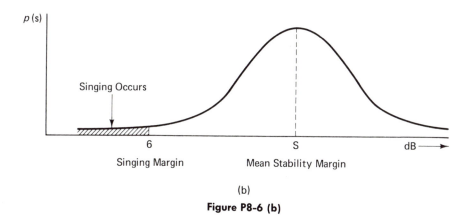

(b)

Figure P8-6 (b)

In general, the quantity $2T + 2B_s$ is symbolized by S and known as the *mean stability margin.*

If the distribution on the stability margin is Gaussian with a standard deviation of σ_s, how many standard deviations should the mean stability margin, S, be above the singing margin of 6 dB if 99.9% of the time the circuit should not sing? Refer to Fig. P8-6(b).

7. Due to an impedance mismatch in a transmission system, the reflected energy or echo can be troublesome if delays exceed 10 ms. The echopath at-

tenuation is equal to $B_E + 2T$ dB as observed from Fig. P8-6(a), where B_E is an average value over the audio range of the balance return loss rather than the minimum value. For a one-way propagation time of 50 ms, representing approximately a 6000 mi connection, an echo path attenuation of 30.9 dB is needed if 50% of the subjects are to find the connection satisfactory. If the subscriber tolerance-to-echo follows a Gaussian distribution with a standard deviation of 2.5 dB, how many people will be satisfied by a loss of 33.4 dB? For longer distances, the attenuation necessary becomes excessive and echo suppressors are needed.

8. List the four major types of transient noise on a cable radio system, and illustrate with waveforms.

9. a. What are the major causes of phase jitter?
   b. Of what significance is the P/AR test?

10. Calculate the peak-to-peak and percent of jitter when a single interfering tone of −40 dBm0 is present on a −18 dBm0 carrier.

9

Pulse Transmission Over Bandlimited Systems

9-1 Introduction

Since the signal spectrum of a pulse is large, bandlimiting can have a drastic effect on the shape of the received pulse. In practice, every channel is band-limited, resulting in severe degradation of the detected signal unless appropriate precautions are observed. This chapter will consider the effects of filtering on the time response of a pulse.

9-2 Bandwidth Limiting of a Pulse Train

To gain a better understanding of the detrimental effects of band limiting on a signal, consider a periodic pulse train of ¼ ms pulse duration and 1 ms period applied to a low-pass filter. Assume the filter is ideal, the skirt has an instant cut-off, and the time delay for the various frequencies is a constant. As an example, look at a 4000 b/s bit stream, where each digit represents a binary state of 1 or 0.

The Fourier series of the pulse stream, shown in Fig. 9-1(a) and explained in Appendix A, is given by:

$$v(t) = \frac{At_p}{T} + \frac{2At_p}{T} \sum_{n=1}^{\infty} \text{sinc} \frac{nt_p}{T} \cos \frac{n2\pi}{T} t \qquad (9\text{-}1)$$

where the sine terms are absent due to the even symmetry. The amplitudes for the various frequency components are:

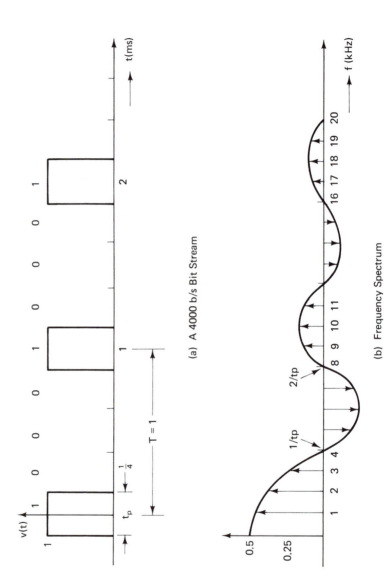

(a) A 4000 b/s Bit Stream

(b) Frequency Spectrum

Figure 9-1 Frequency Spectrum of a 4000 b/s Bit Stream having a fundamental frequency of 1 kHz

Peak Amplitude

$DC\left(\dfrac{At_p}{T}\right)$ 0.25

Harmonics $\left(\dfrac{1}{2}\,\dfrac{\sin n\pi}{\dfrac{4}{n\pi}}\right)$

	Peak Amplitude
Fundamental (n = 1)	0.450
n = 2	0.318
n = 3	0.150
n = 4	0
n = 5	−0.090
n = 6	−0.106
n = 7	−0.064

The waveform thus consists of harmonics of 1 kHz fading off at a rate of:

$$\frac{\sin \dfrac{n\pi t_p}{T}}{\dfrac{n\pi t_p}{T}}$$

By permitting only a finite number of the harmonics to be present, as determined by the cutoff frequency of the filter illustrated in Fig. 9-2, the transmitted pulse undergoes varying degrees of amplitude distortion. This is illustrated in Fig. 9-3 where the bandwidth of the low-pass filter of Fig. 9-2 is gradually increased. In Fig. 9-3(a), only the dc and fundamental frequencies are transmitted. In Fig. 9-3(f) all harmonics up to and including the sixth are passed unaffected.

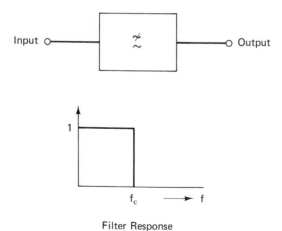

Filter Response

Figure 9-2 Ideal Low Pass Filter

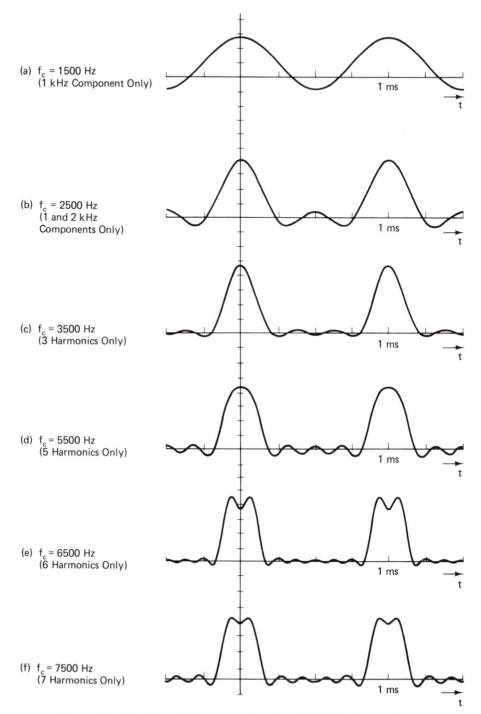

(a) f_c = 1500 Hz
(1 kHz Component Only)

(b) f_c = 2500 Hz
(1 and 2 kHz Components Only)

(c) f_c = 3500 Hz
(3 Harmonics Only)

(d) f_c = 5500 Hz
(5 Harmonics Only)

(e) f_c = 6500 Hz
(6 Harmonics Only)

(f) f_c = 7500 Hz
(7 Harmonics Only)

Figure 9-3 The Effect of Filtering on a Periodic 4000 b/s Bit Stream Having a Fundamental Frequency of 1 kHz.

By employing a nonperiodic bit stream of the same bit rate, the frequency spectrum becomes spread out or continuous rather than discrete. The null points, however, continue to occur at integral intervals of n/t_p.

Note from Fig. 9-3 that when the bandwidth of the channel is around half the pulse rate, i.e., 2 kHz, significant ringing of the pulse leading and trailing tails occurs. This appears as extraneous energy in adjacent keying time slots, which tends to interfere with the reception of the signal in the adjacent keying intervals. This phenomenon is defined as intersymbol interference and steps must be taken to minimize this interference.

9-3   The Unit Impulse

The unit impulse is often used to obtain the frequency response of a network. The unit impulse, $\delta(t)$, is also called the Dirac impulse or delta function. It is a function with the following two properties:

(1) It has value when its argument is zero, otherwise it has a zero value; i.e.,

$$\delta(t) = \begin{cases} 0 & t \neq 0 \\ \infty & t = 0 \end{cases} \qquad (9\text{-}2(a))$$

(2) The area or integral of the function over all time is unity; i.e.,

$$\int_{-\infty}^{\infty} \delta(t) \, dt = 1 \qquad (9\text{-}2(b))$$

More precisely, it can be stated that $\delta(t)$ has the property

$$\int_{-\infty}^{\infty} f(t) \, \delta(t - t_0) \, dt = f(t_0) \qquad (9\text{-}2(c))$$

where $f(t)$ is a function continuous at t_O.

In other words, an impulse $\delta(t - t_0)$ is located at $t = t_0$ and is zero everywhere else. It has a unit area concentrated at the point t_0. One such function is illustrated in Fig. 9-4(a) and (c).

The graphical representation of $A\delta(t - t_0)$, or an impulse of weight A, is shown in Fig. 9-5(a).

The frequency spectrum of such an impulse can be found by taking the Fourier Transform. The result is a continuous spectrum of amplitude A, as shown in Fig. 9-5(b). Applying such an impulse to a network readily provides the frequency response of the system, since the input effectively contains all frequencies of equal amplitude.

As an initial study of pulse propagation through a bandlimited transmission path, a very peculiar impulse wave train is usually used. It consists of a periodic string of impulses spaced by the sampling period T. The inverse of T,

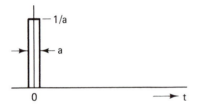
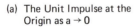

(a) The Unit Impulse at the
Origin as $a \to 0$

(b) Graphical Representation of the
Unit Impulse at the Origin

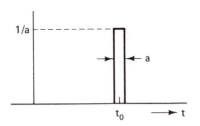

(c) The Unit Impulse at
$t = t_0$ as $a \to 0$

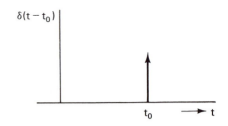

(d) Graphical Representation of the
Unit Impulse at $t = t_0$

Figure 9-4 The Unit Impulse

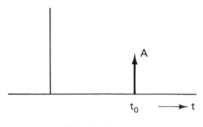

(a) $A \, \delta t(t - t_0)$

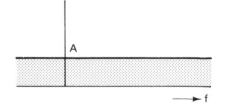

(b) Frequency Spectrum

Figure 9-5 Graphical Representation of $A\delta (t - t_o)$ and Its Frequency Spectrum

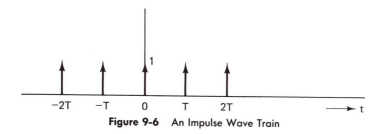

Figure 9-6   An Impulse Wave Train

or $1/T$, is called the sampling frequency. Such an impulse wave train is illustrated in Fig. 9-6 and can be mathematically expressed as

$$\delta(t - nT) \qquad (9\text{-}3)$$

where $\qquad$ n is an integer

This wavetrain will be used to examine the effects of bandpass filters on pulses.

9-4   Frequency Response of an Ideal or Brickwall Filter

Visualize the system shown in Fig. 9-7. In this system, an impulse input of $\delta(t)$ is applied to a transmission channel which has a frequency response corresponding to that of an ideal, or brickwall, low-pass filter of bandwidth B. Although such a filter, and therefore such a response, is unrealizable due to the infinite roll-off of the filter skirts, the ideal gives the approximate response that would be realistically observed when a short pulse is applied to a bandlimited system.

As can be seen from Fig. 9-7(c), the result of bandlimiting an impulse signal is ringing, or time spreading, of the pulse.

The output waveform is given by the expression

$$h(t) = 2B \text{ sinc } 2Bt = 2B \frac{\sin 2\pi\, Bt}{2\pi\, Bt} \qquad (9\text{-}4)$$

From this expression, it can be observed that the ringing tails decay at a rate of $1/t$ and that the amplitude passes through zero at points where $t = n/2B$ where n is an integer, since the sin function is zero whenever $2\pi Bt = n\pi$.

If a wavetrain of pulses is supplied to a bandlimited system, the tails of several pulses can overlap, thus interfering with the major pulse of interest at the detector. For the case illustrated in Fig. 9-8, ignoring any time delays through the system, spurious responses from the third and fourth impulses also appear at the sampling moment $t = 0$. This interference is commonly termed inter-symbol interference, (ISI). If by some method the rate of decay of the tail could be increased from the $1/t$ rate, the ISI would be reduced.

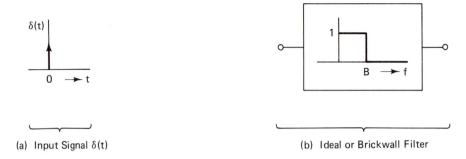

(a) Input Signal δ(t)

(b) Ideal or Brickwall Filter

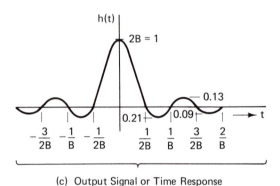

(c) Output Signal or Time Response

Figure 9-7 Time Response of an Ideal or Brickwall Filter Due to Bandlimiting

At sampling time $t = T$, the tails of the several pulses appear within the sampling slot, which results in an observed voltage instead of none at all. This ISI can seriously degrade the noise immunity of the system since the background ISI makes the system tolerant to a smaller amount of noise.

The effects of ISI can be minimized by:

1. Properly timing the pulse train and sampling intervals as they relate to the system bandwidth.
2. Pulse and channel shaping.

By timing the pulse train so that pulses occur or do not occur at intervals of $T = 1/2B$, and are detected at the same rate, the tails of the neighboring pulses are all at zero volts and interference does not occur. At the appropriate sampling instants, only the desired pulse or no pulse will be present. This is illustrated in Fig. 9-9 where the sampling times occur at multiples of $T = 1/f_s = 1/2B$, or at a rate of f_s, which is twice the transmission bandwidth. This is known as the Nyquist rate.

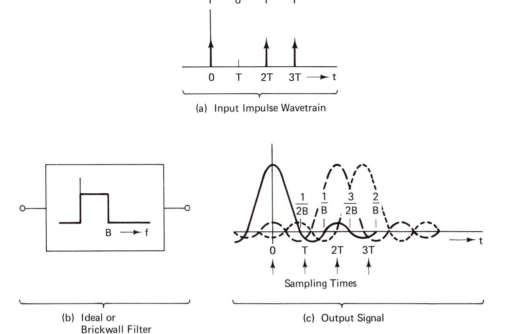

Figure 9-8   Time Response of a 1011 Train after Bandlimiting

$$\text{pulse rate} = 2B \qquad (9\text{-}5)$$

where B is the low-pass channel bandwidth.

Thus, if the sampling at the receiver is performed at the correct instants, or at every T where $T = 1/f_s = 1/2B$, then no ISI occurs.

Stating this somewhat differently, the bandwidth B must be equal to one-half the pulse rate, or

$$B = f_s/2 \qquad (9\text{-}6)$$

This puts very severe restrictions on the sampling times at the receiver. A small time displacement would result in substantial spurious amplitudes at the sampling times. Extremely accurate and stable oscillators, which deviate no more than about 50° in phase from the incoming signal, must be used. Thus, rather expensive equipment is required. Drift due to temperature variations also constitutes a real problem.

Pilot tones can be used to generate the sampling frequency, but noise in the system can cause timing jitter in the actual pilot. In addition, pilot tones require the use of a separate, but parallel, transmission path in order to avoid crosstalk, and phase jitter can be introduced as a result of variations in gains of the parallel path.

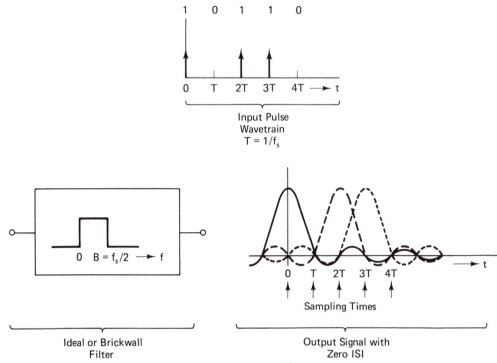

Figure 9-9 Time Response of a 1011 Impulse Train after Bandlimiting, with Pulse Rate Equal to Twice the System Bandwidth ($f_s = 2B$)

The most common solution is to obtain the timing information from the received pulse stream itself. As long as recurring pulses are received, the signal can control an oscillator. Figure 9-9 shows that the bandlimiting of the system prevents the sampling frequency from being transmitted. That is, since the pulse rate is $1/T = f_s$, the bandwidth limiting the ideal low-pass filter with a cutoff frequency of $f_s/2$ will not pass the sampling frequency f_s. To obtain the timing from the pulse stream, the signal is full-wave rectified, which results in doubling the first harmonic of the signal. Figure 9-10 shows the waveforms of a typical received signal before and after rectification. The rectified signal contains a spectral component at the pulse rate of $1/T$ or f_s which may be extracted by means of a sharply-tuned bandpass filter.

In reality, the tails of sample pulses that have a finite width do not vanish at multiples of T, even when $B = f_s/2$. This only occurs as the pulse width is forced to zero. If pulse widths are not too large, the ISI may still be acceptable. The response of an idealized low-pass filter with no delay to a rectangular pulse of width t_p is given by Eq. (9-7). Figure 9-11 shows the response for different pulse durations when $B = 4$ kHz.

$$g(t) = \frac{t_p}{\pi} \int_0^{2\pi B} \frac{\sin \omega t_p/2}{\omega t_p/2} \cos \omega t \, d\omega \qquad (9\text{-}7)$$

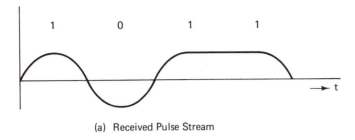

(a) Received Pulse Stream

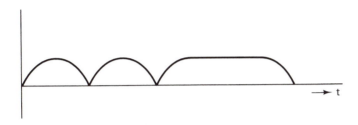

(b) Rectified Pulse Stream

Figure 9-10 Timing Extraction Employing Rectification

Table 9-1 gives the relative percentage of the sample $g(0)$'s amplitude that appears at the adjacent sample instants of $\pm T$. When the pulse duration is $t_p = 0.3\ T$, the percentage error is 0.76.

Because substantial undershoots and overshoots can appear at the adjacent sampling instants as a result of time shift in the signal and non-zero pulse durations, other system frequency characteristics have been developed to reduce these problems. Furthermore, the ideal filter is unattainable in practice. Two solutions in present use that have been found acceptable are the raised cosine response and partial response encoding. These will be briefly considered in Secs. 9-7 and 9-8.

Table 9-1   Relative Percentages of Pulse Present in Terms of the
Maximum Amplitude, at Adjacent Sampling Points

Pulse Duration	% error at $t = \pm T$
0.2T	0.33
0.3T	0.76
0.5T	2.1
T	8.7

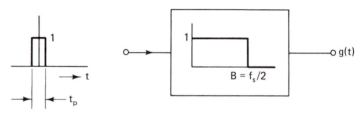

(a) Circuit Configuration

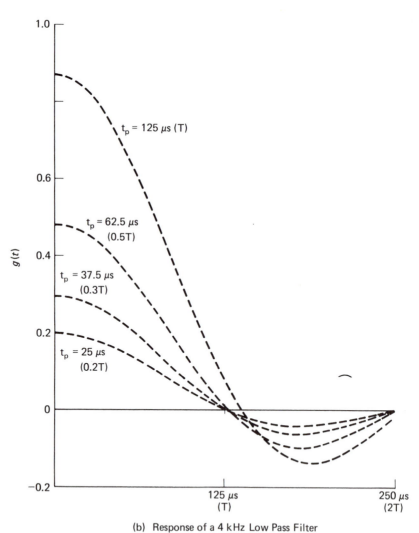

(b) Response of a 4 kHz Low Pass Filter

Figure 9-11 Effect of Filtering on a Pulse of Finite Duration

9-5   The Eye Diagram

In order to determine the effects of random noise, jitter, and so forth on the inter-symbol interference of an incoming pulse stream, the eye diagram is used. It is obtained by applying all the possible pulse combinations to the vertical amplifier of an oscilloscope. About two or three pulse durations are displayed along the horizontal axis. Figure 9-12(a) shows an ideal eye pattern of a three-level system with no noise. With noise, a pattern such as Fig. 9-12(b) or (c) is obtained.

The term "eye diagram" comes from the fact that some areas should appear clear of any signal tracks, making it look very much like an eye. Eye diagram gives an indication of how much noise or error can be tolerated and where the threshold or comparison levels should be centered. The vertical distance between the threshold and the noisy signal tracks indicate approximately how much additional noise can be tolerated, whereas the horizontal spacing gives an indication of the phase error or jitter that can be tolerated.

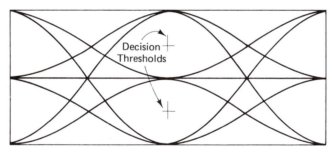

(a) Ideal Three-Level System

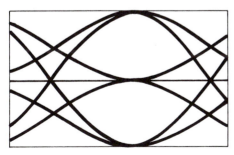

(b) Slightly Noisy System

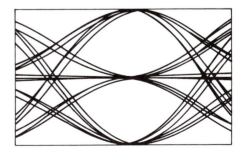

(c) Noisy System

Figure 9-12   Eye Diagrams

If multi-amplitude pulses are used, eye areas would be observed for each sampling instant. The eye centers represent the optimum threshold points for the detector. The ratio of the actual opening to the maximum possible opening represents the degradation caused by the inter-symbol interference in the absence of noise.

For an eye pattern of a two-level system, such as that shown in Fig. 9-13, the peak ISI is about 30%. The equivalent degradation in the S/N rate is 20 log $(1 - 0.3) = 3.1$ dB. This number must be added to the ideal S/N ratio obtained from Fig. 8-12 to determine the actual required S/N. Thus, if the probability of error is 10^{-8}, the peak signal/rms noise ratio (from Fig. 8-12) is ideally about 15 dB. The required peak signal/rms noise ratio must be increased to $15 + 3.1 = 18.1$ dB.

The horizontal opening in the eye can also undergo serious degradation due to variations in the extracted timing, signals, threshold misalignment at the detector, and input pulse shape. Noise and crosstalk can, for instance, cause jitter in the received clock pulse. A percent jitter measurement can be made on the received data by monitoring the short-term movement of the data edges in relationship to a stable clock reference. It would be given by the ratio $\Delta t/T$ of Fig. 9-14.

9-6   Information Capacity

In Chap. 7, the information capacity for a series of uniformly separated pulses was obtained. Looking back at Eq. (*7-14*), n/T represents the time duration between pulse centers for a series of impulse type pulses that are uniformly separated by time intervals of T. Using this equation, the channel capacity for an m-level system similar to that shown in Fig. 9-15 can be given by

$$C = \frac{1}{T} \log_2 m \text{ b/s} \qquad (9\text{-}8)$$

If the pulse rate is at the Nyquist rate, or is twice the system bandwidth, then, according to Eq. (9-5), $1/T = 2B$ and the channel capacity becomes

$$C = 2B \log_2 m \text{ b/s} \qquad (9\text{-}9)$$

From this expression, observe that in order to increase the information capacity, either the bandwidth or the number of signal levels must be increased. If the number of signal levels is 2^n, where $n = 1$ for binary systems, multi-level systems have, in principle, n-times the channel capacity of binary systems at the expense of an increased number of levels. This implies greater sensitivity to noise and more errors.

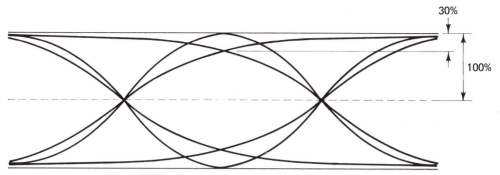

Figure 9-13   Eye Pattern of a Two-Level System

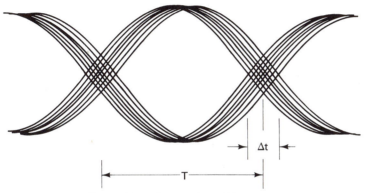

Figure 9-14   Percent Jitter Measurement

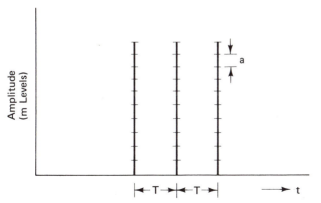

Figure 9-15   *m*-Level Impulse Train

If the message consists of bipolar pulses of *m* levels, the average signal power is related to the "a" level spacing and to the number of levels by Eq. (*8-7*). From this expression, the relationship of *m* to the average signal power and the step size can be obtained. Thus,

$$m = \sqrt{1 + \frac{12S}{a^2}} \qquad (9\text{-}10)$$

Replacing *m* in Eq. (*9-9*) by Eq. (*9-10*) produces

$$C = B \log_2 \left(1 + \frac{12S}{a^2}\right) \text{ b/s} \qquad (9\text{-}11)$$

The relative importance of bandwidth and average signal power to information capacity is indicated in Eq. (*9-11*). If the system capacity is to double, for example, either the bandwidth can be doubled or the argument of logarithm $(1 + 12S/a^2)$ can be squared. For the latter case, the argument must be increased by

$$\left(1 + \frac{12S}{a^2}\right)^2 - \left(1 + \frac{12S}{a^2}\right) = \frac{12S}{a^2} + \frac{144S^2}{a^4} \qquad (9\text{-}12)$$

Equation (*9-11*) can be developed a bit further to relate the capacity to the signal-to-noise ratio. First, relate the level spacing to the mean-noise power in the system by the expression

$$a^2 = k^2 N \qquad (9\text{-}13)$$

where k^2 is a constant of suitable choice to give a low error rate
        N is the mean noise power

Quantization noise is not included since it has been accounted for in Eq. (*9-10*).

By keeping the distance between the levels suitably large, the chances of error can be kept small, as shown in Sec. 8-3. For a given level of spacing a mean-noise power of known noise distribution, it is possible to evaluate the probability of error.

Substituting Eq. (*9-13*) into Eq. (*9-11*) produces the following result:

$$C = B \log_2 \left[1 + \frac{12S}{k^2 N}\right] \text{ b/s} \qquad (9\text{-}14)$$

Given the constant *k* for a system and the S/N ratio, it is thus possible to obtain the channel capacity.

Work done by Shannon* on information theory illustrates that over a white, bandlimited Gaussian channel, it should be possible to attain a channel capacity of

* Shannon, C. E. *A Mathematical Theory of Communication*. BSTJ, vol. 27, pp 379–623, 1948.
Shannon, C. E. *Communication in the Presence of Noise*. Proc. IRE, vol. 37, p 10, 1949.

$$C = B \log_2 \left(1 + \frac{S}{N}\right) \text{ b/s} \qquad (9\text{-}15)$$

where B is the channel bandwidth

                         S is the signal power

and N is the total noise within the channel.

In practice, this goal is far from obtainable because of the extremely complex encoding required and the long time delays. Instead, a $12/k^2$ value of $1/7$ is more typical, which represents an increase in power of $10 \log 7$, or 8.4 dB over that required for Shannon's case.

Assume time division multiplexing is used to insert pulses during the single sampling interval of T shown in Fig. 9-15. If n pulses or channels are inserted, the total channel capacity will go up to

$$C = 2nB \log_2 m \text{ b/s} \qquad (9\text{-}16)$$

This increase in the number of pulses transmitted increases both the required channel capacity and the effective transmission bandwidth (nB).

9-7   The Raised Cosine Channel Response

Impulses are unattainable in practice and the ideal, or brickwall, filter cannot be built. Thus, when designing a digital communication channel, appropriate adjustments must be made to the impulse type signals and filters discussed in Sec. 9-4.

The desired end result is a filter response which is realizable and which produces minimum overshoots and undershoots in its ringing and which passes through zero at adjacent sampling times. When the combination of pulse shape, channel, and equalizer have an overall odd symmetric transfer function about $B/2$, as illustrated in Fig. 9-16, the response to an impulse results in small overshoots and undershoots and passes through zero at the sampling instances of $nT = n/B$. By selecting a suitable response, such as the solid curve in Fig. 9-16, the filter becomes realizable since there are no abrupt changes and the slopes are finite. In such a model, the tails of neighboring pulses do not interfere with the main pulse. The transmission channel can compensate for any deviation of the pulse from the impulse function.

One acceptable solution to the problem of designing a communication channel is the raised-cosine response, which is shown in Fig. 9-17(a). Although such a response is not strictly realizable, it can be more closely approximated than the brickwall response. It has a gradual roll-off of up to about 1.5 times the Nyquist rate, but cuts off rapidly at twice the Nyquist rate. The raised-cosine frequency response is given by

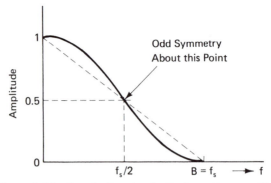

Figure 9-16 Transfer Function for Realizing Minimum ISI

$$H(f) = \frac{1}{2}\left[1 + \cos\left(\frac{\pi f}{B}\right)\right] \quad f < B \qquad (9\text{-}17)$$
$$= 0 \qquad\qquad\qquad f > B$$

and shown in Fig. 9-17(a).

An impulse applied to such a filter gives an output, or corresponding impulse, response similar to that shown in Fig. 9-17(b).

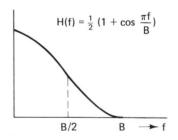

(a) Frequency Response

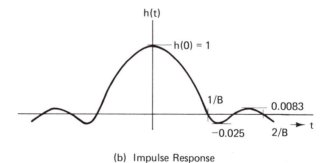

(b) Impulse Response

Figure 9-17 Raised-Cosine Channel Response

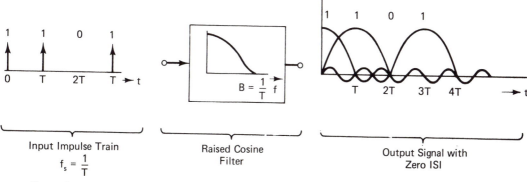

Figure 9-18   Time Response of a 1101-Impulse Train after Bandlimiting with a Raised Cosine Filter in which Pulse Rate Equals Filter Bandwidth ($f_s = B$)

ISI will not occur if the pulse train is timed such that at every time interval $T = 1/B$, the tails of all neighboring pulses are at zero volts. Only the desired pulse or no pulse will be present at the appropriate sampling instants, as shown in Fig. 9-18.

The bandwidth of such a transmission channel is doubled. (Comparing it to the brickwall case, it goes to f_s, or twice the analog signal bandwidth called for in the Sampling Theorem.) The undershoot and overshoot errors, however, are drastically reduced. The slope of the ringing curve is also much gentler near the sampling times, which permits more jitter, or room for time displacement, than the sin x/x type of ringing present in the case of the brickwall filter.

9-8   Partial Response (Correlative) Techniques

In order to meet present-day high speed data requirements, a multilevel system which uses more than two signal states is necessary. If each level is represented by groups of m binary digits, where the total number of signal levels is 2^m, then the speed capability of the m system is m times the speed of the binary system. Increasing the speed capability requires adding more levels, which makes the system more prone to error because of the increased sensitivity to noise.

The horizontal eye opening also deteriorates as more levels are added because the number of possible transitions between levels increases. The vertical eye opening remains unaffected, however.

Figure 9-19 illustrates this for a four-level system where three eyes exist—one between each pair of adjacent levels. Although the vertical eye is fully open, the horizontal eye is reduced over the binary case at the slicing level. The chief contributor to this deterioration is the transition between the extreme levels of 0 and 3. Transition between adjacent levels cause no deterioration.

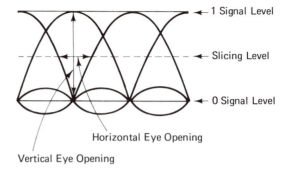

(a) Theoretical Binary
Impulse Eye Pattern

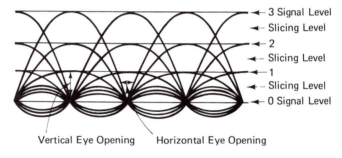

(b) Theoretical 4 Level
Impulse Eye Pattern

Figure 9-19 Eye Patterns of Binary and Multilevel Impulses

Some of the classes of partial response systems discussed in this section permit transitions to occur only between adjacent levels, and thus keep the inter-symbol interference lower than in multilevel systems. The level-coded partial response systems still incur noise penalties when compared to the binary systems, much like the multilevel system described in Problem 8-3. Just as in the multilevel case, partial response systems have bit rates which exceed the Nyquist rate.

Because there is no correlation between the levels of a multilevel system, or, rather, because the past history of an encoded signal has no effect on its present state, each combiration of *m* binary digits is associated with a unique or particular level. At the receiver end, each group of *m* binary digits represents the same level.

In partial response encoding, each signal level depends not only on the present input state but also on the previous input states. That is, there is a correlation between the levels. For this reason, partial response is frequently called correlative. At the receiver, however, each level is still treated independently from the rest. For example, at the transmitter in the encoder, each MARK (or SPACE) is assigned one of several permissible levels. The assignment depends upon the past history and present state of the signal. At the decoder, each level is unique associated with a MARK or SPACE which is independent of the past history of the waveform. Because distinctive patterns are transmitted, violation of these patterns at the receiver results in errors which can be readily detected. Redundant digits do not need to be added to the input since binary data at the transmitter provide for error detection.

Duobinary Technique

One technique for transmitting data is the duobinary. In context, the term duo indicates the doubling of the bit rate over the straight binary system. It is based on an end-to-end system transfer function of

$$|H(f)| = \begin{cases} 2 \cos \pi Tf & f \le \dfrac{1}{2T} \\ 0 & f > \dfrac{1}{2T} \end{cases} \tag{9-18}$$

where T represents the time duration between applied impulses. The B bandwidth of the system is thus $1/2T$ or half the signaling rate. Alternatively, this transfer function can be expressed as

$$2 \cos \pi fT = |1 + e^{-j2\pi fT}| \tag{9-19}$$

since

$$\cos \frac{\theta}{2} = \sqrt{\frac{1 + \cos\theta}{2}}$$

The impulse response of such a system consists of the sum of two sinc functions separated by T, as illustrated in Fig. 9-20. As can be seen, the tails of the resulting output pulse are greatly reduced due to the cancellation of the individual sinc function tails. However, the resulting pulse is 1.5 times as wide as the impulse response of a brickwall filter.

Consider the response for a binary input of 11, which is represented by $\delta(t)$ and $\delta(t - T)$. The input consists of two impulses separated by a time of T. The output, ignoring any time delays in the system, consists of $h(t)$ and $h(t - T)$ where $h(t - T)$ is a delayed version of the $h(t)$ function illustrated in Fig. 9-20. Each of these functions can, in turn, be represented by two sinc type pulses, numbered 1 for $h(t)$ and 2 for $h(t - T)$ in Fig. 9-21. The difficulty in drawing the response can be reduced by drawing the maximum amplitudes of the overlapping sinc functions and assuming that the tails cancel, as shown in the approximate waveform.

Figure 9-22 shows the response to a series of MARK and SPACE impulses. Note that for a full response on the positive side, two or more consecutive MARKS or 1's are needed. For a full response on the negative side, two or more consecutive SPACES or 0's are needed. To obtain 0, the mark and space

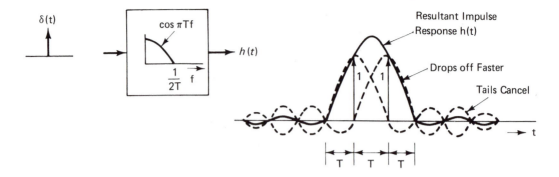

Note: The Impulse Response Consists of Two
Sinc Pulses Marked by a 1.

Figure 9-20   Impulse Response of a Duobinary System

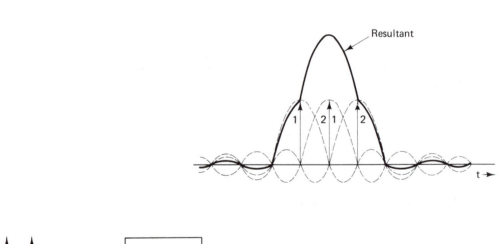

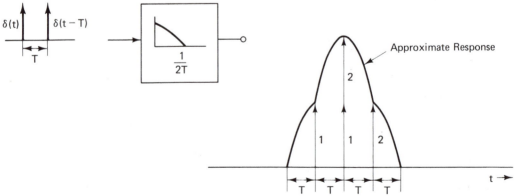

Figure 9-21   Response of a Duobinary System to Two Consecutive Impulses

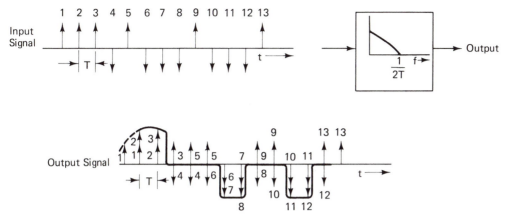

Figure 9-22 Response of a Duobinary System to an Input Data Stream

must alternate. Note that a three-level system now exists—that is, + 2, 0 and − 2—and that transitions can occur only between adjacent levels for successive signaling elements. In other words, there must be at least one interval at 0 when going from a − 2 to a + 2 level.

In addition, if an impulse is at either the + 2 or − 2 level, it must be at the 0 level for an even number of baud intervals before returning to the same + 2 or − 2 level. If the impulse is at the 0 level for an odd number of baud intervals, it must be switching from + 2 to − 2 or from − 2 to + 2.

The signaling rate of the duobinary system is still at the Nyquist rate, that is the bandwidth is equal to $1/2T$, or one-half the signaling rate. Signal rates slightly exceeding the Nyquist rate can be used and reasonable results still obtained.

Figure 9-20 or Equation 9-19 shows that the duobinary response can be produced by using a digital delay circuit and brickwall filter. This is illustrated in Fig. 9-23(a). The effective addition of the delayed pulse introduces a predictable amount of inter-symbol interference at the transmitter. At the receiver, it must be subtracted out again as the decoder in Fig 9-23(b) illustrates.

The various approximate waveforms for a typical data stream are illustrated in Fig. 9-24. Delays, other than those which occur in the digital delay circuits, are omitted.

One undesirable feature of the duobinary waveform is that the signal has three levels at each sampling instant and any level can be interpreted as a MARK or SPACE, depending upon the past history of the signal. It would be much more convenient to recognize the zero level as either a MARK or a SPACE and not have to keep track of the alterations. In addition, if an error were detected, immediate appropriate action could be taken instead of allowing the error to propagate into the next bit because of the delay circuit action. To solve these problems, a special precoder can be inserted before the encoder

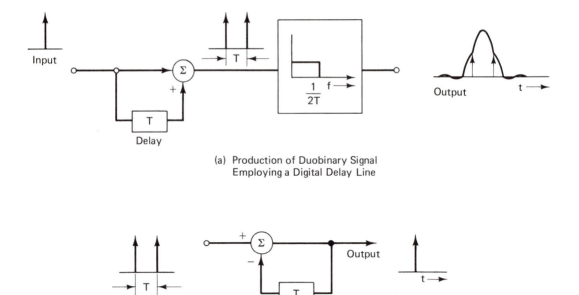

(a) Production of Duobinary Signal
Employing a Digital Delay Line

(b) Duobinary Decoder

Figure 9-23 Encoding and Decoding of a Duobinary Signal

shown in Fig. 9-23(a), and a special post decoder can be inserted after the decoder shown in Fig. 9-23(b). These insertions are illustrated in Fig. 9-25.

The exclusive -OR gates perform Modulo 2 addition, which means the output goes low when the two inputs are the same and high if the inputs are different. Figure 9-26 illustrates the waveforms at the lettered points in Fig. 9-25 for the data given. Notice that for the duobinary waveform (D) with pre-encoding, the amplitude levels can be directly interpreted without referring to the past history. The ± 2 levels can be interpreted as 0 and the 0 level interpreted as 1.

An example of a system that uses the duobinary technique is the RD3 radio system. It produces two 45.5M b/s partial-data channels, one channel of which is in phase quadrature with the other. Block diagrams of the transmitter and receiver are illustrated in Fig. 9-27. The bandpass filter (BPF) selects the summed frequency components from the multiplier sections. The traveling wave tube (TWT) is a broadband power amplifier in common use in microwave and satellite systems. The partial-response filter circuit now has a bandwidth double that of the Nyquist, which is to be expected in a double sideband system.

The RD3 transmits a bandwidth of about 42 MHz rather than the 45.5 MHz suggested in Fig. 9-27(a). This slightly reduces the sample timing toler-

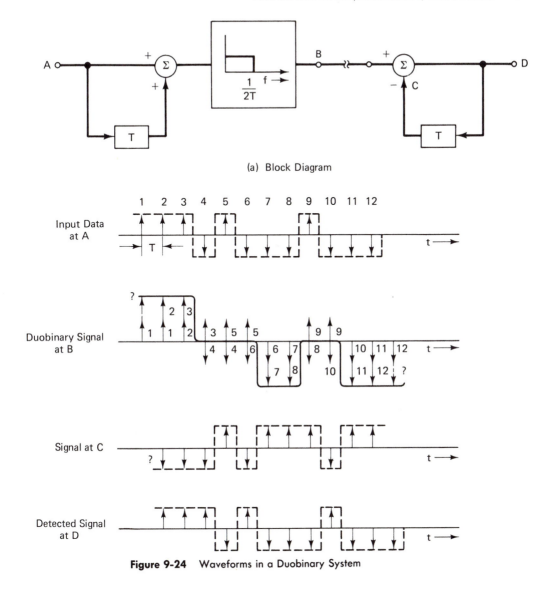

(a) Block Diagram

Figure 9-24 Waveforms in a Duobinary System

ance and causes a small S/N penalty. The RD3 system permits 1344 channels on the radiated bandwidth of a normal heavy microwave route. If the antenna is cross polarized, another 1344 channels, or a total of 2688 channels, are permitted. This is well above the capacity of normal analog microwaves.

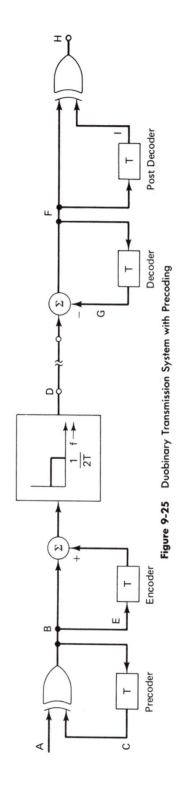

Figure 9-25 Duobinary Transmission System with Precoding

400

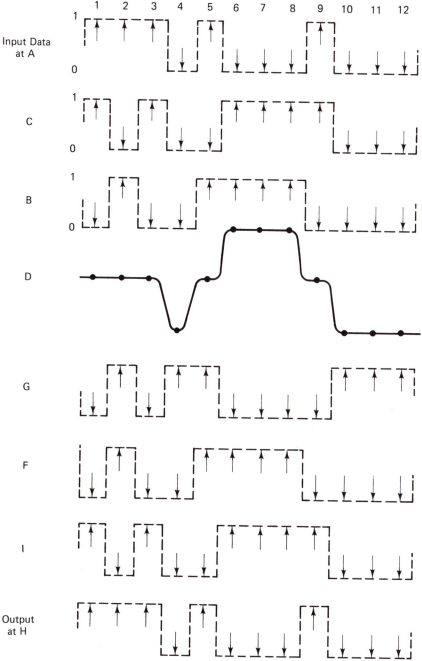

Figure 9-26 Waveforms in the Duobinary System with Precoding (See Fig. 9-25)

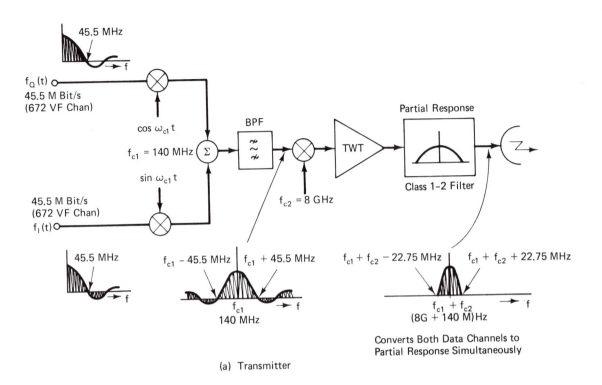

(a) Transmitter

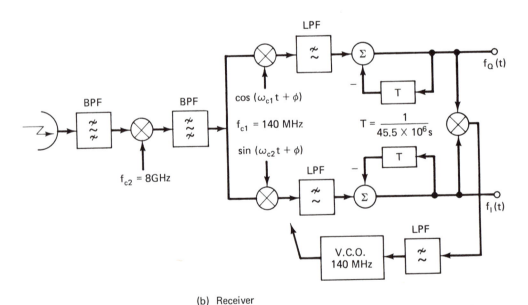

(b) Receiver

Figure 9-27 Block Diagrams of an RD3 Radio System Employing Duobinary Encoding

The various correlative techniques have been categorized into the classes shown in Table 9-2. In this table, the bandwidth of the system is given by F, which is the equivalent of B in our nomenclature. When operating at the Nyquist rate, $F = B = 1/2T$. The k coefficients represent the peak amplitudes of the output sinc responses to an applied impulse. The binary, or 1-1, class has the brickwall type of transfer function. The 1-2 class is the duobinary system just discussed. The 4-3 class, often called the modified duobinary technique, is being used more frequently and will be elaborated on in the next section.

Modified Duobinary Technique (Class 4-3)

As stated in Table 9-2, the modified duobinary system has a transfer function equal to:

$$|H(f)| = \begin{cases} 2 \sin 2\pi fT & f \leq \dfrac{1}{2T} \\ 0 & f > \dfrac{1}{2T} \end{cases} \qquad (9\text{-}20)$$

where T represents the time between adjacent applied impulses. The bandwidth B of the system is $1/2T$, which is equal to ½ the signaling rate. This also is called the Nyquist bandwidth. Alternatively, this transfer function can be expressed as

$$2 \sin 2\pi fT = |1 - e^{-j4\pi fT}| \qquad (9\text{-}21)$$

The impulse response of a system with such a transfer function consists of the sum of *two* sinc functions of oposite polarities that are spaced by a distance of $2T$. (See the k_1, k_2, and k_3 coefficients of Table 9-2). Figure 9-28 shows the two sinc functions and the result, which is the sum of the two sinc functions. As before, the tails of the two sinc functions tend to cancel. Ignoring these tails, the response to a series of MARK and SPACE impulses is shown in Fig. 9-29.

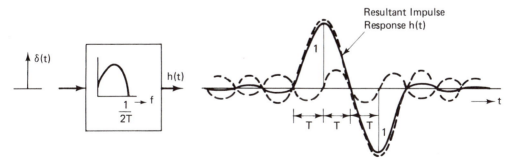

Note: The Impulse Response Consists of Two Sinc Pulses
(of Oppostie Sign) and Separated by 2T. These are
Marked by 1's.

Figure 9-28   Impulse Response of a Modified Duobinary Encoder

Table 9-2 C. R. Kretzmer, "Generalization of a Technique for Binary Data Communications," © 1966 IEEE. Reprinted, with permission, from *IEEE Transactions on Communication Technology*, vol. 14, p. 67, February 1966.

Class	k_1	k_2	k_3	k_4	k_5	h(t)	H(f) [0 < f < F]	No. of Rec. Levels
Binary (Ideal)	1						1	2
1 (n = 2)	1	1					$2\cos\dfrac{\pi f}{2F}$	3
2 (n = 3)	1	2	1				$4\cos^2\dfrac{\pi f}{2F}$	5
3 (n = 3)	2	1	−1				$2 + \cos\dfrac{\pi}{F}f - \cos\dfrac{2\pi f}{F}$ $+ j\left[\sin\dfrac{\pi}{F}f - \sin\dfrac{2\pi f}{F}\right]$	5
4 (n = 3)	1	0	−1				$2\sin\pi\dfrac{f}{F}$	3
5 (n = 5)	−1	0	2	0	−1		$4\sin^2\pi\dfrac{f}{F}$	5

*Entry in Figure is Area Under Curve.

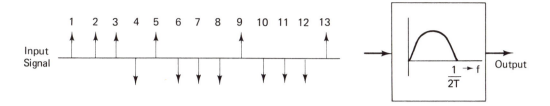

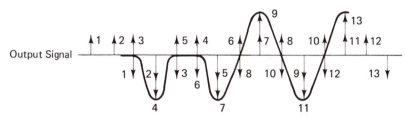

Figure 9-29   Modified Duobinary Pulse Pattern

It can be concluded from Fig. 9-29 that the waveforms have three distinguishable levels—$+2$, 0, and -2—and no dc component at the sampling instants. This is a definite advantage for transmission systems which cannot carry dc.

Similar to the duobinary system, the modified duobinary signal can be produced with an ideal filter and delay circuits as illustrated in Fig. 9-30. Furthermore, the signal can be precoded and later post decoded at the receiver. This is illustrated in Fig. 9-31, where a 1 at the coder output is represented by both the $+2$ and -2 levels and a 0 by the 0 level.

Because the modified duobinary filter characteristic is difficult to approximate in practice, it can be more readily implemented by cascading a digital delay network with a duobinary filter. From Eq. (*9-21*), the transfer function can be rewritten as

$$1 - e^{-j4\pi fT} = (1 - e^{-j2\pi ft})(1 + e^{-j2\pi ft})$$

$$(9\text{-}22)$$

The term, $(1 + e^{-j2\pi ft})$, can be represented by the duobinary filter (see Eq. (9-19)). The term, $1 - e^{-j2\pi fT}$, can be represented by a network with a digital delay circuit, as illustrated in Fig. 9-32.

GTE Lenkurt Inc. uses this modified duobinary technique in a system to produce a single sideband digital radio. A block diagram of an SSB system that uses modified duobinary encoding is shown in Fig. 9-33.

The modified duobinary system gives 4 bits per Hz of bandwidth (BW) compared to the RD3 system's 2 bits per Hz of BW. Both systems use quadrature modulation to obtain these bit rates.

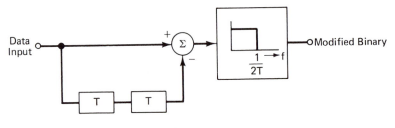

(a) Modified Binary Encoder

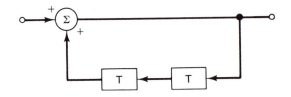

(b) Modified Binary Decoder

Figure 9-30 Transmission and Detection of a Modified Duobinary Signal

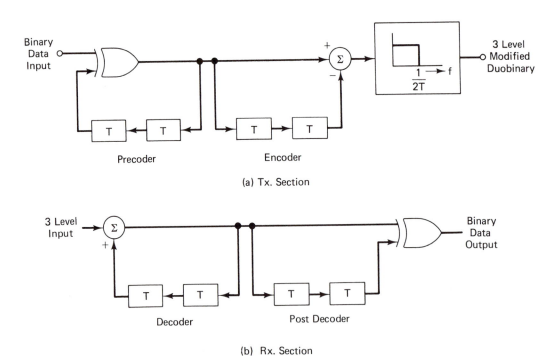

(a) Tx. Section

(b) Rx. Section

Figure 9-31 Modified Duobinary Transmission System with Precoding

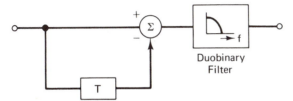

Figure 9-32 Generation of Modified Duobinary Signal (Based on a drawing in Kamile and Feher, *Digital Communications: Microwave Applications*, © 1981, p. 151. Reprinted by permission of Prentice-Hall, Inc., Englewood Cliffs, N.J.)

As a chapter review, consider the transmission of a hi-fidelity signal with a frequency range of 20 to 15,000 Hz and a dynamic range of 100 dB. Assume that the human ear can detect changes of 0.5 dB in intensity level. Now compute the bandwidths for each of the transmission techniques discussed in this chapter. Assume encoding at the baseband level.

1. In the original analog form, a bandwidth of 20 to 15,000 Hz is required, or $B = 15$ kHz.

2. In PAM form, a sampling rate of twice the highest signal frequency is required, or 30 k samples per second. For the ideal or brickwall filter transmission response, the required bandwidth is $B = 15$ kHz. Of course, such a response cannot be achieved in practice.

3. For the raised cosine transmission response in PAM form, the required bandwidth would be $2 \times B$, or 30 kHz.

4. For a binary PCM-encoded signal, distinguishing between 100dB/0.5dB, or 200 amplitude levels, requires 8 bits, or binary pulses, since $2^8 = 256$. Therefore, the bit rate is $8 \times 30,000 = 240$k b/s. The brickwall response requires 120 kHz of bandwidth, or 2 b/Hz of bandwidth. The raised cosine response requires 240k Hz of bandwidth, or 1 b/Hz of bandwidth.

5. For a ternary PCM code transmitted over a channel with a raised cosine response, each sample requires 5 bits, or $3^5 = 243$. This represents a bandwidth of $5 \times 30,000$, or 150 kHz. One of the drawbacks of the ternary system is that it is more susceptible to noise than the binary encoded system.

6. For the PCM signal that is encoded in the duobinary or modified duobinary format, the required bandwidth is still the Nyquist bandwidth of 120 kHz. However, the Nyquist rate can be exceeded by up to 43% in marginal operation to give 2.86 b/Hz of bandwidth.

7. For the PCM signal encoded upwards in frequency to a DSB duobinary format, a bandwidth of 240 kHz is required, since the bandwidth is doubled due to the presence of both sidebands. For the PCM signal encoded upwards into an SSB modified duobinary format, a bandwidth of 120 kHz is required since only one sideband is present.

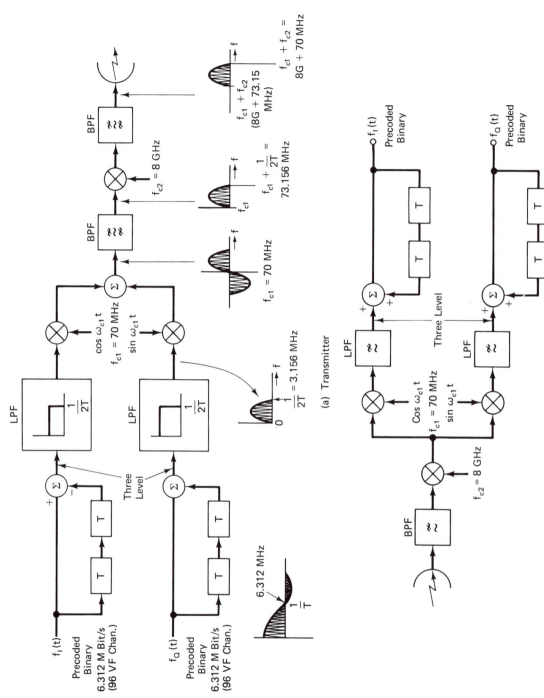

Figure 9-33 Block Diagram of an SSB Radio System Using Modified Duobinary Encoding

Problems

1. a. What are the frequency and time characteristics of the impulse function $\delta(t - t_0)$?
    b. Why is the impulse function a convenient testing pulse for determining the frequency response of a network?

2. Sketch the time response of an impulse function $\delta(t - T)$ which passes through a low-pass brickwall filter with a cut-off frequency of f_c.

3. a. Define Inter-Symbol Interference (ISI).
    b. If a delayed impulse is added to the impulse of Problem 2, what would the spacing be if zero ISI is desired?
    c. Discuss methods of reducing ISI.

4. a. What is the Nyquist rate of a channel?
    b. Determine the Nyquist rate of a channel with the characteristics outlined in Problem 2.

6. a. Give two methods of generating the sampling frequency at the receiver. Why is this timing so critical?
    b. Explain why the sampling frequency cannot be directly extracted from the incoming wavetrain to the receiver when operating at the Nyquist rate.

7. a. Explain why real-life input sample pulses with a finite width still experience ISI, even when the sampling frequency is equal to twice the low-pass filter bandwidth.
    b. Sketch the time response of a pulse of duration 0.3T that passes through a bandpass filter of cut-off frequency $1/2T$. Give the amplitude at the time $t = T$.

8. a. Give a hook-up circuit that would produce an eye diagram.
    b. What information can the eye diagram give?
    c. An eye pattern of a two-level system indicates a closure of 40% due to ISI. If the probability of error is 10^{-7}, determine the required peak signal/rms noise ratio.
    d. If Δt and T of Fig. 9-14 are 1 μs and 10 μs respectively, what is the percent jitter?

9. a. Determine the channel capacity of a system which carries a signal with a peak voltge of 1 and pulse spacings of 1 ms. Assume that amplitude variations of 3.9 mV can be detected. If the channel is operated at the Nyquist rate, what is its bandwidth if an ideal brickwall response is assumed?
    b. Each sample or pulse in (a) is converted into an 8-bit signal. Now what is the channel capacity? If the channel is operated at the Nyquist rate, what is its bandwidth if an ideal response is assumed?
    c. What is the chief disadvantage of the method given in (a) over that in (b) in regards to noise immunity?

10. a. A bipolar pulse stream of 1 W average power with step sizes of 1 mV is transmitted down a 4 kHz bandwidth channel. What bit rate can be sent down such a channel?
    b. If the bit rate is doubled, what is the increase in the average signal power? Assume no change in bandwidth. Is this reasonable in practice?
    c. If the bit rate of (a) is doubled by increasing the bandwidth, what is the new bandwidth?

11. State Shannon's information rate. Why is it so difficult for real systems to approach this theoretical rate?

12. A single, digitized voice channel requires an ideal bandwidth of 32 kHz. What bandwidth is required if 24 such channels are time-division multiplexed?

13. a. What difficulties are experienced in designing a brickwall filter?
    b. Discuss a feasible pulse-channel-equalizer frequency response which can be approximated in real life and which results in a reasonable ISI. Does the raised-cosine response meet this criterion?
    c. What bandwidth is required for the raised-cosine response in order to achieve minimal ISI? Compare this to the brickwall type of response.

14. For a specified probability of error, the partial response technique can transmit substantially more bits per second per hertz of bandwidth than the Nyquist-type systems that have no memory—that is, that ignore past history. Explain why.

15. a. Draw the impulse response of a duobinary system.
    b. If the bandwidth of a duobinary system is *B,* what should the spacing be between applied impulses to keep the ISI to a minimum?
    c. Sketch the duobinary response for the input sequence 1011100100011.

16. For the same input sequence given in Problem 15(c), sketch the waveforms at points A, B, C, and D for Fig. 9-24(a).

17. a. For the same input sequence given in Problem 15(c), sketch the waveforms at points, A to H, inclusive, for Fig. 9-25.
    b.

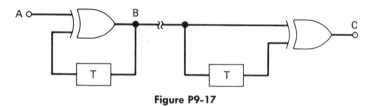

Figure P9-17

Prove that C = A for the precoder/post decoder shown in P9-17.

18.

$f(t) = \cos \omega_1 t \longrightarrow \bigotimes \longrightarrow f_{1m}(t) = \frac{1}{2} [\cos(\omega_2 + \omega_1)t + \cos(\omega_2 - \omega_1)t]$

$\cos \omega_2 t$

$f(t) = \cos \omega_1 t \longrightarrow \bigotimes \longrightarrow f_{2m}(t) = \frac{1}{2} [\sin(\omega_2 + \omega_1)t + \sin(\omega_2 - \omega_1)t]$

$\sin \omega_2 t$

Figure P9-18(a)

When multiplying a signal, $f(t) = \cos \omega_1 t$, by the orthogonal signals $\cos \omega_2 t$ and $\sin \omega_2 t$, the modulated signals expressed in Fig. P9-18(a) are obtained. These modulated signals are then summed and transmitted. Using the following identities:

$$\cos x \cos y = \frac{1}{2} [\cos(x + y) + \cos(x - y)]$$

$$\sin x \cos y = \frac{1}{2} [\sin(x + y) + \sin(x - y)]$$

$$\sin x \sin y = \frac{1}{2} [\cos(x - y) - \cos(x + y)]$$

prove that the original signals can be separated out using the detector (complete with filter) of Fig P9-18(b). Give the expressions for A and B.

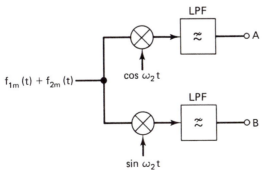

Figure P9-18(b)

19. Referring to Table 9-2, sketch the impulse response and frequency characteristics of a Class 2-3 System. Do the same for a Class 4-3 System. How many levels does each of these systems produce?

20. A hi-fidelity signal which has a 20 Hz to 20 kHz frequency range is to be transmitted, and 512 levels are to be distinguished. Obtain the tranmission bandwidth for the following conditions:

a. Transmission in its original analog form.

b. Transmission in a PAM form via a channel having a brickwall and a raised-cosine response.

c. Transmission in a binary PCM form via a channel having a brickwall and a riased-cosine response.

d. Transmission in a ternary PCM form via a channel having a brickwall and a raised-cosine response.

e. Transmission in a duobinary form.

f. Transmission in a modified duobinary form over an AM and SSB system.

10

Coding for Communications

Since computer systems use binary digits, coding systems have been developed to convert letter and number characters into binary notation. It is usually desirable to use a code which is both recognizable to humans and acceptable to computers. Only 6 bits are required to express numeric and alpha information, which allows for a total of 2^6, or 64, combinations. The extra combinations can be used for punctuation marks, control characters, and other special characters required for data communications.

Special bits, called parity bits, are occasionally added for error detection. In paper tape codes, for example, a check or parity bit is added to the end of a character to create either an even or an odd parity. If even parity is used, a parity bit is added to make the total number of 1 bits even; for odd parity, the sum of 1 bits with parity is odd.

For example, the following table shows the application of *even* parity:

Code Word	Parity Bit
0000	0
1000	1
1010	0
1100	0
0100	1
1011	1

In *odd* parity schemes , the parity bit is used to make up an odd number of ones. For the characters in the example just given, the parity bits would be the opposite of that shown.

Parity can detect, but not correct, single errors. It will miss double errors. For example, if the following code is transmitted in an even parity scheme:

	Data	Parity
	0110	0

but the following code is received:

	Data	Parity
	1110	0

then a parity error exists since an odd number of ones were received with a 0 parity bit. However, if the same code is transmitted but the following code is received:

	Data	Parity
	1010	0

the error will *not* be detected, because the corrrect parity bit was received with an even number of ones.

This chapter will consider some of the more popular binary coding systems and their applications.

10-2   Codes

Baudot Code

One of the first codes developed which is still popular is the Baudot tele-printer code. In this code, each character is represented by five bits of equal length. This would normally allow for $2^5 = 32$ *distinct characters. By using a shift arrangement, much like that of the upper and lower cas*es on a typewriter, the number of distinct characters can be doubled to 64. This is performed by the letters (LTRS) and the figures (FIGS) control characters. Depending upon which one of these two control characters was most recently used determines all of the subsequent series of characters. The LTRS shift indicates that all characters after it are alpha characters. All the succeeding characters after a FIGS shift are numerics and other special characters.

Since six of the possible combinations are used for function codes—LTRS, FIGS, SPACE, CARRIAGE RETURN, LINE FEED and BLANK—58 are left for characters. After taking into account the nine characters of the number system and the 26 characters of the alphabet, this leaves only 23 for special characters.

Because of the code being prefixed by FIGS or LTRS, when a baudot terminal is interfaced to a computer, the software must monitor the FIGS-LTRS status in order to interpret the data correctly. A receiver has no way to tell if an error has occurred, as a parity bit is not used.

The Baudot code is usually transmitted asynchronously, whereby the data bits of each character are preceded by a start bit and followed by a stop interval. The stop interval is 1.42 times the length of the other intervals. Because these start and stop bits add no information to the character, the time taken up by these bits reduces the net data throughput.

The Baudot code is used in telegraph communication on the keyboard printer and paper tape reader.

Binary Coded Decimal (BCD)

One of the earlier codes developed for data processing machines was the Binary Coded Decimal, shown in Table 10-5. Since there are no alpha characters, this code is not suitable for a communications code, only for the storing of numeric information. The BCD code is often hidden in other codes, such as EBCDIC and ASCII.

To represent a numeric greater than 9, other four-bit groupings are used; for instance 49 is represented by the code 0100 1001.

Extended Binary Coded Decimal Interchange Code (EBCDIC)

Another code developed later by IBM as their standard machine language is the EBCDIC code. It is an extension of the BCD code to 8 binary bits, having the capability of accommodating 256 combinations. Table 10-1 illustrates the corresponding binary and hexadecimal equivalents for this code. In the hexadecimal code of base 16, numbers beyond 9 are represented by the letters A, B, C, D, E, and F. This is shown in the first two columns of the table.

The EBCDIC code can handle graphic and control characters. It can be used in data processing applications since it is very compatible with BCD-oriented data processing equipment. This code suffers somewhat from a communications viewpoint as no bits are set aside for parity detection.

American National Standard Code for Information Interchange (ASCII)

A more recent code, which has become the one most extensively used as a communication code, is the ASCII code. (See Table 10-2.) This code is a 7-bit code, with an eighth bit added for parity. The eighth parity bit, can be either odd or even parity, depending on what the user desires. Some terminal vendors are using this eighth bit for creating another 128 special characters from the keyboard rather than for parity.

As an example, using even parity, the letter A would be represented by 01000001. It would be 11000001 for odd parity.

The ASCII code and variations of the code are used by most computer terminal manufacturers. This code is used commonly in computer-to-computer and computer-to-terminal communications.

All characters in either the ASCII or EBCDIC codes can be classified into one of two categories: "Graphic" or "Control."

The Graphic characters are the "printable" alpha-numeric, punctuational and other symbolic characters. Examples are 2, B, !, %, (,), etc.

Table 10-1 EBCDIC Code Chart Courtesy Tektronix

HIGH (B8 B7 B6 B5) \ LOW (B3 B2 B1)	0 (0000)	1 (0001)	2 (0010)	3 (0011)	4 (0100)	5 (0101)	6 (0110)	7 (0111)	8 (1000)	9 (1001)	A (1010)	B (1011)	C (1100)	D (1101)	E (1110)	F (1111)
0 (0000)	NUL	SOH	STX	ETX	PF	HT	LC	DEL		RLF	SMM	VT	FF	CR	SC	SI
1 (0001)	DLE	DC1	DC2	DC3	RES	NL	BS	IL	CAN	EM	CC		ITS	IGS	IRS	IUS
2 (0010)	DS	SOS	FS		BYP	LF	EOB/ETB	ESC/PRE			3M			ENR	ACK	BEL
3 (0011)			SYN		PN	RS	UC	EOT					DC4	NAK		SUB
4 (0100)	SP										¢	.	<	(	+	|
5 (0101)	&										!	$	*	)	;	¬
6 (0110)	−	/									¦	,	%	_	>	?
7 (0111)											:	#	@	'	=	"
8 (1000)		a	b	c	d	e	f	g	h	i						
9 (1001)		j	k	l	m	n	o	p	q	r						
A (1010)			s	t	u	v	w	x	y	z						
B (1011)																
C (1100)	{	A	B	C	D	E	F	G	H	I						
D (1101)	}	J	K	L	M	N	O	P	Q	R						
E (1110)	\		S	T	U	V	W	X	Y	Z						
F (1111)	0	1	2	3	4	5	6	7	8	9						

BINARY — HEX — EBCDIC

Table 10-2   American Standard Code for Information Interchange

Bit Number										
		0	0	0	0	1	1	1	1	
		0	0	1	1	0	0	1	1	
		0	1	0	1	0	1	0	1	
$b_7 b_6 b_5 b_4 b_3 b_2 b_1$ Row	Column	0	1	2	3	4	5	6	7	
0 0 0 0	0	NUL	DLE	SP	Ø	`	P	@	p	
0 0 0 1	1	SOH	DC1	!	1	A	Q	a	q	
0 0 1 0	2	STX	DC2	"	2	B	R	b	r	
0 0 1 1	3	ETX	DC3	#	3	C	S	c	s	
0 1 0 0	4	EOT	DC4	$	4	D	T	d	t	
0 1 0 1	5	ENQ	NAK	%	5	E	U	e	u	
0 1 1 0	6	ACK	SYN	&	6	F	V	f	v	
0 1 1 1	7	BEL	ETB	'	7	G	W	g	w	
1 0 0 0	8	BS	CAN	(	8	H	X	h	x	
1 0 0 1	9	HT	EM	)	9	I	Y	i	y	
1 0 1 0	10	LF	SS	*	:	J	Z	j	z	
1 0 1 1	11	VT	ESC	+	;	K	[	k	{	
1 1 0 0	12	FF	FS	,	<	L	~	l	⌐	
1 1 0 1	13	CR	GS	–	=	M	⌐	m	}	
1 1 1 0	14	SO	RS	.	>	N	^	n	|	
1 1 1 1	15	SI	US	/	?	O	–	o	DEL	

The Control characters perform some function, either for formatting and controlling the functions of the receiving machine, or for maintaining the communications discipline. Examples of format control are BS (Back Space), CR (Carriage Return), FF (Form Feed or Next Page), etc. Examples of communications control are SOH (Start of Header), NAK (Negative Acknowledgement).

Many ASCII terminals and personal computers use a "control key" in conjunction with a second key to perform a control function. That is, the control key labeled CTL or CTRL is held down while a second key labeled with a character from column 4 or 5 is hit. The value transmitted is 40_{16} less than the column 4 or 5 key that was hit.

For example, if the CTL key in conjunction with the key labeled B is hit, the STX control character is transmitted.

$$CTL\ B = B - 40_{16} = 42_{16} - 40_{16} = 02_{16} = STX$$

Although the four device control characters DC1–DC4 have not been defined for specific functions, the characters DC1 and DC3 are often used for flow control. If, for instance, a computer is sending a long string of characters to a printer via a communications line, the small memory or buffer at the printer may become overloaded and overflow, thus losing some data. To prevent this from occurring, the printer sends a Transmission Off (XOFF) character, DC3, back to the computer when the buffer is almost full. Upon receipt of this DC3 control character by the computer, it halts transmission. When the buffer is almost empty, the printer sends a XON character, DC1, to the computer and the computer again resumes transmission. This presumes either a full-duplex channel or reverse channel modems.

Note that the control functions, such as SOH (Start of Heading), can be detected by 0's in the first two bits. If both are not 0's, a character is being transmitted.

In order to transfer information properly, the receiver needs to know what is to be printed and how it is to be printed, and what actions are to be taken under special circumstances. In order to ensure message control, message heading and error detection and correction, control characters or control codes are used. Some of these control characters are explained in Table 10-3.

Hollerith Code

The Hollerith code shown in Table 10-4 is the most widely used code for punched card applications. It is a 12-level code that consists of two parts: zone and data bits. Only one of the three zone bits and one of the nine data bits are on any one code.

Punched cards are frequently used for customer billing and for entering information into a computer for batch processing.

Numeric Codes

Numeric codes represent number values by using binary digits. The most obvious numeric code is the simple binary number system. The following list outlines some special characteristics of numeric codes. Several of the more common numeric codes are tabulated following the list.

1. Weighted Codes. Some numeric codes give a *weight*, or value, to each bit position so that the value of the encoded number is found by adding the weights of the 1's in the code word. For example, look at the 8421 BCD code in Table 10-5. The code for 6_{10} is 0100. The subscript represents the base or radix of the number system.

$$0110 = 0 \times 8 + 1 \times 4 + 1 \times 2 + 0 \times 1$$
$$= 6_{10}$$

   Weighted codes are convenient for analog-to-digital and digital-to-analog conversion.

2. Self-complementing codes. An example of an unweighted code is the BCD Excess 3. This code is obtained by adding 3 to the equivalent of 8421 BCD. No direct weighting exists for this code, but it has the advantage of being self-complementing. If the bits in a code word are complemented (0's become 1's and vice versa) the resulting code word represents the so-called 1's complement. This is parallel to the 9's complement in the decimal system, which is obtained by subtracting the number from 9.

Table 10-3 Control Codes

NULL (null)	All zeros character, used for fill.
SYN (synchronous idle)	Used in synchronous transmission for character synchronization. At least two "sync" characters must precede every transmission.
SOH (Start of Header)	Signifies the start of a heading that will indicate routing information. The following character is the first of the heading.
STX (Start of Text)	Used at the beginning of a sequence of characters which are to be referred to as text.
ETX (End of Text)	Used at the end of text. A message error check character follows ETX and the receiving station must acknowledge the data received.
ETB (End of Block)	Signifies the end of a block of data. More blocks will follow.
EOT (End of Transmission)	Used at end of transmission or end of call. Also appears as the first character in a poll sequence.
ACK, NAK (Acknowledge or Negative Acknowledgment)	Sent by the receiving station to the transmitting station to indicate successful (ACK) or unsuccessful (NAK) reception of a message. The receiving terminal transmits a NAK in response to a "select" message if it is busy.
DLE (Data Link Escape)	Changes the meaning of a limited number of contiguously following characters.
ENQ (Enquiry)	Used as a request for a response from the remote station; typical response may be address or status control of station's buffer. The remote station may wish to "bid" for control of the line.
CAN (Cancel)	Disregard the accompanied data.

In Table 10-5, for example, 0101 Excess 3 represents 2_{10}. The 1's complement is 1010, which represents 7_{10}. Note that the 9's complement of 2_{10} is 7_{10}, i.e., $9_{10}-2_{10}$.

Self-complementing codes are useful in performing arithmetic operations on the digitized data. When wishing to subtract a number for instance, computers usually find its complement and then add. The reader may wish to refer to some basic digital text to review this technique.

3. Exact count codes. One error checking method involves using a code that has groups of bits in which every code word has the same number of 1's. This is called an *exact count* code, and the Biquinary code shown in Table 10-5 is an example of it. In the Biquinary code, each of the two bit

Table 10-4   Hollerith Punched-Card Code

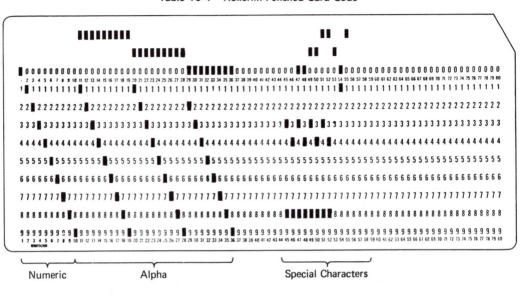

Numeric   Alpha   Special Characters

groups contains a single 1 for every code word. Note that Biquinary is weighted. Other error checking and correcting codes, such as the Hamming Code, are more complex.

4. Unit distance codes. When an adjacent code word differs by only one bit position, the code is called a unit-distance code. The most important example is the Gray code shown in Table 10-5. For example, in going from the decimal 3 (0011) to the adjacent code word of decimal 4 (0110), only one of the bits has changed. This is different from the BCD case where three bits are changed, i.e. 0011 to 0100. The Gray code can be extended beyond four bits, and algorithms have been developed to convert a binary number of any length to the Gray code and vice-versa.

   The Gray code is convenient when positional indications are required—for example, a shaft rotary position. When the shaft encoder is at the binary position 3 (0010) in BCD, if the second most significant bit is changed slightly before the two least significant bits, the output would indicate 7 (0111) in BCD—a highly erroneous number. In the Gray code, however, it would indicate the adjacent position.

In most codes, redundant or superfluous information is added to the original data in order to increase the probability of detecting an error. If enough redundancy is added, it is even possible to correct errors. However, as more redundant bits are added, the efficiency of the coding scheme is decreased. Effi-

Table 10-5 Several Numeric Codes

Decimal	8421 BCD	Excess 3	4221 BCD	Biquinary		Gray
Weights 8421	8421	—	4221	50	43210	—
0	0000	0011	0000	01	00001	0000
1	0001	0100	0001	01	00010	0001
2	0010	0101	0010	01	00100	0011
3	0011	0110	0011	01	01000	0010
4	0100	0111	1000	01	10000	0110
5	0101	1000	0111	10	00001	0111
6	0110	1001	1100	10	00010	0101
7	0111	1010	1101	10	00100	0100
8	1000	1011	1110	10	01000	1100
9	1001	1100	1111	10	10000	1101
10	—	—	—	—		1111
11	—	—	—	—		1110
12	—	—	—	—		1010
13	—	—	—	—		1011
14	—	—	—	—		1001
15	—	—	—	—		1000

ciency is defined as the ratio of information bits to total bits. Thus, for the Baudot teletypewriter code described earlier in this section, the system efficiency is

$$\eta = \frac{\text{Information pulses/character}}{\text{Transmission pulses/character}} \qquad (10\text{-}1)$$

$$= \frac{5}{7.42} = 76.4\%$$

For the seven-bit ASCII code, which has an eighth bit for parity

$$\eta = \frac{7}{8} = 87.5\%$$

10-3   Error Detection and Correction

Error control methods fall into two basic categories: (1) error detection with retransmission, and (2) forward-acting error correction.

The first method involves use of a backward channel in the system. When an error is detected, a retransmission request (ARQ) is sent back to the transmitter.

The second method not only detects error but, by a proper coding technique, also corrects the error via the receiving circuitry. Forward error correction is the only viable technique in situations where one transmitting source is connected to many receivers. (If each receiver, upon detecting an error, requested the transmitter to retransmit, the information throughput in the system would be exceptionally small because of all the time tied up in retransmitting.) Forward error correction is used on some space probes, where retransmission is often impossible. The long time delays would require excessive storage facilities in the probe's electronic circuitry.

To detect errors, a parity bit can be added to the bits describing the character. Even parity is obtained when one bit is added to the characters that have an odd number of 1 bits. If a bit is inverted upon transmission, the combination will result in an odd number of 1 bits, which indicates an error. Double errors will not be detected, as even parity will still be obtained at the detector.

This form of parity check, where a parity check is made on each character, is often called the Vertical Redundancy Check (VRC). It is often applied to the ASCII code since each character has only seven bits. It is not employed with the EBCDIC code since each character already consists of eight bits.

By further increasing the redundancy, more errors can be detected. The negative effect is that system efficiency is reduced. Efficiency, in this sense, is defined as the transmission of the largest number of characters regardless of errors. One code, the 3-out-of-7 binary code, for instance, allows only those sequences which have four 0's and three 1's to be transmitted.

The IA3, which is widely used in radio telegraphy, is an example of such a 3:4 constant ratio code. The IA3 code detects odd errors but not all even errors (for example, when a 1 replaces a 0 and a 0 a 1). The code allows 35 combinations out of 2^7, or 128, possibilities for an efficiency rate of $35/128 = 27\%$.

By using large blocks divided into many characters, it is possible with both row and column parity to detect not only all single but also most double and triple errors. In addition, all single errors can be corrected. When checking message blocks, a complete character known as a "Block Check Character" (BCC) may be appended to each block of information.

This BCC is formed at the transmitter by performing an "exclusive or" operation on all of the characters. All of the characters after the STX code, including the ETB or ETX code, are included in this operation. The resulting BCC is transmitted to the receiving station immediately following the ETB or ETX code character as illustrated in Fig. 10-1. The receiving station also performs the "exclusive or" on the message bits as they are received and compares this result with the BCC received at the end of the message. If these are identical, the received message is assumed to be error free. This method of block

Figure 10-1 Message BCC

Table 10-6 Block Check Character

CHARACTERS	STX	N	A	I	T	I	S	O	K	ETB.	BCC. ←Column Parity Bits
b_1	0	0	1	1	0	1	1	1	1	1	1
b_2	1	1	0	0	0	0	1	1	1	1	0
b_3	0	1	0	0	1	0	0	1	0	1	0
b_4	0	1	0	1	0	1	0	1	1	0	1
b_5	0	0	0	0	1	0	1	0	0	1	1
b_6	0	0	0	0	0	0	0	0	0	0	0
b_7	0	1	1	1	1	1	1	1	1	0	0
b_8	1	0	0	1	1	1	0	1	0	0	1

Message Bits (b_1–b_7) — Character Parity / Row Parity Bits (b_8)

(a) Transmitted Message Block

	N	A	I	~	I	S	O	K		
b_1	0	0	1	1	0	1	1	1	1	1	1 ←Column Parity Check
b_2	1	1	0	0	0	0	1	1	1	1	0
b_3	0	1	0	0	1	0	0	1	0	1	0
b_4	0	1	0	1	(1)	1	0	1	1	0	1 X
b_5	0	0	0	0	1	0	1	0	0	1	1
b_6	0	0	0	0	0	0	0	0	0	0	0
b_7	0	1	1	1	1	1	1	1	1	0	0
b_8	1	0	0	1	1	1	0	1	0	0	1

Row Parity Check (b_8)

X (under column ~)

(b) Detected Message Block

X Indicates an Error in the Parity Check

checking, often called a "Longitudinal Redundancy Check (LRC)," is used with the ASCII code only. In Table 10-6(a), for instance, a block of information that begins with a STX (start of text) character is encoded in ASCII with even parity. The ETB (end of transmitted block) character indicates that the character following is the block check character.

When a single error occurs, as in the case of the 1 in the received message of Table 10-6(b), an error occurs in the row and column parity checks. By checking for even parity in each row and column, it is possible to locate which bit is in error.

In low-speed applications (teletypewriters, for example), the message is often sent in duplicate and compared at the receiver to see if a match occurs. If the messages match, it is assumed that an error has not occurred. If there is no match, a request is sent to the sending station for a retransmission of the signal.

In high-speed synchronous applications, such as computer-to-computer data transfer, data is frequently transmitted in blocks with bits added for parity checking. The receiving station performs parity checks. If sent and received information matches, it is assumed that no error has occurred, and a positive acknowledgement (ACK) signal is sent to the transmitter terminal. If parity does not check, a negative acknowledgement (NAK) is sent to the transmitter, requesting a retransmission of the erroneous block.

For any particular application, there is an optimum block length for net data throughput (NDT), or the rate at which data bits arrive at the receiver end. If block lengths are extremely long there is a high probability of error and therefore retransmission requests are very likely. On the other hand, if block lengths are extremely short, the NDT suffers because a great deal of time is spent on signal control, or ACK and NAK. A typical NDT curve is shown in Fig. 10-2.

Another powerful error check method is the Cyclic Redundancy Check (CRC) discussed in depth in the following section. It is always employed with the EBCDIC code and at times with the ASCII code.

Hamming Distance

In detecting errors, the list of code words is restricted. If a message is received which is not on the restricted list, it must be in error.

An important property of a code is its Hamming distance, which is equal to the number of bits by which two code words differ. Thus, the Hamming distance between the words 1 0 0 1 0 0 and 1 0 1 0 1 0 is three, since the second, third, and fourth bits, counting from the right, are different.

A Hamming distance of one between code words, such as in the Baudot code, does not permit error detection since all combinations are valid. In the ASCII code with even parity and a Hamming distance of two, single errors can be detected since a change in any one bit produces an error code. If a code has a Hamming distance of three, any single error will be closer to its original code word than that of another code word. For instance, if the valid codes of a three bit word are 000 and 111—for a Hamming distance of 3—a change in one bit will still more closely resemble the valid code word and the bit in error can be corrected.

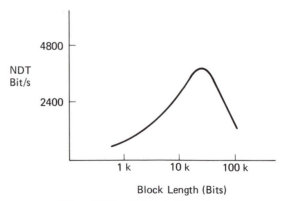

Figure 10-2 Net Data Throughput

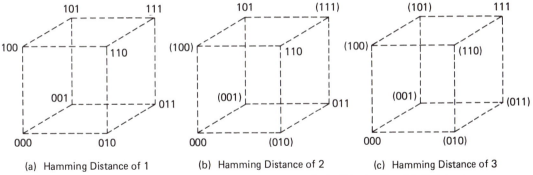

(a) Hamming Distance of 1   (b) Hamming Distance of 2   (c) Hamming Distance of 3

Figure 10-3 Geometric Models of Three-Digit Code Words. (Invalid Code Wrods in Parentheses.)

This concept is readily illustrated by the use of the geometrical models of Fig. 10-3. The vertices of the three dimensional cubes represent the valid or invalid code words. In Cube a, the Hamming distance is 1 and all the code words are valid. An error in any code word resembles another valid combination, thus not allowing for any detection of errors.

In Cube b, the Hamming distance is 2, and error detection is possible since any single error will result in an invalid code word. Since the distance from the invalid combination to adjacent valid code words are identical, no error correction is possible.

In Cube c, where the Hamming distance is three, both single and double error detection is possible. Single error correction is now also possible, since the resulting invalid code word will be closer to its original valid code word than the value of any other valid code word.

10-4   Cyclic Codes

In forward-error correction schemes, bits are added to the data in a way that allows for error detection and correction by the receiver. One such scheme is the combined row-column parity check discussed previously. This technique, however, is not foolproof since it is possible to have double bit errors in rows occur simultaneously with double bit errors in columns, which results in undetected errors. This is of particular concern in communication transmission systems where noise tends to be impulsive and bursts of errors result.

The cyclic codes are one set of codes which are very effective in detecting error bursts. The use of a feedback mechanism, by which any transmitted code word is very much dependent upon the previous history of the message, makes it highly unlikely for an error burst to produce an acceptable or undetected received, encoded message. The hardware required to perform cyclic coding is minimal, but the code efficiency is somewhat poor, as several check bits must be added to the message bits. The cyclic redundancy check code known as the CRC-12 has a 99.955% chance of detecting error bursts of 12 bits in length.

Cyclic redundancy codes add several bits (up to about 25) to the shifted message encoded bits. The total data, or message, block is usually considered one lengthy binary polynomial, $P(x)$. At the transmission end, this message block is divided by a fixed binary polynomial, called the generator polynomial $G(x)$, which results in a polynomial plus a remainder, $R(x)$. This remainder, known as the Block Check Character (BCC), is added to the message block to form the encoded transmitted block.

At the receiver end, the received data block is divided by the same generator polynomial, $G(x)$. If there is a remainder, an error has occurred. Either error correction is performed, or retransmission is requested.

The serial data bits of the various code words in CRC's are usually represented as polynomials for the sake of convenience. They are written in the form $x^n + x^{n-1} + \ldots + x^2 + x^1 + x^0$, where x^n represents the highest order bit present and x^0 represents the least significant bit. For example, the binary word 10101 is represented by $x^4 + x^2 + 1$. The degree of the polynomial is one less than the number of bits in the code word.

The addition or subtraction used in calculating the polynomial is called Modulo 2 addition and is performed by exclusive-OR gates. In this arithmetic

$$
\begin{aligned}
0 \oplus 0 &= 0 \\
0 \oplus 1 &= 1 \\
1 \oplus 0 &= 1 \\
1 \oplus 1 &= 0
\end{aligned}
\qquad (10\text{-}2)
$$

Subtraction is identical to addition. Adding two polynomials, for example, results in

$$
\begin{array}{r}
x^4 \qquad\; + x^2 + x \\
x^3 + x^2 + x \\
\hline
x^4 + x^3 \qquad\qquad
\end{array}
$$

Since multiplication is nothing but continued self-addition, when the two polynomials above are multiplied, the result is

$$
\begin{array}{r}
x^4 + x^2 + x \\
x^3 + x^2 + x \\
\hline
x^7 \qquad\; + x^5 + x^4 \\
x^6 \qquad\quad + x^4 + x^3 \\
x^5 \qquad\quad + x^3 + x^2 \\
\hline
x^7 + x^6 \qquad\qquad\qquad\quad + x^2
\end{array}
$$

Dividing $x^5 + x^3 + x$ by $x^2 + x + 1$ results in

$$
\begin{array}{r}
x^3 + x^2 + x \\
x^2 + x + 1 \,\big)\, \overline{x^5 \qquad + x^3 \qquad\quad + x} \\
x^5 + x^4 + x^3 \\
\hline
x^4 \\
x^4 + x^3 + x^2 \\
\hline
x^3 + x^2 + x \\
x^3 + x^2 + x \\
\hline
\end{array}
$$

If a remainder is present, it can always be expressed to a degree one less than the degree of the divisor.

The code words mentioned above can be defined as follows:

$M(x)$ Originating message polynomial consisting of m bits

$G(x)$ Generator polynomial

$T(x)$ Transmitted polynomial consisting of t bits

$E(x)$ Error polynomial

The encoded message, or transmitted polynomial, is obtained in several steps. First, the originating message, $M(x)$, is shifted c bits to the left, where c is the number of check bits to be added. The value of c is equivalent to the degree of the generator polynomial. The shifted message can be analytically expressed as $M(x)x^c$.

This shifted message is next divided by the generator polynomial, $G(x)$. The remainder of this division operation is added to the shifted message to generate the transmitted polynomial, $T(x)$. This polynomial can always be exactly divided by $G(x)$. Thus,

$$\frac{M(x)x^c}{G(x)} = Q(x) + \frac{R(x)}{G(x)} \qquad (10\text{-}3)$$

where $Q(x)$ is some quotient and $R(x)$ is the remainder, which has a degree one less than the generator polynomial. Multiplying through by $G(x)$

$$M(x)x^c = Q(x)G(x) + R(x) \qquad (10\text{-}4)$$

Since subtraction is identical to addition, then

$$T(x) = M(x)x^c + R(x) = Q(x)G(x) \qquad (10\text{-}5)$$

Thus, it is proven that $T(x)$ can always be exactly divided by $G(x)$.

Consider the following example:

1. Let Message polynomial $M(x) = 101101$ ($x^5 + x^3 + x^2 + 1$), and Generator polynomial $G(x) = 11001$ ($x^4 + x^3 + 1$). $G(x)$ contains 5 bits and is of degree 4. Therefore, the $M(x)$ must be shifted by $c = 4$ bits. The following shifted polynomial is obtained by multiplying $M(x)$ by x^4:

$$M(x) \cdot x^4 = (x^5 + x^3 + x^2 + 1)(x^4) = x^9 + x^7 + x^6 + x^4$$

2. Dividing this product by the generator polynomial $G(x)$ gives

$$
\begin{array}{r}
x^5 + x^4 + x^2 \\
\hline
x^4 + x^3 + 1 \,\big|\; x^9 + \quad\; + x^7 + x^6 \quad\quad\; + x^4 \\
x^9 + x^8 \quad\quad\quad\quad\; + x^5 \\
\hline
x^8 + x^7 + x^6 + x^5 + x^4 \\
x^8 + x^7 \quad\quad\quad\; + x^4 \\
\hline
x^6 + x^5 \\
x^6 + x^5 \quad\quad\; + x^2 \\
\hline
+ x^2
\end{array}
$$

remainder →

The remainder, $R(x) = x^2$, is added to $M(x)x^4$ to give the transmitted polynomial

$$x^9 + x^7 + x^6 + x^4 + x^2$$

Thus, the transmitted code word is

$$1011010100$$

If there is no error at the receiver, this message will produce no error when divided by the same generator polynomial.

As a check

$$
\begin{array}{r}
x^5 + x^4 + x^2 \\
x^4 + x^3 + 1 \;|\; \overline{x^9 \qquad\quad + x^7 + x^6 \qquad\qquad + x^4 \qquad\quad + x^2} \\
\underline{x^9 + x^8 \qquad\qquad\qquad + x^5} \\
x^8 + x^7 + x^6 + x^5 + x^4 \\
\underline{x^8 + x^7 \qquad\qquad\qquad + x^4} \\
x^6 + x^5 \qquad\quad + x^2 \\
\underline{x^6 + x^5 \qquad\quad + x^2} \\
0
\end{array}
$$

no remainder

To implement this particular code, the feedback shift register shown in Fig. 10-4 is used. The number of shift register gates required is equivalent to the order of $G(x)$ and c.

After initially clearing the shift register, the message, or dividend, is entered. The highest power of x corresponds to the first digit which arrives. Since no feedback occurs until a 1 dividend digit arrives at the far or left end, the first four operations are merely simple shifts. Whenever a 1 digit arrives at the left

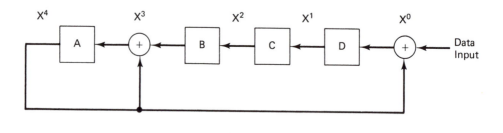

Shift Register

Exclusive OR

Figure 10-4   Circuit to Obtain Remainder Bit with $G(x) = x^4 + x^3 + 1$

end, a 1 is subtracted from the second and fifth digits on the next shift. This is identical to the process just outlined. The message bits are followed by c 0's. What is left in the register forms the remainder, which is appended to form the transmitted message. For the message and generator polynomial previously given, the sequences in the shift register are given in Table 10-7.

The full encoder is illustrated in Fig. 10-5. The delay of 4 shifts (c shifts, in general) is necessary to allow computation of the remainder. Initially the switch is in position 1, which allows the remainder to be calculated after $M(x)x^c$ is divided by $G(x)$. After the message is shifted by adding 4, or c, 0's, the switch is moved to position 2. This allows the remainder, which is located in the register, to be added to the message bits. The identical circuit can be used for decoding. After the received message $R(x)$ has been added or shifted in, the register contains the remainder. If an error has occurred, this remainder will be non-zero.

Often, higher-order generator polynomials are used. One example is the CRC-16 code that has a generator polynomial of $G(x) = x^{16} + x^{15} + x^2 + 1$. It can be implemented by the shift register shown in Fig. 10-6.

Table 10-7   Shift Register Sequence wWith $G(x) = x^4 + x^3 + 1$

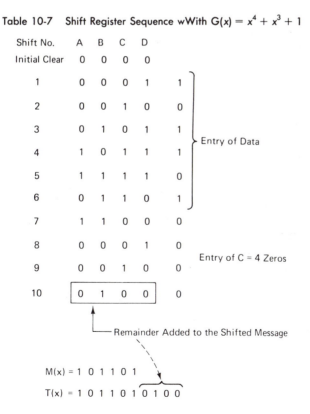

Shift No.	A	B	C	D	
Initial Clear	0	0	0	0	
1	0	0	0	1	1
2	0	0	1	0	0
3	0	1	0	1	1
4	1	0	1	1	1
5	1	1	1	1	0
6	0	1	1	0	1
7	1	1	0	0	0
8	0	0	0	1	0
9	0	0	1	0	0
10	0	1	0	0	0

Entry of Data

Entry of C = 4 Zeros

Remainder Added to the Shifted Message

$M(x) = 1\ 0\ 1\ 1\ 0\ 1$

$T(x) = 1\ 0\ 1\ 1\ 0\ 1\ 0\ 1\ 0\ 0$

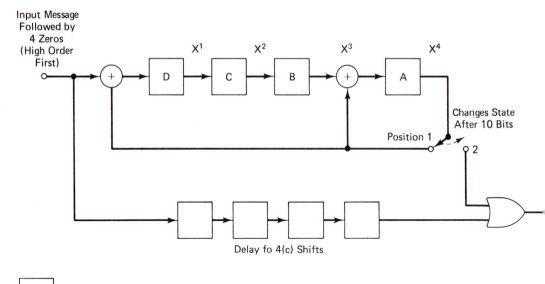

Figure 10-5 Complete Encoder for the Cyclic Code $G(x) = x^4 + x^3 + 1$

Hamming Codes

As discussed in the previous section, the ability to detect errors is related to the Hamming distance between code words. This, in turn, is related to the ability to correct errors. Table 10-8 outlines these relationships.

As an example, consider a case in which the Hamming distance is 3. If a single error occurs, correction can be attempted by trying to change each bit. If the wrong bit is changed, a double error will occur, which is detectable. Only the erroneous bit, when changed, will result in a correct combination.

One of the first cyclic codes devised for error correction was developed by R. W. Hamming of Bell Laboratory. The Hamming code can correct for single errors and detect double errors in a block.

To develop a Hamming code for single-error correction, assume that c check bits are added to the message, M, to form the transmitted word, T, of t bits. With c check bits, 2^c possible error patterns can be indicated. If one of these combinations indicates no error, then $2^c - 1$ combinations remain to indicate where the error is located. Therefore, the length of the transmitted code word, t, must be equal to or less than $2^c - 1$, or

$$t + 1 \leq 2^c \qquad (10\text{-}6)$$

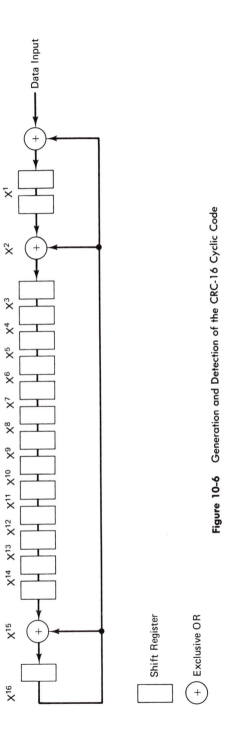

Figure 10-6 Generation and Detection of the CRC-16 Cyclic Code

Table 10-8

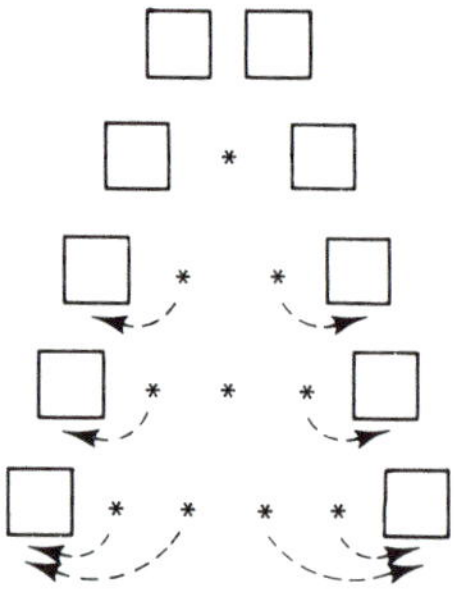

Hamming Distance	Model	Errors Detected	Errors Corrected
1		0	0
2		1	0
3		2	1
4		3	1
5		4	2
d Even		d–1	$\dfrac{d-2}{2}$
d Odd		d–1	$\dfrac{d-1}{2}$

☐ Legal Code Word

* Illegal Code Word

➤ ⌐ Error Correction

For convenience, transmit the maximum allowable code word, or

$$t + 1 = 2^c \qquad (10\text{-}7)$$

If the message consists of m bits, then

$$t = m + c \qquad (10\text{-}8)$$

$$m + c + 1 = 2^c$$

$$m = 2^c - c - 1 \qquad (10\text{-}9)$$

Table 10-9 gives the number of check bits required for a message m bits long.

Table 10-9   Number of Check Bits Required for Single-Error Correction

c (check bit)	m (message length)
2	1
3	4
4	11
5	26
6	57
7	120

Table 10-10

m_3	m_2	m_1	c_2	m_0	c_1	c_0	bit notation
7	6	5	4	3	2	1	bit location

As an example of how such a scheme is implemented, number the transmitted code word of 7-bit length from right to left. From Table 10-9, $m = 4$ and $c = 3$. In the 7-bit code shown in Table 10-10, the check bits, or parity bits, are c_2, c_1, c_0, and the message bits are m_3, m_2, m_1 and m_0. (The subscripts give the weighting of a particular bit.) Now consider the rationale for the location of the check bits.

Upon detection, the check and message bits are checked for parity. If *binary weighting* is applied to the three parity checks of p_2, p_1, and p_0, and p_0 is the LSB, the following will be obtained for even parity:

$$p_0 = 1 \text{ when the error is in location } 1, 3, 5, 7$$
$$p_1 = 1 \text{ when the error is in location } 2, 3, 6, 7$$
$$p_2 = 1 \text{ when the error is in location } 4, 5, 6, 7$$

If each parity is required to check one unique check bit in addition to the three message bits, then by observing the above, the c_0 check bit should be in location 1, c_1 check bit in location 2, and c_2 check bit in location 4. The other locations are excluded because the bits in these locations are repeated.

At the encoder, even parity for a binary weighting on the check bits is obtained when

$$c_0 \oplus m_0 \oplus m_1 \oplus m_3 = 0$$
$$c_1 \oplus m_0 \oplus m_2 \oplus m_3 = 0$$
$$c_2 \oplus m_1 \oplus m_2 \oplus m_3 = 0$$

Since subtraction is the same as addition

$$c_0 = m_0 \oplus m_1 \oplus m_3$$
$$c_1 = m_0 \oplus m_2 \oplus m_3 \qquad (10\text{-}10)$$
$$c_2 = m_1 \oplus m_2 \oplus m_3$$

Determine the Hamming code word which corresponds to the message m_3, m_2, m_1, $m_0 = 1101$. Use even parity.

Example 10-1

$$c_0 = 1 \oplus 0 \oplus 1 = 0$$
$$c_1 = 1 \oplus 1 \oplus 1 = 1$$
$$c_2 = 0 \oplus 1 \oplus 1 = 0$$

The coded word, T, is 1100110.

Assume that a single error has occurred during the transmission and that the received message is

1	1	0	1	1	1	0
m_3	m_2	m_1	c_2	m_0	c_1	c_0

Checking for even parity:

$$p_0 = c_0 \oplus m_0 \oplus m_1 \oplus m_3 = 0 \oplus 1 \oplus 0 \oplus 1 = 0$$
$$p_1 = c_1 \oplus m_0 \oplus m_2 \oplus m_3 = 1 \oplus 1 \oplus 1 \oplus 1 = 0$$
$$p_2 = c_2 \oplus m_1 \oplus m_2 \oplus m_3 = 1 \oplus 0 \oplus 1 \oplus 1 = 1$$

Thus, bit 4 ($p_2 p_1 p_0 = 100$) is in error and the correct word is 1 1 0 0 1 1 0.

Odd parity, can be used as well as even parity. For longer messages, the check bits are located in positions which have a power of 2; that is,

$$2^0, 2^1, 2^2, 2^3, 2^4$$

Although the error detection and correction schemes outlined are very efficient, they are often incompatible with the types of codes required in a typical communication system. For instance, to assure synchronization of the receiver at all times, the encoded signal should undergo transitions continuously. In addition, due to hybrid transformers in a cable system, dc cannot be transmitted and, therefore, the number of 1's and 0's cannot be balanced out to give a 0 dc level. These considerations conflict with the efficient coding schemes just described.

10-5 Line Codes

As just discussed, ideal line codes have their own desired properties. The following list outlines major considerations related to these properties.

1. All cable systems, and most other communication systems, will not allow dc transmission due to either transformer action or, more likely, some low frequency, cut-off restriction. In the case of NRZ (non-return-to-zero) coding (see Fig. 10-7), this restriction forces all the positive and negative elements to have equal probabilities. For this reason, there should not be long strings of element sequences that maintain the same polarity.

2. Because any code adds redundancy, the code efficiency should be as high as possible. Increasing the element rate, or increasing the number of distinguishable levels, requires either an improvement in the S/N ratio or an increase in the bandwidth, both of which are costly.

3. In order to obtain synchronization or timing at the receiver, the receiver usually locks in on the carrier being received. The derivation of the timing can be obtained only if the line signal undergoes a sufficient number of 0 crossings; that is, the transmitted signal should always undergo transitions.

4. In order to minimize crosstalk between channels, the amount of energy in the signal at low frequencies should be small.

Some of the line codes which take into account these considerations are the Alternate Mark Inversion (AMI) code, also known as the Bipolar, and the High Density Bipolar code (HDB3).

Binary 0 1 1 0 0 0 0 1 0 1 0 0 0 0 1 1 0 0 1 1 0 0 0 0 0

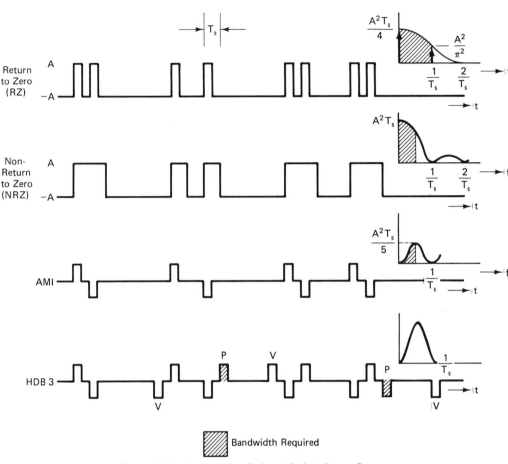

Figure 10-7 Typical Line Codes with their Power Spectra

The AMI code is a three-level code (+, −, and 0). The 0 level represents the binary zero, and the alternative + or − represents the binary 1. Figure 10-7 shows the implementation of the AMI code, whereby the polarity of the alternate pulse representing marks are inverted. AMI suffers from excessive redundancy, but it readily detects single errors. A single error causes either an extra pulse to appear on the line or a pulse to be removed. In either case, the result is two successive pulses with the same polarity.

AMI has a significant advantage in immunity to crosstalk when compared to the NRZ or RZ code schemes. This can be observed from their frequency spectrums as shown in Fig. 10-7.

If a long sequence of 0's is transmitted, no transition can occur, and the AMI code can cause a loss of synchronization at the receiver. To overcome this difficulty, AMI is modified to the High Density Bipolar code (HDB). The HDB3 will allow a maximum of three 0's to occur. After that, a violation pulse is added. The rules for transforming from AMI to HDB3 are:

1. Every fourth 0 in a sequence of 0's is converted to a violation pulse V. Successive V pulses are of opposite polarity.

2. The number of pulses between successive V pulses must be odd to ensure that each V pulse is opposite in sign to the preceding V pulse. If the insertion of a V pulse does not result in a bipolar violation, as in the case of an even number of pulses between successive V pulses, an additional pulse (P) is inserted at the location of the first 0.

Decoding is achieved by replacing every violation pulse V and the three digits preceding V or 0's. The HBD code is quite redundant and has a low noise immunity.

Manchester Coding

Local Area Networks (LAN) such as Ethernet and Cheapernet are increasingly using the digital biphase or Manchester code for signal transmission over the network. It uses the phase of a square wave signal to indicate a "1" or a "0" as shown in Fig. 10-8. A 1 is encoded as a low-to-high transition and a 0 as a high-to-low transition. Each bit has a transition at its center, thus providing a good reference for timing recovery.

For two like adjacent bits, a bit boundary is present that corrects the line polarity for encoding the second bit. Between unlike bits, the bit boundary transitions are absent, leaving only bit-center transitions. The preamble in the beginning of a message packet consists of an 8-byte field of alternating 1's and 0's, permitting the receiver to be synchronized on the bit-center transitions.

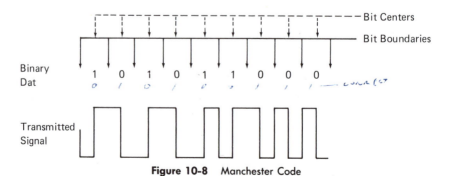

Figure 10-8 Manchester Code

If the pulses are positive and negative around zero volts, the pattern contains no dc. If the zero voltage level resides elsewhere, the Manchester-encoded signal will have a constant dc component. This feature is used in the IEEE-802.3 protocol for detection of collisions. When a collision occurs, i.e., when another transmitter begins to deliver a signal to the network, the current-mode drivers cause the dc voltage on the line to double as the two currents will add. Collisions are detected by monitoring the dc voltage on the cable, averaged over several bits.

10-6 Channel Throughput and Efficiency

Given a channel with a certain bandwidth and bit error rate, there is an optimum operating condition that will provide for a maximum net data throughput (NDT). The NDT is a measure of the usable characters detected at the receiver. It distinguishes between the total number of bits in the text, including all redundant bits, and the number of correct bits detected in a specified period of time.

Consider as an example the technique of requesting retransmission (ARQ) of a block of information after an error is detected. After a block of information is sent, the transmitting terminal stops and waits for a reply from the receiving terminal regarding whether or not the block was received with no error. If the block is received with no detected error, a positive acknowledgement (ACK) is sent to the transmitter and the transmitter is free to send a new block of information. If an error is detected, a negative acknowledgement (NAK) is returned, requesting a retransmission of the block. In the case of half-duplex transmission, where a channel is operating in both directions but not simultaneously, this technique results in a certain dead time, which reduces the NDT.

Figure 10-9 illustrates the procedure which uses a block-by-block data transfer scheme. In this method, a positive or negative acknowledgement is expected after a message block is transmitted. Special time-out, or shut-down, procedures follow if no response is received. If an ACK signal is received, the next block of information is transmitted; a NAK signal induces a retransmission of the error block.

The dead time, T, between the end of the transmission of one block and the beginning of the transmission of the next block may include:

1. *Modem turn-around times.* (Generally in the case of half-duplex operation.) This delay varies with the type of modem. The most significant time delay occurs between the Request to Send signal in the Data Terminal Equipment (DTE) and the Ready for Sending response in the Data Communication Equipment (DCE).

   Modems with adaptive equalizers also have a period of adjustment during which a short, predetermined pulse pattern is transmitted. This

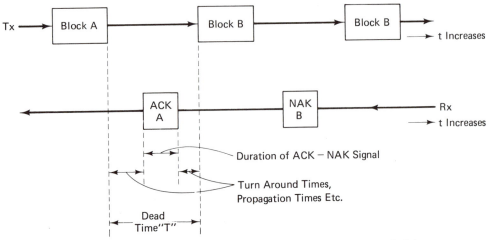

Figure 10-9 Illustration of Dead Time and Turn-Around Time on a Block-by-Block Data Transfer Scheme

period, which can take up to 100 ms, must be included in the modem turn-around time.

2. *Loop propagation time of the system.* Typical propagation times are:

> Unloaded cable = .02 ms/km
>
> polyethylene coaxial line = 5 μs/km
>
> radio/microwave propagation = 3.33 μs/km

A link to a geostationary satellite placed 35,800 km above the equator at a slant height of 41,200 km (5° elevation) would require a propagation time of 275 ms from the earth station to the satellite and back again. Thus, the total delay for a satellite loop between two earth stations is around 550 ms, depending upon the exact location of the earth stations. Additional propagation delays occur in the modem filters, equalizer circuits, and other reactive circuits. These additional delays add a few milliseconds to the total delay.

3. *Transmitter and receiver delays.* The DTE at the receive end must check for errors and assemble the message after receiving the last character. It then returns an acknowledgement that must be interpreted by the DTE at the transmit end for appropriate action. These delays are minimal and usually ignored.

4. *Length of the return message.* This message may vary from 1 bit to 32 bits. Depending upon the bit rate of the return channel, this delay may or may not need to be taken into account. On a 75 baud backward channel, for instance, 32 bits would take 32 b/75 b/s, or 426 ms.

5. *Removal of echo suppressors.* Echo suppressors are often added on long transmission paths to attenuate echos from the far end. Since it takes about 150 ms to remove an echo suppressor, a turn-around time of 300 ms per block can be expected. On full-duplex operation, these echo suppressors are removed during the starting-up procedure and remain OFF for the duration of the hook-up.

A typical curve showing how the NDT varies with block length is illustrated in Fig. 10-2.

Small block lengths reduce the NDT because the relative percentage of dead time compared to available transmission time is large, even though the probability of error in any individual block is extremely small. On the other hand, extremely long block lengths have a high probability of error which results in a high likelihood of retransmission requests, which results in a reduction of NDT.

Data rates can be increased by using a circuit with a reverse, or secondary, channel in a duplex system. In such a system, a primary channel with a high signaling capacity and a secondary channel that flows in the opposite direction are present simultaneously. (In a full-duplex channel, the signaling rate capability is the same in both directions.) In data transmission, the information block is put on the primary channel in the forward direction, and the ACK or NAK signals are placed in the secondary or reverse channel.

One half-duplex system with a backward channel for error control is the 1200-baud modem, which fits the CCITT standard for the general switched telephone network. It transmits data through the use of a 2100 Hz tone that represents a SPACE, or binary 0, and a 1300 Hz tone that represents a MARK, or binary 1. This method of modulation is called frequency shift keying (FSK). The backward channel (MARK, 390 Hz; SPACE, 450 Hz), is used for error control and can handle up to 75 baud. Fig. 10-10 illustrates the frequency

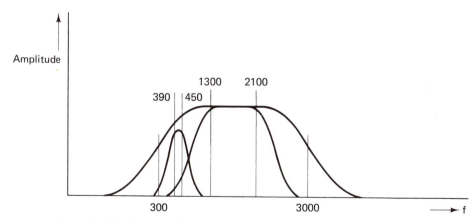

Figure 10-10   Frequency Spectrum Utilization of a 1200-Baud Half-Duplex System with Backward Channel

spectrum used by both the forward transmitted data and the reverse channel.

In an error-free circuit, the throughput efficiency of the data transmission system can be calculated by obtaining the ratio of the time taken in transmitting a data block (B/R) to the sum of the data block times and dead time *T*. This is given by:

$$\eta = \frac{\dfrac{B}{R}}{\dfrac{B}{R} + T} \times 100\% \qquad (10\text{-}11)$$

where η = the throughput efficiency
B = the block length in bits
T = the total dead time
R = the input data rate in b/s

Equation *(10-11)* ignores the losses due to parity and synchronization bits sent with each data block. If desired, these can be taken into account separately.

Example 10-2

Obtain the throughput efficiency of the 1200-baud half-duplex system shown in Fig. 10-10. Assume the following conditions:

$B = 1,000$ bits
Length of backward or return message = 16 bits
Length of path employing polyethylene cable = 800 km
Modem turn-around time = 200 ms
Accumulation of modem delay times, Tx and Rx-delays = 12 ms
For this system $R = 1200$ b/s
Time taken for the 16 bits of the return message at 75 b/s = $^{16}/_{75} = 213$ ms
Time taken for propagation delay (loop delay) = $5\mu s/km \times 2 \times 800 = 8$ ms
Time taken for modem turn-arounds = 400 ms
Time taken for modem Rx + Tx delay = 24 ms
Thus $T = (213 + 8 + 400 + 24)$ ms = 645 ms

$$\eta = \frac{\dfrac{1000}{1200} \times 1000}{\dfrac{1000}{1200} \times 1000 + 645} = 56.4\%$$

If errors are present, efficiency is further reduced by a factor of $1 - P(e)$, when $P(e)$ is the block error probability. If one block in 100 is in error, then $P(e) = .01$ and the throughput efficiency is

$$56.4 \, (1 - .01) = 55.8\%$$

Consider a full-duplex, four-wire system in which a continuous transmit-and-receive channel is available. This system avoids any modem turn-around problems. Assume the same conditions presented in Example 10-2. Also assume a return channel of 1200 b/s. Since there are no turn-around problems, the following T and throughput efficiency is obtained:

Example 10-3

$$T = \left(\frac{16}{1200} \times 1000 + 8 + 24 \right) \text{ms} = 45.3 \text{ ms}$$

$$\eta = \frac{\frac{1000}{1200} \times 1000}{\frac{1000}{1200} \times 1000 + 45.3} = 94.8\%$$

This marks a significant increase in efficiency compared to the system with modem turn-around time involvement.

By increasing the data signaling rate, the time taken to transmit each block, (B/R), will be reduced and T will become proportionally larger. This will result in a reduced throughput efficiency. To compensate for this, it would be desirable to use longer blocks for the higher data rates. Longer blocks, however, have a greater probability of error. For this reason, some optimum block length such as that illustrated in Fig. 10-2 can be obtained for any particular system.

With this block-by-block transmission, the transmission efficiency is relatively poor for long dead times. For example, for a half-duplex satellite interconnection with a total loop delay of 550 ms, the efficiency of Example 10-2 is reduced to 41%.

An alternative method that gives greater throughout efficiency is a continuous mode of transmission whereby data blocks are transmitted without interruption, unless a negative acknowledgement is observed. When a NAK is transmitted back to the transmitter, the error block is retransmitted. This technique, which requires a full display system, is illustrated in Fig. 10-11.

The continuous block transfer system avoids the dead time of the block-by-block scheme but requires much more storage, or buffering. In block-by-block transmission, only one block at a time needs to be stored at either end. The next block is not sent until the first block has been correctly received. In continuous block transmission, when one of the blocks is not received correctly, all of the subsequent blocks must be stored until the error block is acknowledged. Such a system can run out of buffer space very quickly, particularly if several blocks are in error or if one of the input lines is error prone. This can cause the entire system to slow down or cut off, which can also happen when equipment problems or severe signal fading occurs on one line. A method of resolving this problem by limiting the storage at each end to 7 blocks is discussed in the next chapter.

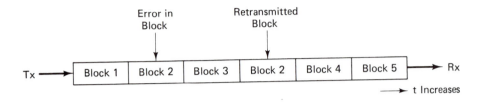

Figure 10-11 Continuous Block Transfer Scheme

Problems

1. Indicate which of the following codes are weighted and which are not weighted.
a. 8421 BCD
b. BCD Excess 3
c. 4221 BCD
d. Biquinary
e. Gray Code

2. What is the weighting of the following codes?

a. Decimal	Code	b. Decimal	Code
0	0000	0	00000
1	0001	1	00001
2	0011	2	00011
3	0100	3	00111
4	0101	4	01111
5	0111	5	10000
6	1000	6	11000
7	1001	7	11100
8	1011	8	11110
9	1100	9	11111

3. Which of the codes mentioned in Problems 1 and 2 are self-complementing?

4. Fill in the parity column in each of the following tables to add the indicated parity bit to each word.

a. Code	Even Parity	b. Code	Odd Parity
01001		1000001	
01100		1000010	
10000		1000011	
11111		1000100	
01100		1000101	
10100		1000110	
11101		1000111	
01110		1001000	
00000		1001001	
10101		1001010	

5. Briefly explain what each of the following terms means, and name a code which exemplifies each term.
a. Exact Count Code
b. Unit Distance Code

6. The following message is to be transmitted in ASCII with odd parity for the character check (row, or horizontal, parity bit) and with even parity for the block check character (column, or vertical, parity check). Give the transmitted message block.

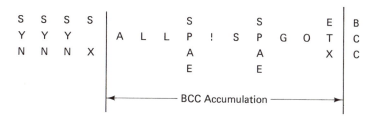

7. Why is the net data throughput lower when the block length is exceptionally short or long?

8. Why are cyclic codes effective in detecting error bursts?

9. The message 1001001010 is to be transmitted in a cyclic code with a generator polynomial of $G(x) = x^2 + 1$.
a. How many check bits does the encoded message contain?
b. Obtain the transmitted code word.
c. Draw the encoding arrangement to obtain the remainder bits.
d. The decoder has the same arrangement as the encoder. After the received word is clocked into the decoder input, what should be the content of the register stores?

10. Sketch the encoding/decoding arrangement for the 16-bit generator polynomial $x^{16} + x^{12} + x^5 + 1$. (CCITT recommendation is V4.)

11. a. What is the "Hamming distance"?
    b. For a Hamming distance of 5, how many errors can be detected? How many can be corrected?

12. a. Determine the Hamming coded word for the message $m_3, m_2, m_1, m_0 = 1010$. Employ even parity and apply binary weighting to the three check bits.
    b. If the received message is 1110010, what is the correct transmitted word, assuming that only a single error may occur?
    c. Regardless of the message length, where should the check bits be located?

13. a. List the properties a line code should have.
    b. Describe how well the AMI code meets the conditions of (a).

14. a. Describe why the continuous block transfer scheme has a better throughput efficiency than the block-by block data transfer scheme. List one disadvantage of the continuous block transfer scheme.

15. List several of the chief contributors to the dead time in the block-by-block transfer scheme.

16. a. Obtain the throughput efficiency of a half-duplex system under the following conditions:

    Block length (B) = 5000 bits.

    Forward (and backward) data rates (R) = 2400 b/s

    Length of return message = 16 bits

    Length of single satellite loop = 82,400 km

    Modem turn around time = 100 ms

    Accumulation of modem delay times, Tx and Rx delays = 12 ms

    b. If the data signaling rates of (a) are increased to 4800 b/s, what happens to the throughput efficiency?
    c. Based on (b), what should happen to the block length as the signaling rate is increased? Since the block error probability rises as the block length is increased, is there some optimum block length? Explain.

17. Why is a greater throughput efficiency obtained in a full-duplex mode than in a half-duplex mode in block-by-block transmission?

18. Encode the following binary data stream into the RZ, NRZ, AMI, HDB3 and Manchester codes. Assume an initial positive going pulse for the first "1" in the AMI and HDB3 codes.

    1 1 0 0 0 0 1 0 0 0 0 0 0

11

Network Protocol*

* Chapters 11 and 12 were authored by Tom McGovern, who is presently with the Computer Systems group at the Northern Alberta Institute of Technology in Edmonton.

11-1   Introduction

Network is a broad term similar in meaning to *system*. Both are used to describe organized complexity. A network implies size, remoteness, and interrelationship. Data Communication Networks are concerned principally with the transfer of information between computer terminals. As the number of terminals and computers in a network increases, problems concerned with the design, efficiency, cost, and control of such a network become paramount.

To begin with, a network may be described as an interrelated group of nodes. These nodes may be terminals or junctions. Terminals are any devices connected to the network, including computers and computer terminals, for the purpose of sending or receiving information. Junctions perform such functions as completing a link between two terminals (switching), deciding in which order the various information transfers should take place (scheduling), and allocating temporary storage of information along the route of its travel (buffering). Figure 11-1 illustrates the basic components of a network and the interrelationship of the nodes.

11-2   Switching

The need for sophistication in protocols developed as networks grew in complexity. The first connections between terminals were made point-to-point, as shown in Fig. 11-2. In point-to-point connection, each terminal is connected to any other terminal by an individual and permanent line. The obvious disadvantages of this system include cost and underutilization of lines; that is, the user pays for the line whether it is being used or not.

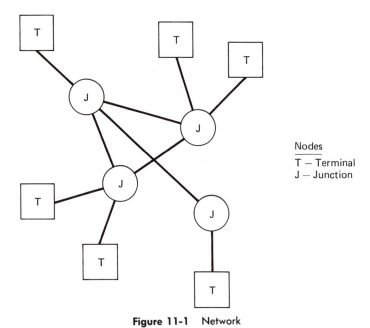

Nodes

T — Terminal
J — Junction

Figure 11-1   Network

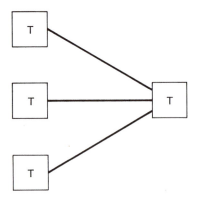

T — Terminal

Figure 11-2   Point-to-Point Connection

Circuit and Multipoint Switching

One solution found to the drawbacks of point-to-point switching was circuit switching, which is shown in Fig. 11-3. In this system, a switching center is created to provide a path between two terminals for the duration of the communication. The user pays only for the time connected and for sharing the lines and switching centers with other users.

An alternative improvement over point-to-point connection is a multipoint, or multidrop, network that facilitates a one-to-and-from-many terminal connection. This is shown in Fig. 11-4. One terminal controls the data transfer between itself and all other terminals by means of polling.

Both circuit switching and multipoint networks have their drawbacks. Circuit switching times can be high and therefore present a problem for interactive systems. Most users of a remote inquiry system would expect an almost immediate response—for example, in checking customer credit ratings. Even a delay of a few seconds, when experienced on a regular basis, becomes intolerable.

In multipoint systems, polling implies a programming overhead; that is, time must be spent checking each terminal to find out if there is a requirement for data transfer from a particular terminal. This is accomplished by the execution of a program in the controlling terminal. The cost of polling is normally distributed among the terminal users. Another problem of multipoint systems is underutilization of terminals due to the fact that there is only one connecting line.

Packets

Two recent developments in computer communication systems that herald the development of new methods of information handling are the emergence of worldwide data networks and the use of packet switching.

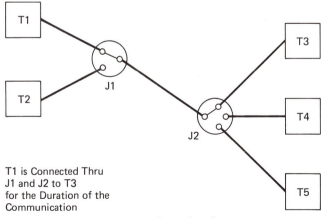

T1 is Connected Thru
J1 and J2 to T3
for the Duration of the
Communication

Figure 11-3   Circuit Switching

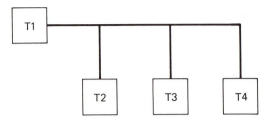

T1 is the Controlling Terminal

Figure 11-4 Multipoint Network

The establishment of standards within countries and across continents is essential to the development of such data processing applications as electronic funds transfer. The network of the future must be accessible to a wide variety of users, and it must perform its own housekeeping functions; i.e., it must be intelligent and use its resources efficiently.

The second development, packet switching, is a refinement of message switching that involves transferring messages between terminals as units of information (See Fig. 11-5). The message travels through the network with a header that precedes the message and identifies the destination. Subsequent messages may take entirely different routes.

In packet switching, information is transferred between terminals in discrete, variable-length units called packets. Fig. 11-6 shows that each packet consists of:

a. a header that contains control data and the destination address

b. the message or data field that is transparent to the network.

A Frame Check Sequence (FCS) is attached to the packet for error control at the local level.

Transparency is a term frequently used on computer systems to describe a lack of awareness. For example, the user is often unaware of how much the operating system does to satisfy a request made in a high-level language. It can be said that the operating system is "transparent" to the user. In this regard, it can be claimed that the network and the software which drives the network have no interest in the actual information being transmitted. The network is concerned about the source, the destination, the quantity, and the accuracy of the information, but the information itself is transparent to the network.

Packets are transmitted via specialized computers which interpret header information, check for accuracy, and route the data. Packet switching, as shown in Fig. 11-7, does not physically link two terminals. (Compare this with circuit switching.) The network facilities are charged primarily on the basis of quantity of information rather than connection time. Packet switching systems are becoming less sensitive to distance with respect to charging than they were in the past.

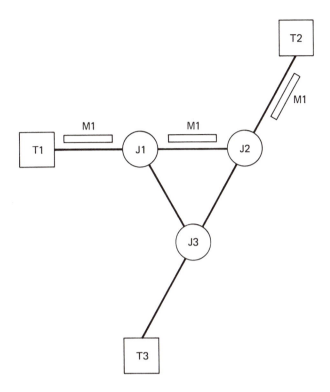

Message M1 is Transmitted from Terminal T1
to Terminal T2 via Junctions J1 and J2

Figure 11-5   Message Switching

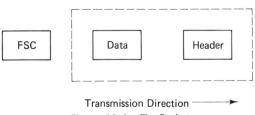

Transmission Direction ⟶

Figure 11-6   The Packet

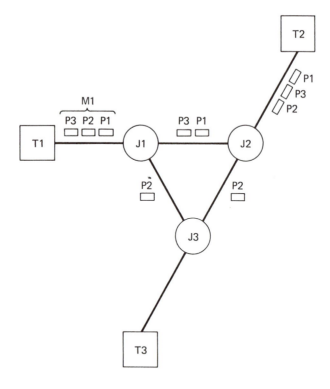

Message M1 Becomes Packets P1, P2, P3
P1 and P3 are Transmitted from T1 to T2 via J1, J2
P2 is Transmitted from T1 to T2 via J1, J3, J2
The Packets are Returned to their Original Sequence
at Terminal T2

Figure 11-7   Packet Switching

11-3   Protocol Levels

As networks become more complex and wide-reaching and as the requirements for interlinking grow, the need for international protocol standards grows proportionately.

Protocol is defined as a "set of rules." Most familiarly used in a military or diplomatic sense, it is now applied to data communication systems. There are protocols for making connections, transferring messages, and performing enough other functions that they may be considered as a hierarchy of levels or layers. The establishment of a connection between two terminals would obey the lowest level of protocol. The transfer of a file of information to solve a specific problem would follow the protocol of a higher level on the hierarchy, and so on.

Protocol may be defined in terms of its level of "closeness" to the hardware in the same way that computer languages are categorized as high or low level. The International Standards Organization (ISO) has proposed a seven-level architecture called the Reference Model of Open Systems Interconnection, which may be remembered by the palindromic acronym ISO-OSI (Fig. 11-8). The ISO-OSI is a set of guidelines, not a well defined standard.

The seven levels of a complete network are defined as follows:

Level 1 is the physical level. Protocols at this level involve such parameters as the signal voltage swing and bit duration, whether transmission is sim-

Layered Level	Function	Example Approaches	
1. Physical	Actual means of a bit transmission across a physical medium.	EIA RS 232 EIA RS 422 EIA RS 449 CCITT X.21	
2. Data Link	Enables logical sequences of messages to be reliably exchanged across a single physical data link.	ISO—HDLC ANSI—ADCCP IBM—SDLC, BSC DEC—DDCMP CCITT X.25	
3. Network Control Level	Intranetwork operations, addressing and routing. Switching control terminal connections	CCITT X.25, X.21 ANSI X281	O F T E N
4. Data Flow Control	End-to-end control and information exchange. Reliability and error control.		M E R G E D
5. Session Control	Support of session dialog if programs and services are available.	ARPANET FTP DECNET DAP	
6. Presentation Control	Compactation, encryption, peripheral device coding and formatting.	System defined requirements	
7. Process Control	Application and system activities control.	User defined steps	

Figure 11-8 ISO-OSI

plex, half duplex, or full duplex, and how connections are established at each end. The Electronic Industries Association's RS-232C and RS-449 standards are examples of level 1 protocols.

Level 2 is the data link level. At this level, outgoing messages are assembled into frames, and acknowledgements (if called for at higher levels) are awaited following each message transmission. Outgoing frames include a destination address at the link level, and, if the higher levels require it, a source address as well, plus a trailer containing an error-detecting or error-correcting code. The data portion of the frame is whatever comes down to this level from level 3, without reference to its significance. Correct operation at this level assures reliable transmission of each message. Three major forms of data link protocol are presently in use. They are:

1. IBM's Binary Synchronous Communications Protocol (BISYNC) which uses special characters for control of message transmission. See Fig. 11-9.

2. DEC's Digital Data Communication Message Protocol (DDCMP). A generalized method of data link that includes message numbering information and a field to keep track of the number of characters in the message. See Fig. 11-10.

3. X.25 Protocol. Developed by the International Consultative Committee for Telephony and Telegraphy (CCITT) to define the rules governing the use of a public packet switching network. Developed jointly by France, Japan, the United States, the United Kingdom, and Canada. In addition, all nine countries of the European Common Market adopted X.25 as the basis for the proposed packet-switching Euronet Network. All transmissions between stations are organized in blocks of information, or frames, with a standard basic structure. See Fig. 11-11.

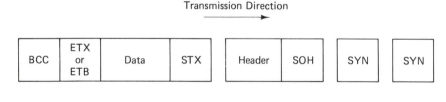

Transmission Direction

| BCC | ETX or ETB | Data | STX | Header | SOH | SYN | SYN |

SYN — Synchronisation Character
SOH — Start of Header
STX — Start of Text (Data)
ETX — End of Text
ETB — End of Text Block
BCC — Block Check Character

Figure 11-9 BISYNC (IBM's Binary Synchronous Communications Protocol)

CRC 2 (16)	Data	CRC 1 (16)	Address (8)	Sequencer (8)	Response (8)	Flag (2)	Count (14)	Class (8)	SYN (8)	SYN (8)

The Bracketed Numbers Show the Allocation
in Bits for Each Control Field for a Total of
12 Bytes (96 Bits) of Control Information.
The Maximum Number of Bytes in the Data
Field is 16,383

Figure 11-10   DDCMP (Digital Equipment Corporation's Digital Data Communications Message Protocol)

Level 3 is the network level. At this level, outgoing messages are divided into packets. Incoming packets are assembled into messages for the higher levels, and routing defines the destination of the packet and indicates the order of transmission. (The packets are not necessarily received in the same order in which they were sent when a packet network is used.) The header usually includes a source address. Level 2 constructs the frame containing the packet's data and header.

Level 4 is the transport level. This may be the busiest of all the architectural levels. Its protocol establishes network connections for a given transmission—for example, whether several parallel paths will be required for high throughput, whether several paths can be multiplexed onto a single connection

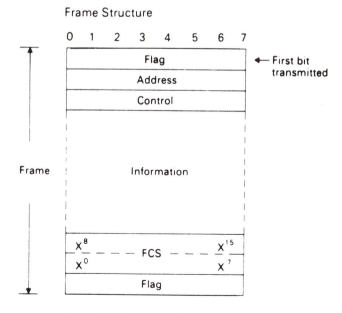

Figure 11-11   Frame Structure

to reduce the cost of transmission, or whether the transmission should be broadcast. This is the lowest level of strictly end-to-end communication, where the involvement or even the existence of intervening nodes is ignored.

Level 5 is the session level, at which the user establishes the system-to-system connection. It controls logging on and off, user identification and billing, and session management. For example, on a data base management system, a failure of a transmitting node during a transaction would be calamitous, because it would leave the data base in an inconsistent state. Level 5 organizes message transmissions in such a way as to minimize the probability of such a mishap—perhaps by buffering the user's inputs and sending them all in a group, more quickly than they could be sent under control of a higher level.

Level 6 is the presentation level. It controls functions that the user requests often, and that therefore warrant general treatment. Such functions include library routines, encryption, and code conversion.

Level 7 is the application level, the one seen by individual users. At this level network transparency is maintained, hiding the physical distribution of resources from the human user, partitioning a problem among several machines in distributed-processing applications, and providing access to distributed data bases that seem, to the user, to be concentrated in his CRT terminal.

11-4 Level 1

The physical level represents the lowest level of interconnection. To interconnect data terminal equipment (DTE) such as a computer, terminal, word processor, or printer, to data communication equipment (DCE) commonly known as a modem or data set, an interface standard is required. The EIA RS-232C, the EIA RS-449, and the CCITT X.21 are three prominent physical level standards.

11-4.1 The EIA RS-232C

This standard is used extensively in North America and contains detailed specification for a 25-pin connector in terms of:

electrical signal characteristics

mechanical characteristics

a functional description of the interchange circuits specifications for particular applications

The most significant standards for establishing the protocol for communication between a terminal and a modem are the signals defined for the interchange circuits, the most important of which are:

Request to Send: a request from the DTE to the DCE for permission to transmit data

Clear to Send (CTS): a signal from the DCE informing the DTE that it can start transmitting data

Transmitted Data: carries the data to be modulated

Received Data: carries the demodulated data from the DCE to the DTE

Data Carrier Detect (DCD): sent by the DCE to inform the DTE that it is ready to demodulate data

Recall that the two main problems associated with serial transmission are the conversion between serial and parallel data and mutual synchronization. The solution to the synchronization problem leads to two major classes of protocol used in computer communication systems, asynchronous protocols and synchronous protocols.

Asynchronous protocols

The proliferation of personal computers has created a demand for a relatively low speed, unsophisticated method of communication between the microcomputer, and a mainframe installation. The most commonly used system to date is an asynchronous full duplex link running at 300–1200 baud.

Fig. 11-12 shows the signal sequence for a half duplex communication at Level 1. The RTS initiates the carrier which turns on DCD and CTS, then the data is transmitted. When transmission is complete, RTS is turned off, which turns off the carrier and CTS. Carrier off turns off DCD.

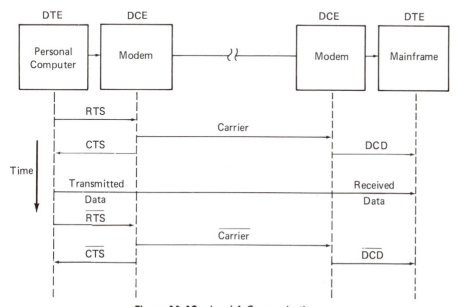

Figure 11-12   Level 1 Communication

In full duplex mode both DTE's are transmitting and receiving so that Transmitted Data and Received Data are active at the same time. This means RTS and DCD are always on.

We have assumed here that there is no dialing or switching involved. On a switched line, manual or autodialing would precede these steps. On a leased line the modem power switch also initiates the communication sequence.

Synchronous Protocols

For large volumes of data transmitted between mainframes, a higher transmission rate reduces line costs. Synchronous transmission is more efficient in this context, particularly in conjunction with a large number of remote sites.

At level 1, the protocol established between the DTE and the DCE is similar to that of the asynchronous system. The main difference is that the data is sent in synchronism with timing signals generated by a clock in the DCE. Synchronous modems communicate with one another to synchronise their clocks. Two additional signals on the RS-232C are thus required:

Transmit Timing: a signal from the DCE to the DTE to "clock" the data being transmitted

Receive Timing: a signal from the DCE to the DTE to enable the incoming data to be sampled at the correct speed. This clock is derived from the carrier by the DCE.

11-4.2   The EIA RS-449

An EIA upgrade of the RS-232C, the RS-449 consists of a 37-pin connector containing additional functions such as diagnostic circuits and a 9-pin connector for secondary channel circuits. The RS-449 has been designed to accommodate a growing variety of communication needs including increased cable length and transmission speeds.

11-4.3   The X.21

X.21 is the Level 1 standard for X.25, the CCITT's packet switching protocol. Instead of each connector pin being assigned a specific function (RS-232C and RS-449), each function is assigned a character stream. This approach has reduced the connector requirement to 15 pins but demands more intelligence of the DTE and the DCE.

11-5   Level 2

Asynchronous Protocols

In the context of personal computer communication with mainframes, a variety of software packages has been developed, of varying degrees of sophistication, which provide the Level 2 requirements for:

a. converting the personal computer to a simple (dumb) terminal

b. transmitting files by block in either direction

c. detecting errors and retransmitting blocks

Sometimes a copy of the package has to be available at each end. Typical examples include Kermit, ASYNC, and Crosstalk.

Kermit

The Kermit protocol was designed at the Columbia University Center for Computing Activities (CUCCA) by Bill Catchings and Frank da Cruz. Its purpose is to transfer sequential files over ordinary serial telecommunication lines. Kermit is a terminal-oriented file transfer protocol which has been implemented on a wide variety of microcomputers and mainframes. Data is transferred in packets, the format of which is shown on Fig. 11-13.

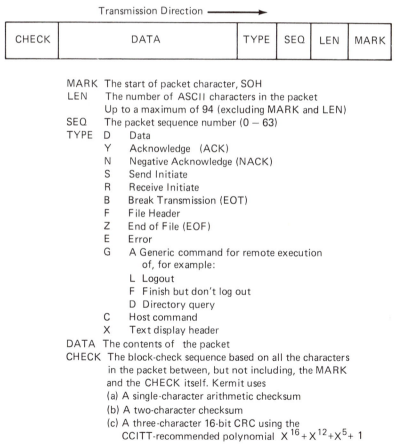

Transmission Direction ⟶

| CHECK | DATA | TYPE | SEQ | LEN | MARK |

MARK The start of packet character, SOH
LEN   The number of ASCII characters in the packet
          Up to a maximum of 94 (excluding MARK and LEN)
SEQ   The packet sequence number (0 – 63)
TYPE D   Data
         Y   Acknowledge (ACK)
         N   Negative Acknowledge (NACK)
         S   Send Initiate
         R   Receive Initiate
         B   Break Transmission (EOT)
         F   File Header
         Z   End of File (EOF)
         E   Error
         G   A Generic command for remote execution
                  of, for example:
              L Logout
              F Finish but don't log out
              D Directory query
         C   Host command
         X   Text display header
DATA The contents of the packet
CHECK The block-check sequence based on all the characters
          in the packet between, but not including, the MARK
          and the CHECK itself. Kermit uses
          (a) A single-character arithmetic checksum
          (b) A two-character checksum
          (c) A three-character 16-bit CRC using the
                CCITT-recommended polynomial $X^{16}+X^{12}+X^5+1$

Figure 11-13 **The Kermit Packet**

Kermit can be used on a variety of systems by using a common subset of their features. Communication is half duplex. A 96-character packet can be accommodated on most hosts without buffering problems. Packets are sent in alternate directions in a handshaking mode. A time-out facility allows transmission to resume after a packet is lost. The transmission code is ASCII and includes prefixing of ASCII control characters.

File Transfer

The simplicity of Kermit's file transfer protocol is based on the fact that each packet must be acknowledged. Fig. 11-14 illustrates the sequence of communication for receiving files based on the packet types listed in Fig. 11-13. Although this flow-chart indicates the packet sequence for receiving several files:

SFDDDDDDDDDDFDDDZFDDDDDDDDZFDDDDDDDDDDDDDDDDDZB

it does not include:

- the ACK packet which acknowledges receipt of a valid packet
- the NACK packet which acknowledges receipt of a bad packet or that a time-out has occurred
- the error packet which is sent by the side that encountered the error and that contains an error message

The flowchart for sending files is very similar but begins with a Receive Initiate packet (R) and incorporates Y, N, and E packets.

Kermit operates so that during a particular transaction the sender is the master and the receiver is the slave. These roles may be reversed for the next transaction. The Kermit at either end of the line may therefore act as either a master or a slave. A simpler way to operate between a mainframe and a microcomputer is to make the mainframe a permanent slave or server. The mainframe Kermit then gets all its instructions from the microcomputer Kermit in the form of a special command packet. The server will even log itself out upon command from the microcomputer Kermit.

Sample Session

Kermit is prompt-oriented. Kermit issues a prompt, the user types a command, and Kermit executes the command and issues another prompt. A typical Kermit session in a microcomputer to mainframe connection consists of:

- invoking the microcomputer's Kermit
- making the connection
- signing on to the mainframe
- invoking the remote Kermit and entering server mode
- returning to the microcomputer's Kermit
- transfer of files
- releasing the remote Kermit

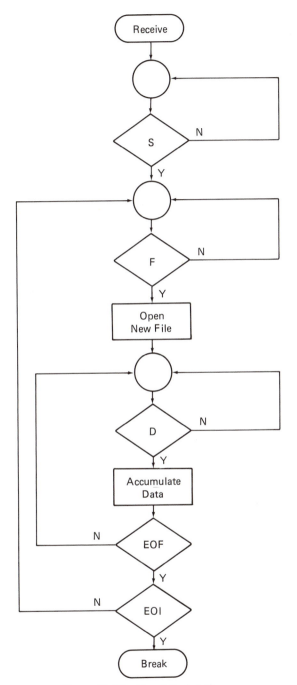

Figure 11-14   Kermit Packet Flow

- signing off the mainframe
- exiting from Kermit

Fig. 11-15 shows a typical session between the CDC CYBER 170 and an IBM PC.

```
Turn on microcomputer and monitor

A > KERMIT
IBM - PC Kermit - MS  V2.26
Type ? for help
Kermit - MS > SET PARITY SPACE
Kermit - MS > SET BAUD 300
Kermit - MS > C
(Connecting to host, type CTRL right square bracket
 to return to PC)
<Return>
<Return>
Normal CYBER log-in procedure
/ATTACH, KERMIT/UN = LIBRARY
/KERMIT
Kermit — 170 > SERVER
<CTRL> ] C
Kermit - MS > GET MYFILE

        Filename: MYFILE
   Kbytes transferred:
            Receiving:
    Number of packets:
    Number of retries:
           Last error:
         Last warning:
Kermit -  MS > FINISH
Kermit - MS > C
/BYE
<CTRL> ] C
Kermit - MS > QUIT
A>
```

Figure 11-15   A Typical Kermit Session (user input underlined)

ASYNC

ASYNC, designed by IE Systems, Inc., has the same objectives as Kermit. It allows you to sign on to a remote host as well as up-and-down load files. File transfer is block oriented with continuous error checking and automatic retries. The format for an ASYNC block is shown in Fig. 11-16. ASYNC can operate in three modes:

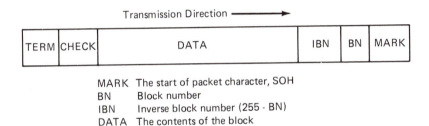

MARK   The start of packet character, SOH
BN   Block number
IBN   Inverse block number (255 - BN)
DATA   The contents of the block
CHECK The block check sequence
TERM   The block termination character (NAK)

Figure 11-16   The ASYNC Block

1. Terminal mode uses the microcomputer as a terminal in full duplex oper-ation.
2. Terminal echo mode treats the microcomputer as a terminal in half duplex operation.
3. Datalink mode is used to transfer files between the microcomputer and another computer (normally a mainframe).

One major difference between ASYNC and Kermit is that ASYNC is menu driven. In fact, there are three ways of communicating with ASYNC:

a. Selecting requirements via a series of menus may be a relatively slow process but is very useful to the first time user.
b. Once familiar with ASYNC the use of a command line which is really an extension of the command used to invoke ASYNC and has the format:

A>ASYNC PS.BB FILENAME.EXT$TELEPHONE

where P identifies primary options:

R   receive a file with error-correcting protocol
S   send a file with error-correcting protocol
T   act as a terminal with full duplex
E   act as a terminal with half duplex
D   disconnect the phone
Esc   return to main memory

where S identifies secondary options:

A   answer phone
C   use CRC-16 error checking instead of checksum
D   disconnect phone on exit
E   revert to terminal echo mode after file transfer
G   get a file from mainframe

H   enable handshaking to prevent overfilling of receiving buffer dur-
     ing a SEND

I   inform receiver of file size

P   put a file to the mainframe

Q   enable quiet mode of transfer

T   revert to terminal mode after file transfer

X   disable direct cursor addressing during file transfer

Note:

   1. You can use only one primary option but up to seven secondary
      options.

   2. Put(P) and Get(G) are used to send raw ASCII files in terminal and
      terminal echo modes (without error correcting)

   3. Send(S) and Receive(R) are used in datalink mode to transmit and
      receive blocked records with automatic error retransmission.

c. A third way of operating ASYNC is by means of a single-screen cursor-
   controlled menu select which is faster than (a) yet supplies more reference information than (b).

      Fig. 11-17 shows the file transmission protocol for sending a file.
   The process is started by a NAK from the mainframe. Notice the time-
   out limit of 80 seconds and the error count which allows for 10 retries before aborting.

   Fig. 11-18 is the flowchart for the corresponding protocol used when receiving a file. Here the time-out limit is 10 seconds although the retry counter is
still set for 10. Note too, the ability of the protocol to accept the most likely possibility of a block being sent twice without incrementing the blockcount.

Sample Session

   Use of the command line in datalink mode to transfer a microcomputer
file to the DECsystem10 is illustrated in Fig. 11-19.

Crosstalk

   Crosstalk is a complete, self-contained data communications program
designed by Microstuf, Inc. for most popular 8- and 16-bit micros and operates
under most versions of CP/M and MS-DOS. Like ASYNC, Crosstalk is
menu-driven and allows sign-on to a remote host as well as up and down loading of files. In addition, Crosstalk has terminal emulation capability for the
DEC VT-100 terminal, the IBM 3101, the Televideo 910/920 Series, and the
Adds Viewpoint. Transfer of error checked files is available between two
Crosstalk compatible systems. Crosstalk automatically detects errors using
CRC and corrects them by retransmission.

Synchronous Protocols

(1) BISYNC. IBM's Binary Synchronous Communications, a more traditional
approach to protocol, is still in common use. The format of a BISYNC message

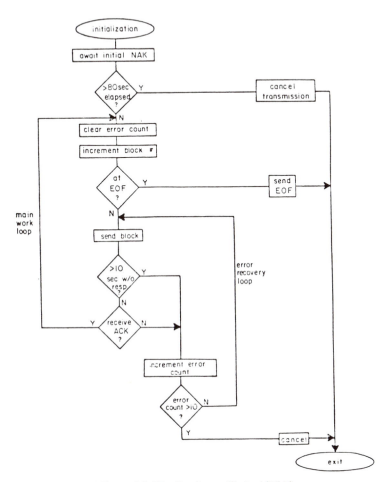

Figure 11-17 Sending a File in ASYNC

is shown in Fig. 11-9.

SYN is a synchronization character that is used to keep the sending and receiving terminals in step. SYN is added by the sender and removed by the receiver.

SOH (Start of Header) tells the receiver that the information following is control information which relates to addressing and sequencing of user data.

STX (Start of Text) identifies the start of the user's information.

There are two kinds of information that may be transferred:

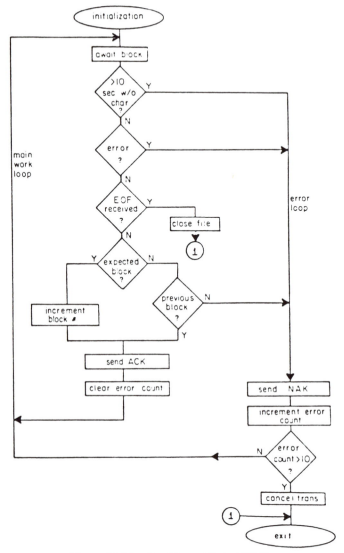

Figure 11-18 Receiving a File in ASYNC

1. Characters from a standard character set, such as ASCII or EBCDIC. The user's information is then terminated by ETX(End of Text) or ETB(End of Text Block). See Fig. 11-20.

2. Data that may contain control characters, from an analog to digital converter, for example. To prevent user data from being "recognized" by the receiver as control characters, BISYNC uses DLE(Data Link Escape). See Fig. 11-21.

```
A > ASYNC T.30                    ; invoke ASYNC
· LOGIN 123,456                   ; Log-in to mainframe
Password:_____
· RUN MICRO
/RECEIVE MYFILE. DAT              ; Tell MICRO to receive
^ V ^ E                           ; Exit to micro
A > ASYNC S. 30 MYFILE.DAT        ; Send file

file transferred
A > ASYNC T.30                    ; Sign off
/EXIT
KJOB
^ V ^ E
A>
```

Figure 11-19   A Typical ASYNC Session (user input underlined).

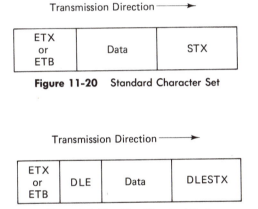

Transmission Direction ⟶

ETX or ETB	Data	STX

Figure 11-20   Standard Character Set

Transmission Direction ⟶

ETX or ETB	DLE	Data	DLESTX

Figure 11-21   Non-Standard Text

When DLE is first encountered by the receiver, the data that follows is treated as user data; that is, it is not tested for control characters. Any control characters present on the data are "transparent" to the receiver. The only problem, of course, is that DLE is also used to signal termination of user's data. If a pattern of bits occurs on the user's data that is equivalent to a DLE, the transmitter adds a second DLE to indicate that the first one is data. The receiver removes the second occurrence of all pairs of DLE's. The end of user's data is detected by means of DLE ETX or DLE ETB.

BCC is a block check character used for detection of errors which have occurred during the transmission of the user's data. BISYNC is limited to serial-synchronous, half-duplex lines. Error checking is by VRC/LRC if the code used is ASCII.

Vertical Redundancy Checking (VRC) is used to check each character in a message by means of a parity bit and usually accompanies Longitudinal Redundancy Checking (LRC). Cyclic Redundancy Checking (CRC) is used for messages coded in EBCDIC.

The BISYNC sequence of operation is illustrated in the flowchart of Fig. 11-22.

The transmitter attempts to make contact with the receiver by sending an enquiry signal(ENQ). The receiver may respond with an ACK(acknowledgement) to say "you may send the data." The data to be sent is handled in blocks and the transmitter keeps track of these by a numbering system. The transmitter initializes the block number to 1 and transmits the first block. The receiver acknowledges receipt(ACK). The transmitter checks to see if all blocks have been sent. If not, it sends the next block until all ten blocks have been sent. After all blocks are sent, the transmitter will terminate the communication by sending an end-of-transmission signal(EOT).

The flowchart in Fig. 11-22 is an oversimplification of the situation, but it does cover initiation, termination, and control of a two-way communication. It contains one obvious flaw. If the receiver sends a NAK signal back to the transmitter indicating an error in the message it received, the flowchart enters an endless loop. This problem can easily be solved by means of a retry counter which would limit the number of times a transmitter would send a particular block of data.

There are many other problems associated with accurate data transmission that have been solved in various ways. For example, to prevent a block of data from being lost completely, two forms of acknowledgement are used, ACK0 for even blocks and ACK1 for odd blocks.

There are contingencies associated with delays incurred in the system. Timeouts are used to create pauses before resending data and reseeking acknowledgement of accurate transfer.

Figure 11-23 is an improved version of the Fig. 11-22 flowchart. It explains the sequential operation of BISYNC protocol and incorporates the retry counter, busy receiver, and timeouts.

(2) X.25. The prime function of Level 2, the data link, is the transfer of data between the user(DTE) and the network(DCE). This transfer must include control transmissions for initiating, sequencing, checking, and terminating the exchange of user information. For this level of protocol, the CCITT X.25 recommends a procedure compatible with the High-Level Data Control(HDLC) procedure standardized by the International Organization for Standardization(ISO). IBM's Synchronous Data Link Control(SDLC) is a variant of HDLC.

The control procedures subscribe to the principles of a new ISO class of procedures for a point-to-point balanced system. The link configuration is a point-to-point channel with two stations. Each station has two functions. The primary function is to manage its own information transfer and recovery. The

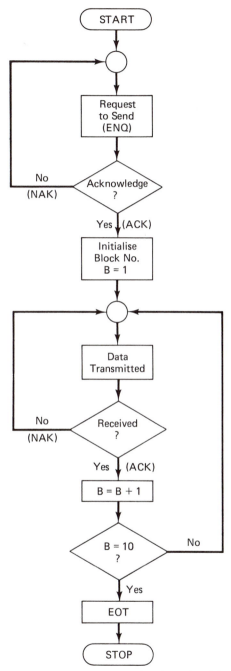

Figure 11-22   BISYNC Flowchart (Simple)

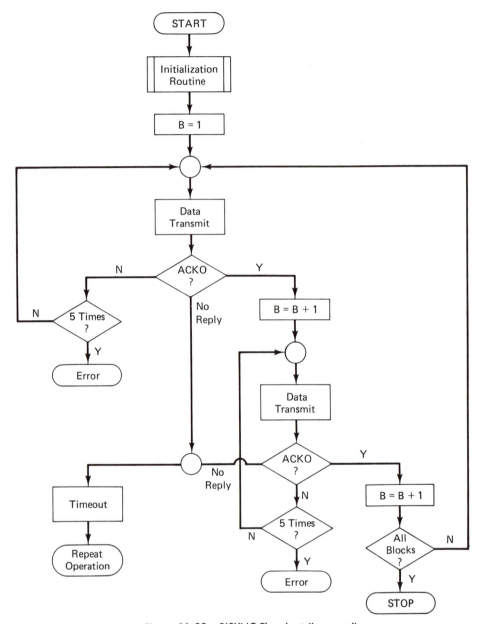

Figure 11-23 BISYNC Flowchart (Improved)

secondary function is to respond to requests from the other stations. In essence, these functions are commands and responses respectively.

The data link procedures are defined in terms of commands and responses and may be thought of as two independent but complementary transmission paths that are superimposed on a single physical circuit. The DTE controls its transmissions to the DCE; the DCE controls its transmissions to the DTE.

Synchronization

To transmit information across a link, the link must be synchronized. This is accomplished by enclosing each frame with unique bit patterns, called flags. Once the receiver recognizes the flag, it knows it is in step with the transmitter and in a position to accept the content of the frame. Similarly, when the receiver recognizes the second flag, it knows that all the information has been received. The receiver is now in a position to check the transmission and act upon the information it contains.

The potential occurrence of a flag bit pattern within the information is prevented by a mechanism similar to that described in the BISYNC protocol. The occurrence of a flag bit pattern in the information is made transparent to the receiving station. The transmitting station examines the frame content between the two flags, which includes the address, control, and FCS(Frame Control Sequence) data, and inserts a 0 after all sequences of five contiguous 1 bits to ensure that a flag sequence is not simulated. The flag bit pattern is 01111110. The receiving station removes any 0 bit which directly follows five 1's.

Commands and Responses

There are three variations on the frame structure of the X.25 data link control protocol:

1. Information Frames, which are used to perform an information transfer, and may also be used to respond to a correctly transmitted information frame.
2. Supervisory Frames, which are used to perform link control functions responding to received information frames.
3. Unnumbered Frames, which provide additional link control functions.

11-6   Level 3

This level of protocol specifies the way in which users establish, maintain, and clear calls in the network. It also specifies the ways in which user's data and control information are structured into packets. The relationship between the user data packet and the frame structure of the X.25 data link control protocol is shown in Fig. 11-24.

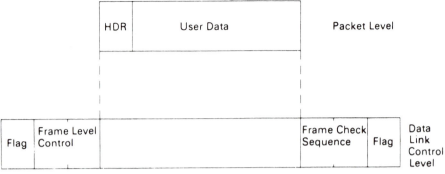

Figure 11-24 Frame/ Packet Relationship

A fundamental concept of packet switching is that of the virtual circuit. Just as circuit switching permits a physical link between two stations, so does a virtual circuit permit a logical bidirectional association between two stations(DTE's). This logical link is assigned only when packets are being transferred. A permanent virtual circuit is a permanent association between two DTE's, much like a private line. A switched virtual circuit is a temporary association between two DTE's and is initiated when a DTE sends a call request packet to the network.

At the packet level, DTE's may establish simultaneous communication with a number of other DTE's in the network. This is accomplished in a single physical circuit by means of packet-interleaved asynchronous time division multiplexing (ATDM). This type of multiplexing is similar to the straightforward time division multiplexing concept of improving the efficiency of information transfer by interleaving messages into the gaps of other messages. See Fig. 11-25(a).

Dynamically allocating the bandwidth to active, virtual-circuit ATDM further improves the transfer rate. This technique is referred to as statistical multiplexing and is illustrated in Fig. 11-25(b). The time previously allocated to non-active terminals is now used on a first come, first served basis.

The interleaved packets are transferred from the DTE's to the DCE's using the frame format of the data link control protocol. Each packet contains a logical channel identifier to relate the packet to a switched or permanent virtual circuit. Figure 11-26 illustrates a typical packet transfer between the DTE and the DCE.

The logical channel identifier(LCI) is used strictly at the local level of the network, or between DTE's and DCE's. The LCI's are selected at each DTE and are used independently of all other DTE's. The LCI's associated with permanent virtual circuits cannot be used for any other purpose. The LCI's used with switched virtual circuits are all free initially and can be used by the DTE to originate new calls or to receive incoming calls from the network.

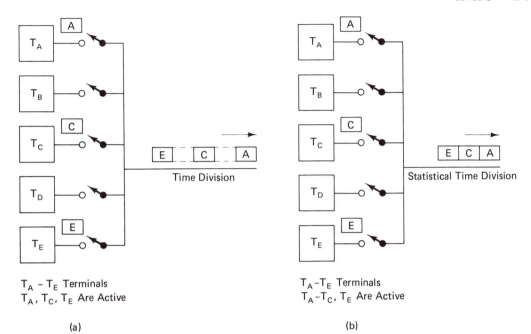

T_A – T_E Terminals
T_A, T_C, T_E Are Active

T_A – T_E Terminals
T_A – T_C, T_E Are Active

(a)   (b)

Figure 11-25   Time Division (a) and Statistical Time Division (b) Multiplexing

Call Establishment and Clearing

There are several packets associated with setting up a communication link between two DTE's. These packets include:

1. A Call Request Packet sent by the calling DTE to the network. This packet includes the logical channel identifier; the address of the DTE called; a facility field for reverse charging, priority, or user restrictions; and user data of 16 bytes.

2. A Call Connected Packet sent to the calling DTE to confirm that the call has been accepted.

3. A Clear Indication Packet sent to the calling DTE to indicate that the call was not established and to provide one of several reasons for this, such as Number Busy.

4. An Incoming Packet received by the called DTE on an available logical channel as a result of a Call Request Packet. (The formats are very similar.)

5. A Call Accepted Packet sent by the called DTE that indicates a positive response to a call request. The network uses the Call Accepted Packet to generate a Call Connected Packet for the calling DTE.

6. A Clear Request Packet sent by either DTE to free the logical channel after data transfer is complete.

A sequence diagram showing the stimulus/response relationship required to establish a call between two DTE's is given in Fig. 11-27.

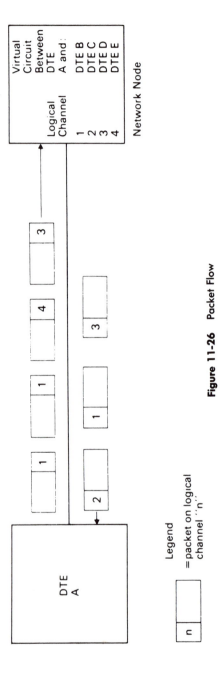

Figure 11-26 Packet Flow

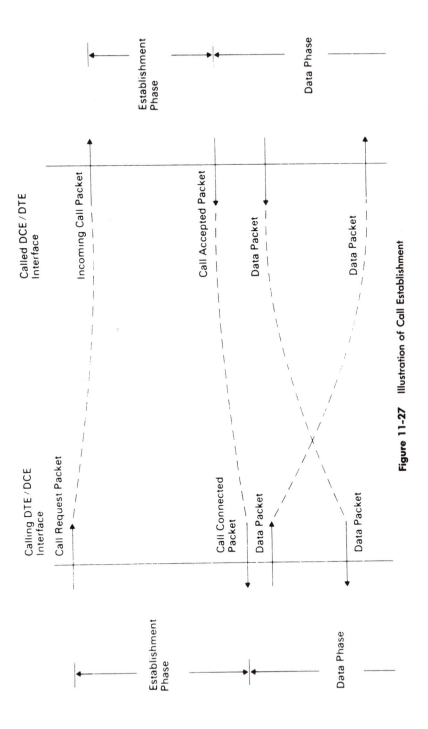

Figure 11-27 Illustration of Call Establishment

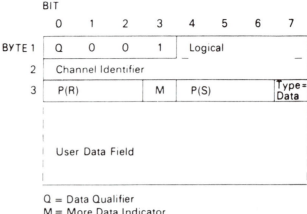

Q = Data Qualifier
M = More Data Indicator

Figure 11-28 Data Packet Format

Data Transfer

Data transfer can take place only after the virtual circuit has been established. The format is shown in Fig. 11-28. $P(S)$ is the packet send-sequence number. (Only Data Packets are numbered.) $P(R)$ is the receive-sequence number used to locally confirm the receipt of packets. M is used in a full data packet to indicate continuation of the data in the next packet. Type-Data distinguishes this format from all the others. The maximum data field length is normally initialized to 256 bytes, but this may be revised by the user.

Figure 11-29 offers another approach to understanding information transfer at the packet level. The following list of events at each logical channel will help in comprehending the diagram.

State 1. A logical channel is in the ready state if there is no call in existence.

State 2. A Call Request Packet sent by a DTE puts the logical channel in the DTE Waiting State.

State 3. An Incoming Call Packet sent by a DTE puts the logical channel in the DCE Waiting State.

State 4. A Call Accepted Packet from a DTE, resulting from an Incoming Call Packet, puts the logical channel in the Data Transfer State. Similarly, a Call Connected Packet from a DCE, resulting from a Call Request, leads to Data Transfer.

State 5. When a DTE and DCE transfer a Call Request Packet and an Incoming Call Packet, respectively, at the same time, a call collision occurs. The DCE will proceed with the request and cancel the incoming call.

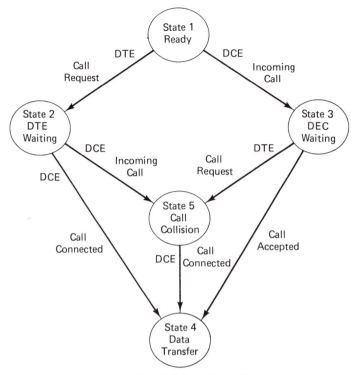

Figure 11-29 Call Setup State Diagram

11-7 Level 4

The transport level provides the facilities that allow end users to transmit across several intervening nodes. Of the upper four levels, the transport level is the one which is relatively well defined; for example, the European Computer Manufacturers' Association defined ECMA-72 transport protocol standard. The services provided by this level include optimizing:

- costs
- quality of service
- multiplexing
- data unit size
- addressing

The major function of the transport level is to free the higher levels from cost-effective and reliability considerations.

11-8   Level 5

The purpose of the session level is to provide the end users with the means of organizing and synchronizing their dialog as well as managing their data exchange. This level provides:

- initiation of the session
- management and structuring of all session requested data transport actions
- termination of the session

11-9   Level 6

The presentation level provides for the representation of selected information. Any function that is requested often enough to justify a permanent place, is held in the presentation level. Three protocols are being developed for the presentation level:

- virtual terminal
- virtual file
- job transfer and manipulation

11-10   Level 7

The application level is the highest level of the ISO-OSI and is normally developed by the user. The user determines the messages used and the actions to be taken on receipt of a message. Other features include:

- identification of partners
- establishment of authority
- network management statistics
- network transparency
- network monitoring

Although an absolute standard for protocol levels is not available, there is a definite trend towards ISO-OSI seven-level architecture. This reference model attempts to provide a conceptual and functional framework which permits the independent development of standards for each of the functional levels.

11-11   Conclusion

Public packet switching networks use point-to-point synchronous circuits. Non-packet switching networks use a physical channel as an increment of dedicated bandwidth. This bandwidth may be permanently leased as a private line, or periodically accessed via circuit switching. Both may result in inefficient use of facilities because the bandwidth is reserved whether or not data is being transmitted.

In packet switching, all user data is converted to discrete, variable-length units called packets. Packets of data from many users are dynamically interleaved over shared network transmission facilities and routed to their destinations. Bandwidth is allocated to a user only when he is actively transmitting data.

Packet switching provides the following benefits:

1. *Universality.* Resource sharing that is built into the network allows both small and large organizations to participate in a nationwide system.

2. *Communication Interconnection Flexibility.* The specification of a Standard Network Access protocol facilitates communications between user systems.

3. *Low Cost.* The ability to share transmission facilities and to share switching and control equipment results in lower communication costs for a large segment of data users.

4. *Accuracy.* Powerful error detection and correction facilities provide users with virtually error-free, end-to-end data transfer.

5. *Reliability.* Standby facilities, alternate network routes, and network monitoring and control result in high network availability.

Even with the universal adoption of X.25, there are still users critical of packet switching technology. The main thrust of criticism comes from the proponents of datagram, a network which transmits disconnected packets without any knowledge or concept of messages, connections, or flow control. The packet assembly and disassembly, sequence numbering, and virtual connection processing are done by computer DTE's. The problem with the datagram system is that it requires a more coordinated and cooperative set of users than is likely to exist in a public environment.

Packet switching is criticized for the following reasons:

1. The interface is complex and was drafted with undue haste.

2. There are flaws in the design.

3. There are standards which are open to different interpretation.

4. X.25 does not provide end-to-end integrity of data flow.

These criticisms are partially answered by CCITT recommendations X.28 and X.29. X.28 applies to the interface between the user and the terminal. X.29 applies to the interface between a remote handler and an application program, or an access method, located in a computer.

The establishment of public data networks on a national and international scale is fraught with technical, economic, and political difficulties. Considerable progress has been made by the CCITT in setting standards to meet customer demands on an international basis. It is not surprising that there are problems which still need to be solved.

New developments create their own set of new problems but the possibilities for improving communication systems are unlimited. Possibilities for the future include incorporating broadcast satellite facilities(packet satellite), sharing one wideband channel among many radio stations(packet radio), and compressing voice transmission by packet switching.

Problems

1. At each level of the ISO-OSI, control information is added to the data transmitted; e.g., an application header at the application level, a transport header at the transport level, etc. What control information is added at the datalink level?

2. What are the main differences between synchronous and asynchronous protocols?

3. Assuming they were being used for control purposes, how would you make the number characters of the EBCDIC code transparent to the receiver?

4. Draw a detailed flowchart for the initialization routine of Fig. 11-23.

5. Using Fig. 11-30, answer the following questions:
   a. Which signal is used by the DTE to indicate clearing?
   b. Which signal is used by the DCE to indicate clearing?
   c. When the logical channel is in the DTE clear request state, which signal does the DTE use to free the logical channel?

Note: After transferring a clear request packet, a DTE may receive other types of packet before the channel is freed by the DCE.

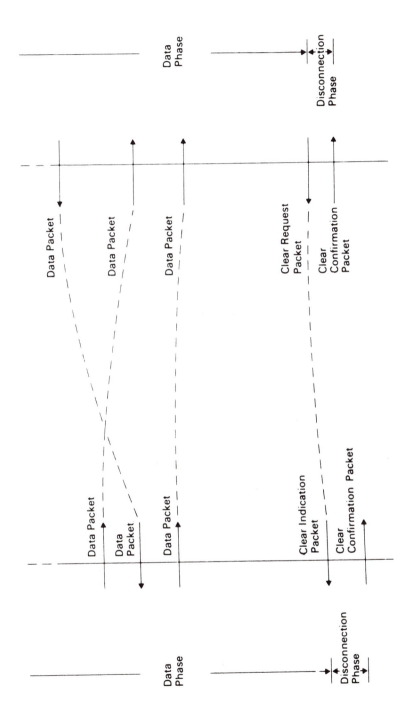

(a) Data Transfer and Clearing

Figure 11-30 (a)   Data Transfer and Clearing

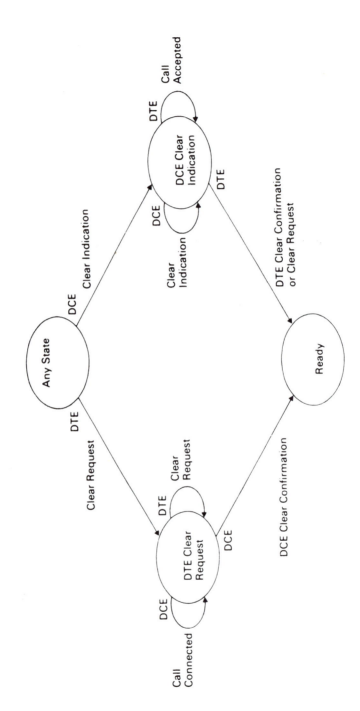

(b) Call Clearing

Figure 11-30 (b) Illustration of Data Transfer and Clearing (a) and Call Clearing (b)

12

Local Area Networks

A Local Area Network (LAN) is a system for interconnecting data communicating components within a relatively confined space. A system is a group of interrelated parts with the focus on the interrelationship. LANs are principally concerned with methods of communication among their components. The components are mutually compatible devices and include microcomputers, disk storage, and printers. The emphasis on compatibility is purely practical. Incompatible microcomputers may be able to share a printer but not programs or data. The term, "a relatively confined space," is used in conjunction with the distinction between LANs and global or distance networks. LANs are most commonly contained within one building but may spread to contiguous buildings. Distance networks usually operate between cities, countries, and continents. LAN's are restricted to interbuilding communication, the maximum distance dependent on the medium used to connect components.

Other significant characteristics of LANs, which do not appear in the definition, include ownership, speed, and availability. LANs are privately owned and therefore not subject to regulation by public bodies or networks. This fact, coupled with the simple, symmetrical topologies used in LANs, facilitates most acceptable speeds of between one and 10 megabits per second. Although a large number of LANs are now commercially available, the variety of technical options is limited. There are three major media, three topologies, and two protocols. Before considering the technology we should look at the purpose of LANs.

Purpose of LANs

This discussion centers around two considerations: the technological role of the LAN and the application spectrum. The initial impetus for LAN development came from the need to interconnect a growing number and variety of intelligent machines in the office. These intelligent machines had been purchased to automate office functions, as well as augment data processing development where in many cases there existed the perception that the data processing center was experiencing enormous backlogs and was unresponsive to the needs of the small user. Initially the purchase of office machines was on an individual need basis and independent of broad organizational planning. Inevitably, some form of coordinated effort was made to consider such functions as compatibility, standardization, and interdepartmental communication. This led quickly to the possibility of interconnection of these systems to improve efficiency, facilitate integration of the applications, and share the resources.

Resource Sharing

The major advantage of LANs is their ability to share equipment, data, and software. The physical resources to be shared are the expensive parts of small systems, viz: file space (disk handlers) and reporting capability (printers). By allowing several workstations (terminals, wordprocessors, personal computers) to share equipment, information costs may be drastically reduced and spread over a number of departments.

The ability of a group of workstations to share data facilitates integration of work in different departments and decreases duplication of effort. A single master copy of a file promotes consistency, accuracy, and reliability of data. Changes to a master file by one user are immediately available to all other users.

Shared software means immediate cost savings. In the educational environment, for example, a typical personal computer lab may have 20 microcomputers. Instead of buying 20 copies of a spreadsheet program, a system of three LAN's of six, seven, and seven workstations requires only three copies of the software. Further benefits include the disk space saving of three copies over 20 copies, and the consistent use of the same version of the software by all users.

The benefits of resource sharing are accompanied by corresponding restrictions. Incompatible microcomputers may share disks and printers but not files and programs. The integration of incompatible devices may be more costly than the installation of a new system. The user can choose between a single vendor system which minimizes compatibility and maintenance problems or the cost benefits and corresponding human communication problems of dealing with more than one vendor.

Personal Computer Networks

With the introduction of personal computers to the office and their interconnection by LANs comes a new potential for increased productivity. Sharing data files on customers, parts, and vendors gives all users fast access to useful, consistent information. Word processing packages facilitate memo, letter, and report writing which can be standardized on the LAN. Spreadsheet and graphics software provide figures and diagrams to enhance reports. By sharing this software, costs are reduced and dispersed.

Office Automation

The interconnection of intelligent machines in the office integrates the job of managing information. Again the emphasis is on increased productivity, this time of secretarial functions such as electronic filing (data collection), scheduling (calendar and reminder files), data base inquiry (customer, parts, vendor), word processing (memo, letter, report), and dissemination (memo, letter, report).

The overlap of functions in these two applications is intentional and highlights the changing role of the computer as a direct aid to managers and the changing role of secretaries to office and information managers.

Process Control

The use of LANs in chemical plants and manufacturing factories in less well-developed. Microcomputers are commonly used to monitor physical data (temperature, pressure, and flow) and control these physical variables with switches and motors. The use of LANs to improve the economics of the situation and provide flexibility and backup, constitutes a new development.

In a production line environment, individual microcomputers can be used to monitor and control a particular stage while at the same time receiving data from the microcomputer at the previous stage, and passing information forward to the next stage.

12-2   LAN Technology

The terminal nodes of a LAN are usually either special- or general-purpose microcomputers. The compatibility of these terminal nodes is crucial to the maximization of resource sharing. The junctions on a LAN are interface boards which are capable of generating and receiving control and data signals for the network. These interface boards are housed in the workstation (microcomputer) and file server (hard disk control unit). In addition, a junction box is used at specific points in the network to physically connect workstations to the network. The most significant component is the line, i.e., the medium of communication. There are three popular media: twisted pair, coaxial cable, and fiber optics. The technology of LANs, as distinct from distance networks, is simple and contains a high degree of symmetry. There are three principal topologies; bus, ring, and star.

LAN protocol, the network's way of controlling traffic, is also distinguishable from distance networks, which normally use some form of polling, in that the workstations have an equal say in controlling the traffic. There are two main protocols or access methods for LANs: Carrier Sense Multiple Access/Collision Detection (CSMA/CD), and token passing.

12-2.1   Transmission Media

The transmission medium is the physical path between nodes of the network. In distance networks, the physical path may be wires, cables, and fibers, as well as the air waves via microwave guide, radio, and satellite. In LANs the choice is much simpler.

Twisted Pair

A twisted pair consists of two insulated wires wound round each other to minimize interference from other pairs in a multiwire cable. This is a traditional form of wiring in telephone systems and for that reason is readily available and inexpensive. Twisted pair cabling is also easy to install. Its main disadvantage is that it is susceptible to electrical interference which restricts it to relatively short distances and low transmission rates. For this reason its popularity as a LAN medium may be short lived.

Coaxial Cable

Coaxial cable (coax) consists of a central insulated wire which carries the signal, surrounded by a fine copper mesh and an outer insulated shield. Coax is the most popular LAN medium because of its high immunity to electrical interference and high transmission rates.

There are two signaling techniques used with coax which have been misleadingly applied to the type of coax used. Baseband coax is defined as 50 ohm (⅜ inch) cable used to transmit a single digital signal at speeds of up to 10 Megabaud. Broadband coax is defined as 75 ohm (½ inch) cable used in the cable television industry and to transmit multiple analog data signals as well as voice and video signals.

Baseband coax is easy to install, moderately expensive, and used in short distance, data only, LANs.

Broadband coax is sensitive, must be tuned, and carefully installed, and is most effective in large scale LANs requiring voice and video as well as data signals.

Fiber Optics

The optical fiber is made of plastic or glass and serves as an extremely high performance transmission medium. Its characteristics are impressive: extremely wide bandwidths (in the gigahertz range), speeds of 1 gigabaud, immunity to electrical interference, extremely compact, and lightweight.

The major problem associated with fiber optics across the communications industry is its incompatibility with all previous development. As a relatively new field, it is expensive to implement. Its compactness and requirement for light sources and detectors make it difficult to implement. The inability to tap a fiber optics signal, a major advantage in terms of security, is a disadvantgage for LANs in terms of future changes and expansion of the network. Although its future potential may be immeasurable, its present use for LANs is limited. Fig. 12-1 shows a table of comparison of transmission media for LANs.

12-2.2 Topology

Topology describes the pattern of connection for network nodes. It determines the layout of communication links between terminal nodes, junctions, and network servers. Topological design algorithms must select links and link capaci-

Medium	LAN cost	Speed	Application
Twisted Paint	low	10Mband	data
Baseband	low	10Mband	data
Broadband	high	20 channels at 5Mband	data, voice, video
Fiber Optics	very high	1Gband	data, voice, video

Figure. 12-1 Comparison of Media

ties on the basis of: message transmission delay, cost, network traffic, and perhaps most importantly future expandability.

Star

In the star topology each node is connected via a point-to-point link to a central control node (Fig. 12-2). All routing of network traffic, from the central node to outlying nodes and between outlying nodes, is performed by the central node. The routing algorithm is a simple matter of table lookup.

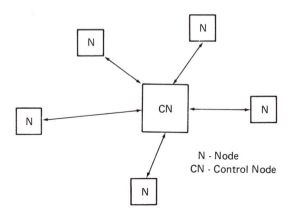

N - Node
CN - Control Node

Figure 12-2 Stat Topology

The star formation works best when the bulk of communication is between the central node and outlying nodes. When the message volume is high among outlying nodes, the central switching feature may cause delays.

The central control is the most complex of the nodes and governs the success, size, and capacity of the network. Star networks require greater emphasis on reliability for this reason.

Ring

Each node is connected to two other nodes by point-to-point links in a closed loop (Fig. 12-3). Transmitted messages travel from node to node round the ring in one direction. Each node must be able to recognize its own address as well as retransmit messages addressed to other nodes. Since the message route is determined by the topology, no routing algorithm is required, messages automatically travel to the next node on the ring. In its place is the concept of circulating a bit pattern, called a token, to facilitate sharing the communications channel. A node gains exclusive use of the channel by "grabbing" the token. It passes the right to access the channel on to the other nodes when it has finished transmitting. This is the basic protocol used in ring topologies. When control is distributed, each node can communicate directly with all other nodes under its own initiative.

Ring networks with centralized control are often referred to as loops (Fig. 12-4). The control node permits the other nodes to transmit messages and acts like the central control node of a star network.

Since ring communication is unidirectional, failure of one node brings the network down. However, simple bypass mechanisms may be employed to minimize downtime.

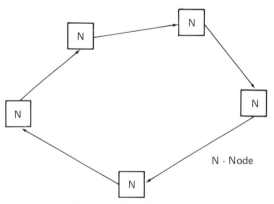

Figure 12-3   Ring Topology

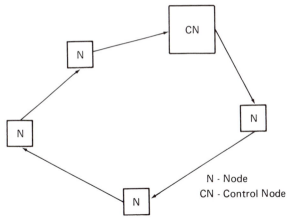

Figure 12-4   Loop Topology

Bus

In bus topology, unlike the star topology, there is no switching required, and no repeating messages as they are passed round the ring. The bus is simply the cable which connects the nodes (Fig. 12-5). As with the ring, the message is "broadcast" by one node to all other nodes and recognized by means of an address by the receiving node. This broadcast of the message propagates the length of the medium. One advantage over the ring is that the delays due to repeating and forwarding the message are eliminated.

Since the nodes are passive the system is failsafe. A faulty node will not affect the other nodes. Bus networks are also easily installed and expanded.

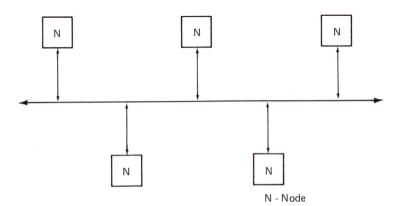

Figure 12-5   Bus Topology

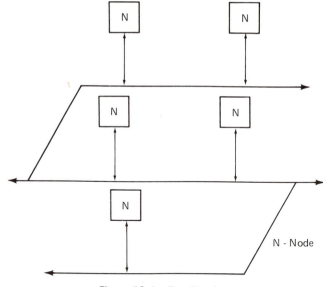

Figure 12-6 Tree Topology

A tree is a generalization of bus topology in which the cable branches at either or both ends, but which offers only one transmission path between any two nodes. The branches coming off either end of the bus never link up (Fig. 12-6). As with the bus, any node "broadcasts" its message which can be picked up by any other node in the network.

At present, the LAN market is dominated by bus and tree topologies in North America, while token passing rings are more popular in Europe.

12-2.3 Protocol

The International Standards Organization's reference model of Open Systems Interconnection (ISO-OSI) discussed in Chapter 11 may also be applied to LANs. Since LANs are primarily concerned with the transmission of information over a physical medium, only the first two levels—physical and data link control—need to be addressed.

The physical level has already been defined as consisting of an interface card coupled to an appropriate medium. The data link control level may be discussed in terms of message format (the frame), error handling (CRC), flow control (buffering), and line control.

Line control may be subdivided into two categories: polling (for networks with a control node) and distributed access methods (for LANs in which each node has equal control). Two distributed access methods have emerged to dominate the LAN market: contention, in which any node has the ability to initiate transfer at any time, and token passing, in which each node must wait its turn.

Contention

The most common form of the contention protocol is Carrier Sense Multiple Access with Collision Detect (CSMA/CD), which is usually associated with bus/tree topology. Carrier Sense is the ability of each node to detect traffic on the channel by "listening." Nodes will not transmit while they "hear" traffic on the channel. Multiple Access lets any node send a message immediately upon sensing that the channel is free of traffic. This eliminates the waiting that is characteristic of non-contention protocols. One problem that may arise is that because of propagation delays across the network, two nodes may detect that the channel is free at the same time, since each will not have detected the other's signal. This causes a collision. Collision Detect is the ability of a transmitting node to "listen" while transmitting, identify a collision, abandon transmitting, wait, and try again. Fig. 12-7 shows the sequence of events for this protocol. CSMA/CD is a highly efficient form of distributed access.

Token Passing

Token passing is most often used with ring topologies, although it can be applied to bus/tree topologies by assigning the nodes logical positions in an ordered sequence with the last member followed by the first—in other words, a logical ring.

Each node in turn, in a predetermined order, receives and passes the right, in the form of a token, to the channel. Tokens are special bit patterns that circulate round the ring when there is no traffic. Possession of the token gives a node exclusive access to the network for transmitting its message. This technique eliminates the possibility of conflict among nodes.

For example, node A holds the token and sends a message specifying node B. All the nodes on the ring check the message as it passes. Each node is responsible for identifying and accepting messages addressed to it, as well as for repeating and passing on messages addressed to other nodes. Node B accepts the message and sends the message to node A to confirm its receipt. When node A receives confirmation that the message arrived and was accepted it must send the empty token round so that another node will have a chance to take over the channel. Fig. 12-8 shows the sequence of events for both the transmitting and receiving nodes.

A variation of token passing, known as slotted access, is found exclusively in ring topology. Instead of a token, the nodes circulate empty data frames, which a node may fill in its turn. Usually, a fixed number of slots or frames of fixed size circulate. Each frame consists of source and destination addresses, parity and control, and data. Note that the slotted access technique limits not only when a node transmits but how much at a time.

12-2.4   LAN Software

The major purpose of LANs is to share resources. To do this we need a way of connecting the workstations—i.e., a medium, a topology, and a protocol. Then

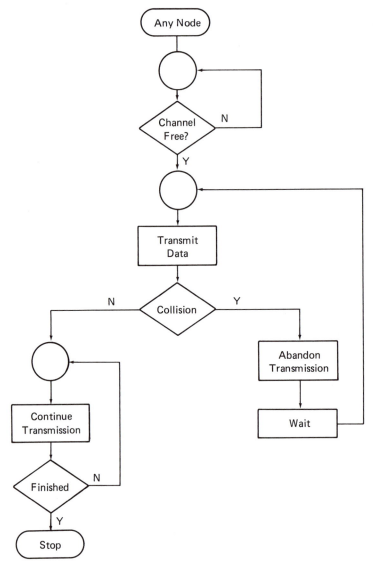

Figure 12-7 Contention Protocol

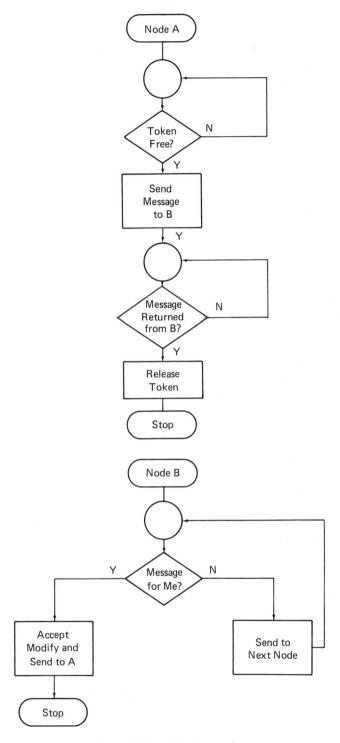

Figure 12-8   Token Protocol

we need to control the sharing of equipment, data, and software; in other words a local area network operating system. This is not as formidable a task as it sounds since microcomputers already have well-developed operating systems such as MS-DOS, CP/M, and UNIX, to name but a few. The problem then becomes one of interfacing network function software with the workstation operating system already available. The LAN software will add its own restrictions to those of the workstation operating system as well as increase file storage available to the workstation, provide for private and shared files, and provide for printer capability at convenient locations.

Most LAN software packages come with modules for logging on and off the network, disk/file sharing, and print spooling.

The logging on and logging off network module may include such considerations as password security, validation of user access to specific files and software, an automatic log-on feature for specific workstations, as well as releasing reserved files on log off. Other features may include password changing, help menus, and error messages for log-on problems.

Sharing files is different from sharing disk space. With simple LAN software it may be possible that each station is allocated a portion of the hard disk for its own use. Sharing data or programs may not be possible. Each workstation takes its own copy of the data or the program. Changes to a file by one workstation will not automatically be passed on to other users. This type of LAN software is termed disk-sharing-software.

Data-sharing-software involves either write protecting a file so that only one workstation may change the information at a time, or with sophisticated LAN software, allowing multiple changes simultaneously. This challenging chore is often transferred to the applications program. For example, multiuser data base management systems handle simultaneous updates of their data bases.

Print spooling is an integral part of printer sharing. It describes the process of copying a file to a temporary storage area on the file server until it can be printed. While the file is in the spooler it is queued; i.e., each file, as it arrives, is given a number and the files printed in the order of the assigned numbers.

12-3   Examples

VictorLAN

The Victor Local Area Network is an adaptation of the Corvus Systems Omninet. It is a system of two to 64 Victor computers, each with its own interface card called a transporter. The transporter is an intelligent interface, containing a microprocessor and its own firmware (permanently stored programs), to reduce the communications load on the workstation's microprocessor. Network junction boxes are used to connect the transporter to the network trunk, and ensure matched impedance and minimum interference. The trunk line cable is a shielded twisted pair.

```
Medium            : Baseband coaxial cable
Topology          : Bus
Protocol          : CSMA/CD
Speed             : 10 M baud
Max. no. of nodes : 1024
Max. trunk length : 2.8 Kilometers
```

Figure 12-9   Victor LAN Characteristics

The VictorLAN fileserver is a 10-megabyte hard disk with 256 kilobytes of RAM and a 1.2-megabyte floppy disk, as well as one parallel and two serial ports. The hard disk stores user files and may be divided into two, four, or eight equal sized sections at installation time. The sections are then accessible as if they were separate physical disk drives. The floppy disk may be used to transfer files between LANs. For example, a serial port on one workstation may be connected via modem to the serial port of a workstation on another LAN, and files transferred between them using an asynchronous protocol such as Kermit (Sec. 11-5). The floppy disk may also be used for archiving. The VictorLAN hardware characteristics are summarized in Fig. 12-9.

VictorLAN resources are assigned to users, not to workstations. Users have access to the same hardware and software regardless of which station they use. Each user may be assigned up to 15 disk volumes which may be distributed among hard disks and floppys on different servers as well as on the workstation being used. Server volumes may be shared among several users or restricted to one user. Printers may also be assigned selectively. The resources associated with each server are managed by the software installed on that server. Files sent to a server for printing are spooled and sequenced automatically.

The VictorLAN Server Operating System, version 2.0, is implemented under Microsoft MS-DOS 2.0. This is transparent to the workstation user who operates under MS-DOS 1.25. In addition to the MS-DOS 1.25 commands which all operate both locally and on the network (except for FORMAT, DCOPY, and CHKDSK which only operate locally), there are 11 network commands, tabulated in Fig. 12-10.

Network Configurations

The simplest possible network consists of a single workstation and a server (Fig. 12-11). The trunk may be extended anywhere along its length, not only from its ends. The cable of course will physically go around corners and up and down walls but must remain as a bus. Fig. 12-12 shows the expansion of the previous network by two nodes.

More typical networks consist of one to 54 workstations and one to 10 servers, each supporting a maximum of three printers. Fig. 12-13 shows a typical configuration. Note that the printer attached to the hard disk station cannot be used by other stations in the network. Servers may be equipped with key-

Command	Function
LOGIN	Logs the user on to network
LOGOUT	Ends session, protects files, allows new log-in
NETPRINT	Sends file to spooler for printing
NETSTAT	Displays network status information
NETUSERS	Displays active users
PASSWORD	Allows password to be changed
PROTECT	Changes status of server file
RESERVE	Temporarily restricts file access
RELEASE	Cancels RESERVE
SHOWLIST	Identifies active printers
STATION	Displays workstation status

Figure 12-10   Victor LAN Commands

boards and displays, both of which are inoperable during network operation but may be used in stand-alone mode.

Ethernet

In 1980, Digital Equipment Corporation, Xerox Corporation, and Intel Corporation joined forces to advance the Ethernet approach to LANs. This resulted in a precise, detailed specification for a baseband LAN. The workstation interface card called the transceiver is connected to the coaxial trunk cable by means of four twisted-pair wires, each of which carries one of the four signals: transmit, receive, collision presence, and power. Ethernet is described as a branching bus topology with a maximum distance of 2.8 kilometers between the two furthest nodes of the network. Up to 1024 nodes can be tapped onto the Ethernet coaxial cable. Hardware characteristics are summarized in Fig. 12-14.

Ethernet provides the local environment for print and file servers to share storage and I/O resources, and terminal servers to convert dumb terminals to intelligent terminals.

One version of the Ethernet implementation for personal computers is that of 3Com Corporation and is called the 3Com Etherseries. Software for this LAN includes Ethershare, which provides sharing of hard disk and printers. Subdivision of hard disk volumes may be public, private, or shared. Etherprint allows all network users to share up to two printers attached to each server.

Configurations

A small Ethernet LAN would consist of a simple bus topology with perhaps three workstations sharing one file server, all within the maximum coaxial cable limit of 500 meters (Fig. 12-15).

A more sophisticated Ethernet LAN would be configured as a tree with a repeater being used to link the segments together. A repeater enables two cable segments to function as if they were one. It amplifies transmission signals and

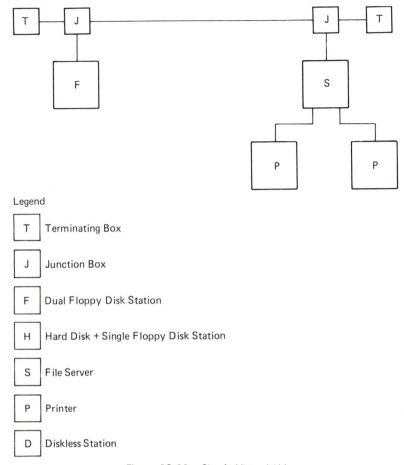

Legend

T	Terminating Box
J	Junction Box
F	Dual Floppy Disk Station
H	Hard Disk + Single Floppy Disk Station
S	File Server
P	Printer
D	Diskless Station

Figure 12-11   Simple Victor LAN

passes data between the coaxial cable segments. A local repeater connects two segments within a limited distance of up to 100 meters (Fig. 12-16).

A large scale Ethernet configuration would include remote as well as local repeaters. A remote repeater connects two coaxial cable segments over a distance of up to 1000 meters. The remote repeater consists of two local repeaters connected by a fiber optic link (Fig. 12-17).

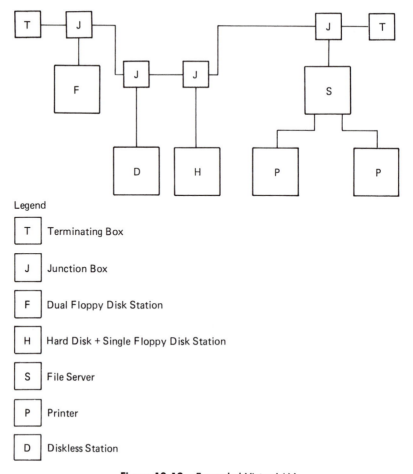

Legend

T   Terminating Box

J   Junction Box

F   Dual Floppy Disk Station

H   Hard Disk + Single Floppy Disk Station

S   File Server

P   Printer

D   Diskless Station

Figure 12-12 Expanded Victor LAN

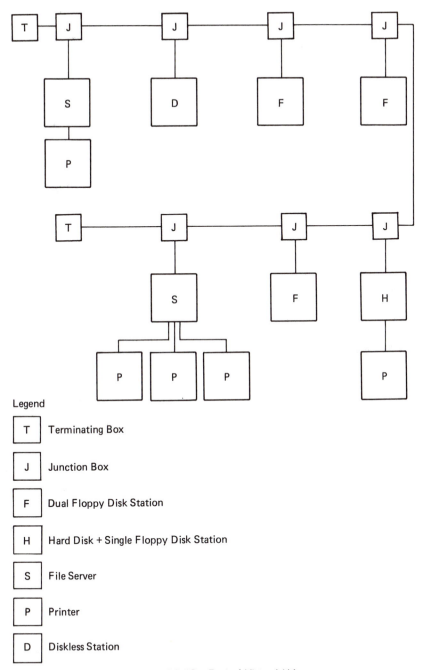

Figure 12-13 Typical Victor LAN

Medium   : Twisted pair
Topology   : Bus
Protocol   : CSMA/CD
Speed   : 1 M baud
Max. no. of nodes : 64
Max. trunk length : 500 meters

Figure 12-14   Ethernet Characteristics

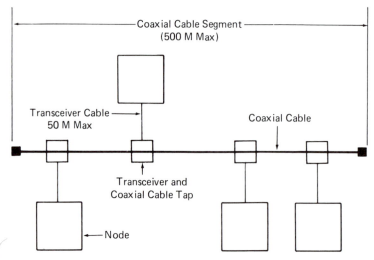

Figure 12-15   Small Ethernet (*Courtesy of Digital Equipment Corporation*)

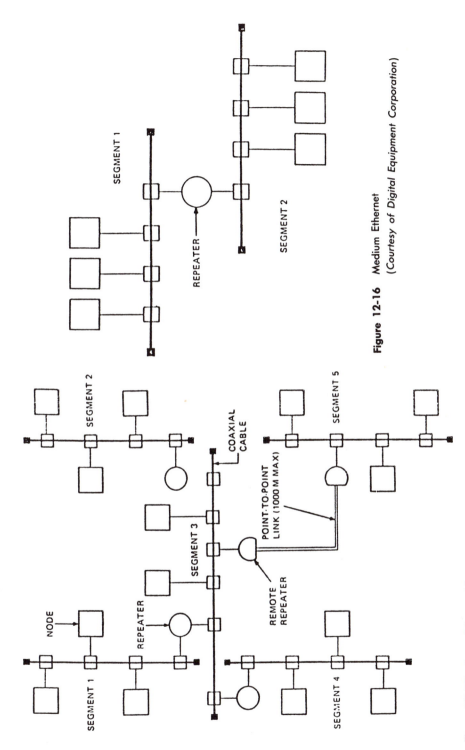

Figure 12-16 Medium Ethernet (*Courtesy of Digital Equipment Corporation*)

Figure 12-17 Large Ethernet (*Courtesy of Digital Equipment Corporation*)

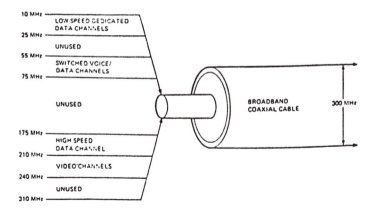

Figure 12-18 Broadband Subchannels

Broadband LANs

Broadband LANs use frequency division multiplexing (FDM) to subdivide the coaxial cable into many channels, each of which can have the capacity of a baseband LAN. Since this is an analog signal technique, the digital data signals must be converted. The main drawback of broadband LANs is the cost of modems and FDMs. The major benefit lies in the fact that the network can transmit data, voice, and video signals simultaneously. The coaxial cable used in broadband LANs is the same as that used in cable television (75ohm). Its capabilities include a bandwidth of 300 megahertz and a transmission rate of up to five megabits per second per channel.

To accommodate transmissions of data, voice, and video the cable is divided into bands that are multiplexed into many subchannels. Fig. 12-18 shows a possible allocation of bands for low speed data, switched voice and data, and video channels, with capacity reserved for future expansion.

One distinction between broadband and baseband systems is that signals in broadband systems must travel in one direction, whereas on baseband the signal travels in both directions. Both use the modified bus or tree topology. In broadband LANs separate send and receive channels allow two-way exchange of information.

On single cable systems, compatible with cable T.V., this is accomplished by halving the capacity of the cable. A Central Retransmission Facility (CRF) remodulates lowband (send) to highband (receive) signals. Fig. 12-19 shows the IEEE 802 working group recommendation for a broadband standard and establishes a 108–162 megahertz guardband between high and low band nodes. Source nodes transmit messages at lower frequencies (10 to 100 megahertz) to the CRF which amplifies and modulates these signals to higher frequencies (170 to 300 megahertz).

In a dual cable system the entire bandwidth is available in both directions. Nodes use one cable for sending and one for receiving, with both signals

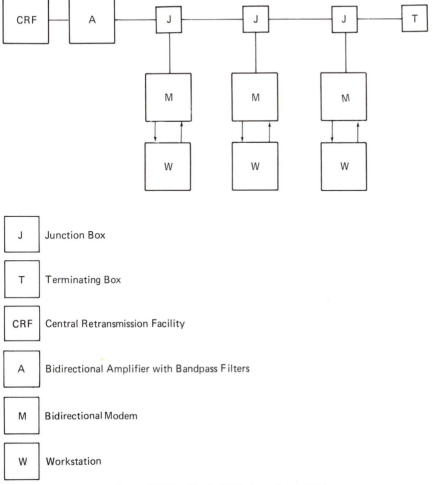

J — Junction Box

T — Terminating Box

CRF — Central Retransmission Facility

A — Bidirectional Amplifier with Bandpass Filters

M — Bidirectional Modem

W — Workstation

Figure 12-19 Single Cable Broadband LAN

occurring at the same frequency. The cables are connected by means of a passive connection called the midcable loop. Fig. 12-20 shows the message flow for a simple dual cable link-up.

The trade-off between the systems is cost versus bandwidth. The single cable broadband LAN is less expensive but supplies only half the bandwidth of the dual cable broadband LAN.

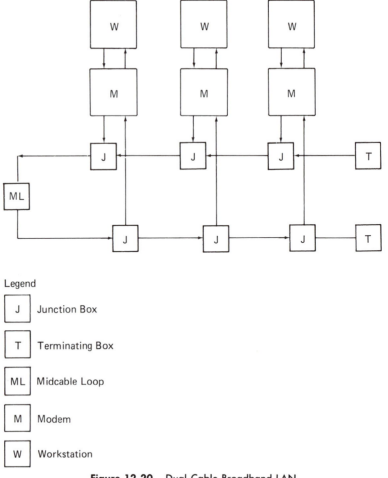

Legend

J	Junction Box
T	Terminating Box
ML	Midcable Loop
M	Modem
W	Workstation

Figure 12-20   Dual Cable Broadband LAN

12-4   Conclusions

LANs provide a viable means of information exchange and resource sharing within a building complex. New LANs are being introduced on the market-place with regularity, yet only a few technologies compete for prominence. Twisted pair and baseband coaxial cable compete for the lower end of the market, most often using bus or tree topology and CSMA/CD protocol. The future may lie with the baseband coaxial cable system initially, but in the long term, with the broadband system which combines the transmission of data, voice, and video in an extremely flexible though expensive system.

Other factors which will have important impact on LAN development include the need for communication between LANs and other computer based systems via a gateway, the competition from Computerized Branch Exchanges (CBXs), and the introduction of Metropolitan networks.

Gateway

A gateway is a device for connecting two systems that use different protocol. A gateway has intelligence and acts as a communications controller and protocol converter. Gateways will be used to connect LANs to a mainframe host computer or a public packet switching network, or to connect dumb/limited intelligence terminals to the LAN.

CBX

CBX provides a viable alternative to the LAN as a way of exchanging both data and voice communication and for resource sharing. Using time division multiplexing and an all digital computer controlled switch, CBX replaces the Private Branch Exchange (PBX) used as the traditional office switchboard. The transmission medium, twisted pair, provides analog voice signals which must be converted at the switch to digital and converted back to analog after being rerouted. Data signals are converted by modem prior to transmission. The CBX is capable of interconnecting a variety of office equipment, including workstations, intelligent copiers, facsimile transceivers, and printers. Fig. 12-21 shows a brief comparison of CBX and LAN systems and identifies some of the factors which may influence user selection.

Metropolitan Networks

Distance networks are used for intercity communication, LANs for interoffice/interbuilding. Many organizations, banks, insurance and finance companies, and retail outlets have enough data traffic within a city to see intracity networking as a means of reducing high communication costs. The need exists within a metropolitan area for a network that supports data, voice, video, and facsimile transmission. These demands are not well catered for by public telephone companies. LANs are also under criticism for having failed to fulfill their potential in terms of resource sharing and savings. It's early days yet. LANs remain a major contender in office automation and have obvious potential in the industrial area. A more reasonable criticism is that LANs are standalone in nature. Their future may lie in the development of sophisticated gateways which will combine LAN's spread across a city into a Metropolitan Network.

Factor	LAN	CBX
Speed	1-10Mband	64Kband
Data	All LANs	Yes
Voice	Broadband	Yes
Video	Broadband	No
Cost	Higher	Lower
Installation	Required	Already there

Fig. 12-21 LAN versus CBX

Problems

1. Identify the main technical difference between LANs and Distance Networks.

2. Which resource, shared by LANs, contributes most to increased productivity in office automation?

3. What would be the main advantage of a bidirectional ring over the unidirectional system of LANs?

4. Complete the following table:

   LAN characteristic   ISO-OSI LEVEL
   bus
   token passing
   fiber optics
   transporter
   network operating system

5. What is the main purpose of the RESERVE command in VictorLan?

6. Identify two characteristics which distinguish Local from Remote repeaters on Ethernet.

7. The CRF in single cable broadband networks and the midcable loop in dual cable broadband networks are sometimes referred to as the "head end." What is the main difference between them in terms of hardware?

APPENDIX A

Fourier Series

Any periodic function with a finite number of axis crossings over some fixed time interval can be represented by an infinite trigonometric series. Such a function $f(t)$ having a period T where $f(t + T) = f(t)$, may be represented by a series called the Fourier series as indicated:

$$f(t) = a_0 + \sum_{n=1}^{\infty} \left(a_n \cos \frac{2\pi nt}{T} + b_n \sin \frac{2\pi nt}{T} \right) \qquad (A\text{-}1)$$

where the coefficients are given by

$$a_0 = \frac{1}{T} \int_{-T/2}^{T/2} f(t)\, dt \qquad (A\text{-}2a)$$

$$a_n = \frac{2}{T} \int_{-T/2}^{T/2} f(t) \cos \frac{2\pi nt}{T} \qquad (A\text{-}2b)$$

$$b_n = \frac{2}{T} \int_{-T/2}^{T/2} f(t) \sin \frac{2\pi nt}{T} \qquad (A\text{-}2c)$$

and where $1/T$ is the fundamental frequency of the periodic function.

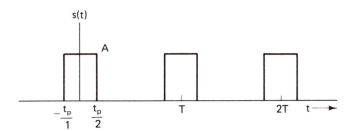

Figure A-1   Rectangular Wave Train

As an illustration, find the Fourier series of a periodic train of pulses of amplitude A and duration t_p as shown in the accompanying figure.

Thus,

$$a_0 = \frac{1}{T} \int_{-t_p/2}^{t_p/2} A\, dt = \frac{A}{T} t \Big|_{-t_p/2}^{t_p/2} = \frac{A t_p}{T}$$

This component represents the dc term in circuit theory.

$$a_n = \frac{2}{T} \int_{-t_p/2}^{t_p/2} A \cos \frac{2\pi nt}{T} \, dt$$

$$= \frac{2A}{T} \frac{T}{2\pi n} \sin \frac{2\pi nt}{T} \Bigg|_{-p/2}^{t_p/2}$$

$$= \frac{2At_p}{T} \frac{\sin \frac{\pi nt_p}{T}}{\frac{\pi nt_p}{T}}$$

The sinc function is defined as

$$\text{sinc } x = \frac{\sin \pi x}{x}$$

Thus

$$a_n = \frac{2At_p}{T} \text{ sinc } \frac{nt_p}{T}$$

Because of even symmetry,

$$b_n = 0$$

Thus the Fourier series for $S(t)$ becomes

$$S(t) = \frac{At_p}{2} + \frac{2At_p}{T} \sum_{n=1}^{\infty} \text{sinc } \frac{nt_p}{T} \cos \frac{2\pi nt}{T} \qquad (A\text{-}3)$$

By expanding the sin function as a MacLaurin's series, the value of the sinc x function can be found at $x = 0$.

$$\text{Thus,} \quad \text{sinc } x = \frac{\sin \pi x}{\pi x} = \frac{\pi x - \frac{(\pi x)^3}{3!} + \frac{(\pi x)^5}{5!}}{\pi x}$$

$$= 1 - \frac{(\pi x)^2}{3!} + \frac{(\pi x)^4}{5!}$$

$$\lim_{x \to 0} \text{sinc } x = 1$$

The Fourier series for $S(t)$ has a dc term of $At_p/2$ and harmonics of $1/T$ having an envelope given by $\frac{2At_p}{\pi}$ sinc nt_p/T. This is illustrated in Fig A-2. The sinc nt_p/T goes through zero whenever $\sin \pi nt_p/T = 0$ or when $\pi nt_p/T = m\pi$. This is equivalent to $n/T = m/t_p$.

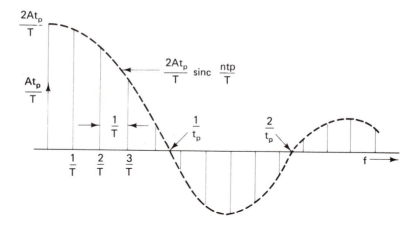

Figure A-2   Fourier Series of a Rectangular Wave Train (Spectrum Graph)

APPENDIX B

Noise Power Ratio (NPR)

The NPR ratio is defined as the ratio (in dB) of the noise in the test channel with all channels loaded with white noise to the noise in the test channel with all channels except the test channel loaded with white noise.

$$\text{NPR} = 10 \log \left(\frac{N_G \times \dfrac{\text{channel slot bandwidth}}{\text{base bandwidth}}}{N} \right)$$

where

N_G = power due to noise generators (mW)
N = noise power due to intermodulation and thermal noise

Dividing and multiplying the term in the brackets by the test tone level of the telephone channel, S

$$\text{NPR} = 10 \log \left(\frac{\dfrac{S}{N} \times \dfrac{1}{\text{base bandwidth}} \times \dfrac{N_G}{S}}{\text{channel slot bandwidth}} \right)$$

$$= 10 \log \frac{S}{N} - 10 \log \left(\frac{\text{base bandwidth}}{\text{channel slot bandwidth}} \right)$$

$$+ \, 10 \log \frac{N_G}{S}$$

therefore

$$10 \log \frac{S}{N} = \text{NPR} - \log \frac{N_G}{S} + 10 \log \left(\frac{\text{base bandwidth}}{\text{channel slot BW}} \right)$$

The term $10 \log N_G/S$ is given in the CCIR-loading equations of (3-21).

APPENDIX C

Group Delay

A traveling wave on a transmission line can be expressed by $\cos(\omega t - \beta x)$*
where:

> ω represents the angular frequency ($\omega = 2\pi f$)
>
> t represents the time
>
> β represents the phase shift constant per unit length of line
>
> x represents the position in the transmission line

Consider two closely spaced angular frequencies of ω_1 and ω_2, each a distance of ω_m (modulating angular velocity) from a carrier angular velocity of ω_c, as illustrated in Fig. C-1, so that

$$\omega_1 = \omega_c - \omega_m$$
$$\omega_2 = \omega_c + \omega_m$$

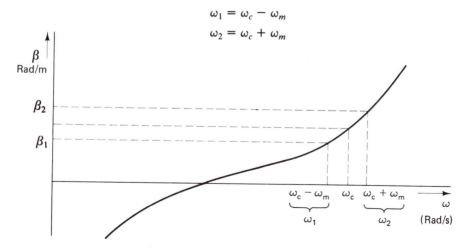

Figure C-1

Let the sum of the two frequencies be represented by equal sinusoidal voltages:

---

* From W. Sinnema, *Electronic Transmission Technology; Lines, Waves and Antennas.* (Englewood Cliffs, N.J.: Prentice-Hall, Inc., 1979), Chapter 3.

$$v = \cos(\omega_1 t - \beta_1 x) + \cos(\omega_2 t - \beta_2 x)$$

$$= 2 \cos \left[\frac{(\omega_1 + \omega_2)}{2} t - \frac{(\beta_1 + \beta_2)}{2} x \right]$$

$$\times \cos \left[\frac{(\omega_2 - \omega_1)}{2} \left[t - \frac{(\beta_2 - \beta_1)}{2} x \right] \right] \qquad (C\text{-}1)$$

by use of the trigonometric identity.

$$\cos x + \cos y = 2 \cos \frac{(x + y)}{2} \cos \frac{(x - y)}{2}$$

The phase constants can be evaluated around ω_c by using the Taylors series. Thus:

$$\beta_1 = \beta_c - \left. \frac{d\beta}{d\omega} \right|_{\omega_c} \omega_m + \frac{1}{2!} \left. \frac{d^2\beta}{d\omega^2} \right|_{\omega_c} \omega_m^2 + \ldots \qquad (C\text{-}2a)$$

$$\beta_2 = \beta_c + \left. \frac{d\beta}{d\omega} \right|_{\omega_c} \omega_m + \frac{1}{2!} \left. \frac{d^2\beta}{d\omega^2} \right|_{\omega_c} \omega_m^2 + \ldots \qquad (C\text{-}2b)$$

Substituting Eq. $(C\text{-}2)$ into $(C\text{-}1)$ and keeping ω_m small,

$$v = 2 \cos \left(\omega_m t - \left. \frac{d\beta}{d\omega} \right|_{\omega_c} \omega_m x \right) \cos(\omega_c t - \beta_c x) \qquad (C\text{-}3)$$

This is interpreted as a carrier traveling wave, $[\cos(\omega_c t - \omega_c x)]$, whose amplitude or envelope is varying with time and position:

$$\cos \left(\omega_m t - \left. \frac{d\beta}{d\omega} \right|_{\omega_c} \omega_m x \right)$$

The phase velocity for the carrier can be determined by imagining an observer following a constant phase point on the wave. That is, $\beta x - \omega t = $ a constant.

Taking the derivative of this expression with respect to time:

$$\frac{d}{dt}(\beta x) - \frac{d}{dt}(\omega t) = \frac{d}{dt} \text{(constant)}$$

$$\beta \frac{dx}{dt} - \omega = 0$$

$$v_p = \frac{dx}{dt} = \frac{\omega}{\beta} \qquad (C\text{-}4)$$

Similarly, the phase velocity, v_g, of the envelope (signal frequency ω_m) is given by $v_g = d\omega/d\beta \big|_{\omega_c}$ and is called group velocity. If a constant phase point $\beta x - \omega T = 0$, for example, continues to be followed, the time delay of the carrier is given by

$$T = \frac{\beta x}{\omega}$$

$$= \frac{b}{\omega} \qquad (C\text{-}5)$$

where b represents the total phase shift for a line of length x. Similarly, the time delay in the envelope, commonly called the envelope or group delay is given by:

$$T_g = \frac{d\beta}{d\omega}\bigg|_{\omega_c} x$$

$$= \frac{db}{d\omega}\bigg|_{\omega_c} \qquad (C\text{-}6)$$

where b represents the total phase shift for a line of length x.

APPENDIX D

The XR-567 Touch-Tone Decoder

The XR-567 consists of a phase-locked loop, a quadrature phase detector, a voltage capacitor, and an output logic circuit. Figure D-1 shows the block diagram of the XR-561. Figure D-2 illustrates its component connections.

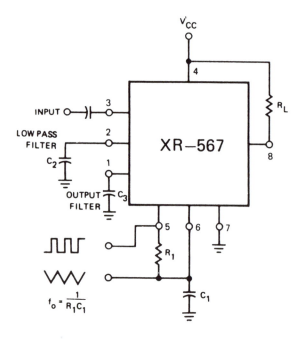

Figure D-1   XR-567 Block Diagram (Courtesy of EXAR Integrated Systems, Inc.)

The nominal free-running frequency of the oscillator within the PLL can be approximated by:

$$f_0 = \frac{1}{R_1 C_1} \qquad \begin{array}{l} R_1 \text{ in ohms } (\Omega) \\ C_1 \text{ in farads } (F) \end{array}$$

If the incoming signal is within the PLL's capture bandwith centered around f_0, the phase-locked loop will be locked in on the incoming tone. This bandwidth, expressed as a percentage of f_0, can be approximated by:

$$\text{BW} = 1070\sqrt{\frac{V_i}{f_0 C_2}} \qquad \begin{array}{l} V_i \text{ in volts rms} \\ C_2 \text{ in microfarads} \end{array}$$

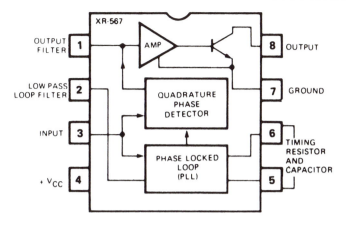

Figure D-2 XR-567 International Block Diagram (Courtesy of EXAR Integrated Systems, Inc.)

where the input voltage V_i must be greater than the threshold voltage of about 20 mV rms.

With the PLL locked on an input one, the quadrature phase detector undergoes an output voltage shift. This is transferred to the output logic circuit, which forces the output "open-collector" power transistor to its saturated value of about 0.5 volts. The load resistance R_L connected to the collector (see Fig. D-3) determines the current level. A typical input and output waveform is shown in Fig. D-4; normally the output is in a high state, but when a tone is applied which lies within the decoder's capture range, the output goes to a low state. The capacitor C_3 in Fig. D-1 determines the output response speed.

To obtain a dual-tone decoder in which two tones are simultaneously

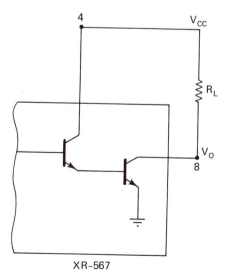

Figure D-3   Output Driver

present, the circuit shown in Fig. D-5 can be used. The R_1C_1 values of each decoder are chosen so that one of the desired tones is selected. When both tones are present, the collector of the tone decoders goes low and the output from the NOR gate goes high.

A touch-tone decoder can be made up of several XR-567's as shown in Fig. 2-13. When appropriate tone pairs are present, the respective NOR gate output goes high.

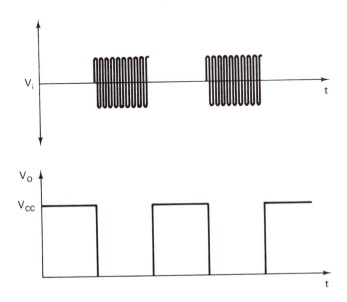

Figure D-4   Response to Input Tone Bursts

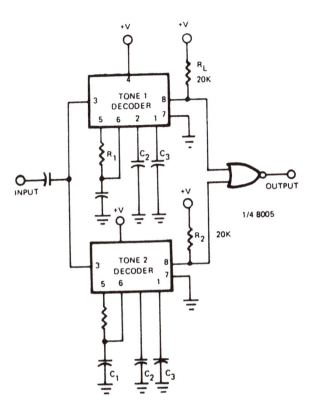

Figure D-5 Two-Tone Detector. (Courtesy of EXAR Integrated Systems, Inc.)

APPENDIX E

The Double-Balanced Ring Modulator

The high amplitude carrier, $\cos \omega_c t$, of the double-balanced ring modulator shown in Fig. E-1 turns the diodes OFF and ON at a carrier frequency rate. The transformer removes the dc component.

When the carrier is positive, diodes A and B act as shorts and the output $v_o(t) = m(t)$, assuming no transformer losses and a 1:1 transformer turns ratio. When the carrier is negative, diodes C and D act as shorts and the modulator reverses the polarity of input signal, $v_o(t) = -m(t)$. If this switching action is represented by the switching function $S(t)$ where $m(t)$ is multiplied by $+1$ when $\cos \omega t > 0$ and by -1 when $\cos \omega t < 0$, Fig. E-2 shows that the output or DSBSC signal can be expressed as:

$$v_o = m(t) \times S(t) \qquad (E\text{-}1)$$

Appendix A shows that the frequency content of the square wave, $S(t)$, consists of all the odd harmonics of $\omega_c/2\pi$ or f_c as $S(t)$ is a square wave with no dc component. It rolls off with the normal sinc response as illustrated in the $S(f)$ plot of Fig. E-2(b).

Since multiplying sinusoids in the time domain gives sum and difference frequencies in the frequency domain, the resultant frequency spectrum of the DSBSC signal consists of sidebands of $m(t)$ around the odd harmonics of the switching function $S(t)$. The harmonics of $S(t)$ are not present at the output because the $m(t)$ applied to the diodes has the dc content removed by the trans-

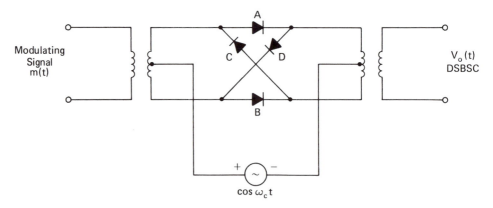

Figure E-1　Double-Balanced Ring Modulator

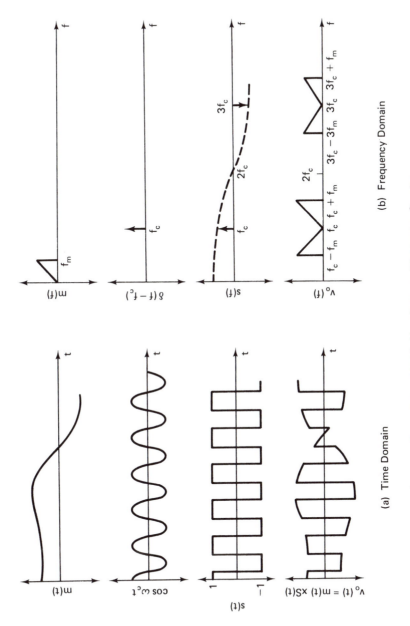

(a) Time Domain

(b) Frequency Domain

Figure E-2 DSBSC Waveforms and Frequency Spectra

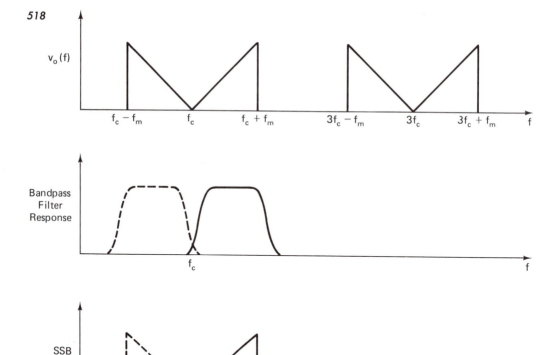

Figure E-3 SSB Frequency Spectrum. The dotted line indicates transmission of the lower sideband whereas the solid line indicates transmission of the upper sideband.

former. When multiplying a sinusoid by a zero, one obtains a null in that component.

To obtain high carrier rejection, small resistors of a few hundred ohms are usually placed in series with each diode in order to obtain equal resistance in all legs when the diodes are forward biased.

SSB is obtained by placing a bandpass filter after the balanced modulator. Fig. E-3 shows the resultant frequency spectrum. This figure also illustrates the difficulties experienced when low frequencies are present, in that the lower frequency components of the desired sideband experience greater attenuation than the mid and high frequency components.

Index